Proceedings of the Steklov Institute of Mathematics

edited by
S. M. Nikol'skiĭ

Number 132 (1973)

Proceedings of the International Conference on Number Theory

(Moscow, September 14-18, 1971)

edited by

I. M. Vinogradov (Editor-in-Chief)
A. A. Karacuba, K. I. Oskolkov
and A. N. Paršin

AMERICAN MATHEMATICAL SOCIETY
Providence, Rhode Island
1975

Академия Наук
Союза Советских Социалистических Республик

ТРУДЫ
ордена Ленина
МАТЕМАТИЧЕСКОГО ИНСТИТУТА
имени В. А. СТЕКЛОВА
CXXXII

Ответственный редактор
академик С. М. НИКОЛСКИЙ

ТРУДЫ МЕЖДУНАРОДНОЙ КОНФЕРЕНЦИИ ПО ТЕОРИИ ЧИСЕЛ
Москва, 14–18 сентября 1971 г.

Редколлегия сборника:
И. М. ВИНОГРАДОВ (главный редактор),
А. А. КАРАЦУБА, К. И. ОСКОЛКОВ, А. Н. ПАРШИН

Издательство "Наука"
Москва 1973

AMS (MOS) subject classifications (1970). Primary 10-02, 10A25, 10B10, 10C15, 10C20, 10C30, 10D05, 10D15, 10E05, 10F35, 10G05, 10H10, 10H25, 10H30, 10H35, 10J05, 10J20, 10K15, 10K20, 10M20, 12A70, 12B10, 12C05, 12D05, 12F20, 14E05, 14G05, 14G13, 14G25, 14H05, 14H25, 14J25, 14K10, 14K15, 14K99, 20G30, 20G40, 30A16, 39A15; Secondary 01A65, 02E99, 10-03, 10A15, 10B45, 10B99, 10D10, 10F30, 10G05, 10H05, 10H35, 10J15, 10K05, 12A05, 12A25, 12A35, 12B15, 14C20, 14D20, 14G05, 14G25, 14J20, 14K30, 20E15, 60F05, 60G50.

Library of Congress Cataloging in Publication Data CIP

International Conference on Number Theory, Moscow, 1971.
Proceedings of the International Conference on Number Theory (Moscow, September 14–18, 1971)

(Proceedings of the Steklov Institute of Mathematics; no. 132 (1973))
Translation of Trudy Mezhdunarodnoĭ konferenfsii po teorii chisel.
Bibliography: p.
1. Numbers, Theory of–Congresses. 2. Algebraic number theory–Congresses. 3. Geometry, Algebraic–Congresses. I. Vinogradov, Ivan Matveevich, 1891– II. Series: Akademiía nauk SSSR. Matematicheskiĭ institut. Proceedings; no. 132.
QA1.A413 no. 132 [QA241] 510'.8s [512'.7]
ISBN 0-8218-3032-5 75-14189

PROCEEDINGS OF THE STEKLOV INSTITUTE OF MATHEMATICS
IN THE ACADEMY OF SCIENCES OF THE USSR

(Труды математического института им. В. А. Стеклова, т. CXXXII, 1973)

TABLE OF CONTENTS[1]

---

(1) In addition to the papers listed, the program of the conference also included the following: T. Tonkov, *On the mean length of finite continued fractions*; A. A. Albert, *On associative division algebras*; B. J. Birch, *Canonical points of elliptic curves.*

ADDRESS OF THE PRESIDENT OF THE ACADEMY OF SCIENCES OF THE USSR ACADEMICIAN M. V. KELDYŠ AT THE OPENING OF THE CONFERENCE

Distinguished colleagues!

Ladies and Gentlemen! Comrades!

On behalf of the Academy of Sciences of the Soviet Union it gives me great pleasure to welcome the participants and guests of the International Conference on the Theory of Numbers. We value the participation in this conference of our guests—outstanding scholars creating their own methods and trends in the theory of Diophantine equations, the geometry of numbers, the theory of transcendental numbers, algebraic number theory and other domains.

The notion of number and elementary geometric representations are the original abstractions, being the cornerstones on which grow the vast edifice of contemporary mathematics, embracing a broad range of scientific problems starting from abstract questions of epistemology and ending with numerous questions directly related to various problems of science and technology. The notion of number led to the creation of algebra; under the influence of problems of natural science it led to the birth of mathematical analysis, and under the influence of difficulties arising from the infinitude of the natural series of numbers and other sets it led to the creation of mathematical logic and to the birth of other domains of contemporary mathematics, which is turn exerted an influence on the development of ideas on number.

The presence in number theory of difficult and profound problems in conjunction with the fact that number is the simplest elementary mathematical notion stimulated the development of new ideas and methods having the greatest significance not only in number theory but also in other parts of mathematics. In our time number theory also enriches mathematics with new concepts, directions and methods, and is closely connected with the investigation of logical foundations, on which the whole edifice of mathematics rests.

Permit me to wish all participants in the conference an interesting and fruitful time.

Research in number theory is a long-standing tradition of our Acadmey of Sciences. An invaluable contribution in this domain was made by menbers of our Academy: Euler, Čebyšev and many others.

The opening day of this conference coincides with the 80th birthday of one of the greatest mathematicians of our time—Academician Ivan Matveevič Vinogradov, whose extensive contribution to number theory is well known to you. I may say with confidence that the works of Academician Vinogradov have had a revolutionary influence on analytic number theory and have changed its appearance.

At the center of his greatest achievements stands the development of the method of trigonometric sums. In 1934 Ivan Matveevič Vinogradov created a new method in analytic number theory. The essentially new, more precise estimates of trigonometric sums established by him permitted him to obtain results far better than earlier ones involving the distribution of fractional parts of polynomials, the solution of Waring's problem, the distribution of primes and a whole series of other problems.

In 1937 I. M. Vinogradov discovered a method of estimating trigonometric sums extended only over prime numbers. This discovery created the possibility of solving additive problems in primes. In the same year he gave a proof of an asymptotic formula for the number of representations of an odd integer as a sum of three primes. Thereby the ternary problem of Goldbach was solved for the first time. Further development of Vinogradov's method led to a number of fundamental results, in particular to the establishment of properties of characters of shifted primes that cannot be deduced even from the extended Riemann hypothesis.

The sphere of application of Vinogradov's method is not restricted merely to number theory, but also exerts an influence on such domains as function theory, mathematical analysis, probability theory and approximate computations. Ivan Matveevič Vinogradov himself constantly worked on the perfection of his method, and found new possibilities for its application.

Soviet mathematicians are sincerely grateful to Academician Vinogradov for the beneficial influence that he exerted on the development of science in our country. Since 1934 he has been continuously the director of the V. A. Steklov Mathematical Institute. Under the leadership of Academician Vinogradov the Mathematical Institute has become one of the most important centers of mathematical thought.

Ivan Matveevič's attentive regard for newcomers and assistants promoted the development of many scientific trends in mathematics, mechanics and theoretical physics. On the basis of scientific groups created in the Mathematical Institute grew many of the scientific establishments playing an important role in the development of science in our country.

On behalf of the conference participants permit me to welcome our illustrious hero and also to present him with the highest government award—the Second Gold Star of Hero of Socialist Labor.

Translated by B. SILVER

Trudy Mat. Inst. Steklov.
132 (1973)

Proc. Steklov Inst. Math.
132 (1973)

EXPONENTIAL SUMS IN THE DEVELOPMENT OF NUMBER THEORY*

K. CHANDRASEKHARAN

The purpose of this article is to survey some of the important results on exponential sums which have influenced the development of the theory of numbers. It is intended to be selective rather than exhaustive, as befits the occasion.(1)

1. Gaussian sums and quadratic reciprocity

The importance of exponential sums in arithmetic was recognized, for the first time, by Gauss (1808, Werke, (ii), p. 11). In his «fourth proof» of the law of quadratic reciprocity, he evaluates a sum of the kind

$$G(m, n)=\sum_{k=1}^{|n|} e^{2\pi i \frac{m}{n} k^2},$$

where m and n are non-zero integers. Such a sum is now known as a Gaussian sum, and can be evaluated in many ways. That the law of quadratic reciprocity is a consequence of a reciprocity formula for Gaussian sums was first recognized by Kronecker (Monatsber. Kgl. Preuss. Akad. Wiss. Berlin, 1880, 686; J. reine und angew. Math., 105, 1889, 267). For non-zero integers m and n, let

$$g(m, n)=\sum_{k=1}^{|n|} e^{\pi i \frac{m}{n} k^2+\pi i m k},$$

so that if m is even, $g(m, n)$ reduces to a Gaussian sum. The reciprocity formula for $g(m, n)$ is as follows:

$$\frac{1}{\sqrt{|n|}} g(m, n)=\frac{1}{\sqrt{|m|}} g(-n, m)\, e^{\frac{\pi i}{4}(1-|mn|)\operatorname{sgn}(mn)}. \tag{1.1}$$

This can be proved by Cauchy's theorem of residues. The proof has points in common with the proof of the functional equation of Riemann's zeta-function ζ, contained in Riemann's posthumous work, which makes no use

---

AMS (MOS) subject classifications (1970). Primary 10-02, 10G05; Secondary 10A15, 10K05, 10H05, 12A05, 10J15.

**Editor's note*. This reprinting of the original paper incorporates suggestions made by the author.

(1) Lecture given at the opening ceremony of the International Conference on Number theory held in Moscow on 14 September 1971, marking the 80-th birthday of Academician I. M. Vinogradov.

of theta-functions. It appears that Riemann, in his lectures, based the transformation theory of theta-functions on the properties of a contour integral which occurs in the proof of (1. 1) (see *C. L. Siegel.* Quellen und Studien zur Geschichte der Mathematik. Astronomie und Physik, **2**, 1932, 45; Ges. Abh., (i), S. 275).

Gaussian sums are related to theta-functions in a simple way. If z is a complex variable, and

$$\vartheta(z) = \sum_{n=-\infty}^{\infty} e^{-\pi n^2 z}, \qquad \operatorname{Re} z > 0, \tag{1. 2}$$

then the «imaginary axis» ($\operatorname{Re} z=0$) is a line of singularities for the analytic function $\vartheta(z)$ defined by (1. 2). If we take $z=\delta+2ir$, where $\delta>0$, r rational, $r=p/q$, $q>0$, $i=\sqrt{-1}$, then we have

$$\lim_{\delta\to 0+} \sqrt{\delta}\,\vartheta\left(\delta+2i\frac{p}{q}\right)=\frac{1}{q}\,G(-p,\ q),$$

where $G(-p,\ q)$ is a Gaussian sum. Because of the transformation formula

$$\sqrt{z}\,\vartheta(z)=\vartheta\left(\frac{1}{z}\right), \qquad \operatorname{Re} z>0,$$

the behaviour of $\vartheta(z)$ near $z=2i\frac{p}{q}$ is reflected by the behaviour of $\vartheta(z')$ near $z'=\frac{q}{2ip}=-\frac{2iq}{4p}$. Hence one can deduce a reciprocity formula between $G(-p,\ q)$ and $G(q,\ 4p)$. It is this idea which Hecke (Vorlesungen über die Theorie der alg. Zahlen, 1923, Kap. VIII) used in his *analytical* approach to the law of quadratic reciprocity in algebraic number fields of arbitrary degree. Mordell had earlier proved (Proc. Lond. Math. Soc., **20**, 1921, 289) a reciprocity formula for Gaussian sums in a quadratic field.

Hecke constructed, for each algebraic number field of degree n, an expression which he called a Gaussian sum, and proved a reciprocity formula for such sums, by considering the boundary-behaviour of theta-functions in n variables. From this he derived a proof of the law of quadratic reciprocity which Hilbert was the first to set up in algebraic number fields by the introduction of his norm-residue symbol. Hilbert's work has been followed up by the contributions of Furtwängler, Takagi, Artin, Hasse and Chevalley, and by the development of class-field theory. One important end-result is Hasse's law of reciprocity for l-th power residues (J. reine und angew. Math., **153**, 1924, 192), which gives, for $l=2$, a more explicit form of the law of quadratic reciprocity than Hecke's, which contains an undetermined eighth root of unity.

Important in the context of exponential sums, however, is Siegel's proof (Göttinger Nachr., 1960, 1; Ges. Abh., (iii), S. 334) of Hasse's formula (explicitly for $l=2$). The proof consists in generalizing the Gauss-Hecke sums (as the sums $g(m,\ n)$ generalize $G(m,\ n)$) and proving a reciprocity formula for them in the manner of Kronecker, coupled with an argument of induction on the degree of the field. The proof avoids the use of theta-functions of several varables, and the use of class-field theory. It is the use, however, of an *explicit formula* involving exponential sums rather than their *estimation*, which plays the key role here.

2. Uniform distribution modulo 1, and Weyl's method

The *estimation* of exponential sums is of crucial importance in a study of the uniform *distribution modulo 1* of an infinite sequence of real numbers (λ_j), $j=1, 2, \ldots$, as was first recognized by Hardy and Littlewood (*G. H. Hardy*. Collected Papers, (i), 119; Proc. Nat. Acad. Sci. USA, **3**, 1916, 84), and, independently, and at about the same time, by Hermann Weyl (Math. Ann., **77**, 1916, 313). The sequence (λ_j) is defined to be uniformly distributed modulo 1, if the sequence (μ_j) of fractional parts of (λ_j), $\mu_j=\{\lambda_j\}$, has the following property: Let a, b be *any* pair of real numbers such that $0 \leqslant a < b \leqslant 1$, and $\varphi_n(a, b)$ the set of μ's such that $a \leqslant \mu_j < b$, $1 \leqslant j \leqslant n$. Then $\frac{\varphi_n(a, b)}{n} \to (b-a)$, as $n \to \infty$, for *every* such pair (a, b). The so-called *Weyl principle* states that a necessary and sufficient condition for (λ_j) to be uniformly distributed modulo 1 is that

$$\sum_{j=1}^{n} e^{2\pi i m\lambda_j} = o(n), \quad \text{as } n \to \infty, \tag{2. 1}$$

for every integer $m \neq 0$. The question of uniform distribution modulo 1 of (λ_j) is thus reduced to the problem of estimating the size of an exponential sum involving λ_j. Weyl further proved that if k is a positive integer, and λ_j a polynomial in j, of degree k, with real coefficients at least one of which (other than the constant term) is irrational, then (2. 1) holds. If, say, $\lambda_j = \alpha j^k + \alpha_1 j^{k-1} + \cdots + \alpha_k$, and α is irrational, then (2. 1) holds for any particular α, uniformly in $\alpha_1, \alpha_2, \ldots, \alpha_k$. This is a consequence of a general method of estimating exponential sums due to Weyl.

Let k be a positive integer, and

$$P(m) = \alpha m^k + \alpha_{k-1} m^{k-1} + \ldots + \alpha_1 m,$$

$\alpha \neq 0$, $\alpha, \alpha_1, \ldots, \alpha_{k-1}$ being real. Let μ be a positive integer, and let

$$S = \sum_{m=1}^{\mu} e^{2\pi i P(m)}. \tag{2. 2}$$

Then, if $k=1$,

$$|S| \leqslant \min(\mu, |\operatorname{cosec} \pi\alpha|), \tag{2. 3}$$

and if $k \geqslant 2$, $K = 2^{k-1}$, then

$$|S|^K \leqslant 2^{2K} \mu^{K-1} + 2^K \mu^{K-k} \sum_{r_1, \ldots, r_{k-1}} \min(\mu, |\operatorname{cosec}(\pi\alpha k! R)|),$$

where $R = r_1 r_2 \ldots r_{k-1}$, and each r_j varies from 1 to $\mu - 1$.

Weyl used this method to prove (Math. Z., **10**, 1921, 88; Ges. Abh., (ii), S. 181) that

$$\zeta(1+it) = O\left(\frac{\log t}{\log\log t}\right), \qquad \text{as } t \to \infty. \tag{2.4}$$

The proof depends on the estimates for sections of the zeta-function, that is for sums of the form $\sum_{n=N}^{N'} n^{-it} = \sum_{n=N}^{N'} e^{-it\log n}$, which result from the estimates for sums of the form $\sum_{n=N}^{N'} e^{iP(n)}$, where $P(n)$ is a polynomial approximation to $\log(1+n)$. One can, in fact, prove (following Landau) that

$$\zeta(s) = O(\log^A t), \quad A > 1, \quad s = \sigma + it,$$

for $\sigma \geqslant 1 - \frac{(\log\log t)^2}{\log t}$, $t \geqslant 3$, and derive from this the result, due to Littlewood, that

$$\zeta(s) \neq 0, \text{ for } \sigma > 1 - \frac{A_1 \log\log t}{\log t}, \qquad t > t_0, \ A_1 > 0. \tag{2.5}$$

By a general analytical argument, one can deduce from this the following estimate for the error-term in the prime number theorem, namely

$$\pi(x) - \operatorname{li}(x) = O\left(xe^{-a\sqrt{\log x \log\log x}}\right), \tag{2.6}$$

for a positive, absolute constant a. Here $\pi(x)$ denotes the number of primes not exceeding x, and for $x > 1$,

$$\operatorname{li}(x) = \lim_{\delta\to 0}\left(\int_0^{1-\delta} + \int_{1+\delta}^{x}\right)\frac{du}{\log u}.$$

It is known, after von Koch (Acta Math., **24**, 1901, 159), that the Riemann hypothesis implies that

$$\pi(x) - \operatorname{li}(x) = O(x^{1/2}\log x). \tag{2.7}$$

To see how effective Weyl's estimation of the sum S is, for $k \geqslant 2$, we have to find out by what factor it differs from the trivial estimate $|S| \leqslant \mu$. Since $|\operatorname{cosec} \pi\alpha| \leqslant \frac{1}{2\|\alpha\|}$, where α is non-integral, and $\|\alpha\|$ denotes the distance between α and the integer nearest to it, it is instructive to consider the case in which α has a rational approximation a/q, $q > 0$, $(a, q) = 1$, such that

$$\left|\alpha - \frac{a}{q}\right| \leqslant \frac{1}{q^2}.$$

In this case, we have (see *L. K. Hua.* Additive Primzahltheorie, 1959, p. 24),

$$S = O\left(\mu^{1+\varepsilon}q^{\varepsilon}\left(\frac{1}{\mu} + \frac{1}{q} + \frac{q}{\mu^k}\right)^{1/K}\right)$$

for $\epsilon > 0$, $K = 2^{k-1}$. If we consider the particular case in which $0 < q \leqslant \mu$, and let *both* μ and q tend to infinity, in a *suitable* manner, then

$$S = O(\mu^{1-\rho}), \quad \text{where} \quad \rho = \frac{1}{2^{k-1}} - \eta, \qquad \eta > 0. \tag{2.8}$$

If k is large, then ρ is small, and the estimate is weak. We shall see that Vinogradov's method gives $\rho = \frac{1}{ck^2 \log k}$ (where c is a numerical constant) which is substantially better for large values of k.

3. Van der Corput's method

Estimates for $\zeta(s)$ comparable to those obtained by Weyl's method can also be obtained by a method discovered by J. G. van der Corput (Math. Ann., 84, 1921, 53; 87, 1922, 39). This consists in first approximating exponential sums $\sum_n e^{2\pi i f(n)}$ by integrals with a sufficiently sharp error-term, and then estimating the integrals under suitable assumptions on the size of f and of its derivatives (see, for instance, *E. C. Titchmarsh*. The zeta-function of Riemann, Oxford, 1951, p. 90).

If f is a real function of a real variable, with a continuous and monotone decreasing derivative f', in an interval (a, b), and $f'(b) = \alpha$, $f'(a) = \beta$, then

$$\sum_{a<n\leqslant b} e^{2\pi i f(n)} = \sum_{\alpha-\eta<\nu<\beta+\eta} \int_a^b e^{2\pi i (f(x)-\nu x)} dx + O(\log(\beta - \alpha + 2)), \tag{3.1}$$

where η is any constant such that $0 < \eta < 1$.

A simple consequence is the following: *If f is twice differentiable, and*

$$0 < \lambda_2 \leqslant |f''(x)| \leqslant h\lambda_2,$$

for $x \in (a, b)$, where $b - a \geqslant 1$, then

$$\sum_{a<n\leqslant b} e^{2\pi i f(n)} = O(h(b-a)\lambda_2^{1/2}) + O(\lambda_2^{-1/2}). \tag{3.2}$$

This approximation can be applied, for instance, to the function $g(x) = f(x+r) - f(x)$ (instead of f) where r is an integer, if f has three continuous derivatives, since $g''(x) = rf'''(\xi)$ for $x < \xi < x + r$. For, if $a < n \leqslant b$, and $0 < q \leqslant b - a$, then

$$\left| \sum_{a<n\leqslant b} e^{2\pi i f(n)} \right| < c\frac{b-a}{q^{1/2}} + c\left(\frac{b-a}{q} \sum_{r=1}^{q-1} \left| \sum_{a<n\leqslant b-r} e^{2\pi i \{f(n+r) - f(n)\}} \right| \right)^{1/2},$$

where c is a numerical constant. Thus one obtains the following result:

If the real function f *of a real variable has a continuous derivative of order three in (a, b), where $b - a \geqslant 1$, and if*

$$0 < \lambda_3 \leqslant |f'''(x)| \leqslant h\lambda_3, \tag{3.3}$$

then

$$\sum_{a<n\leqslant b} e^{2\pi i f(n)} = O\left(h^{1/2}(b-a)\lambda_3^{1/6}\right) + O\left((b-a)^{1/2}\lambda_3^{-1/6}\right).$$

More generally, by an induction argument on k, the order of differentiability of f, one obtains the following inequality (3. 4) due to van der Corput.

If the real function f has a continuous derivative of order k, $k \geqslant 4$, in (a, b), where $b-a \geqslant 1$, and if

$$0 < \lambda_k \leqslant |f^{(k)}(x)| \leqslant h\lambda_k,$$

then

$$\sum_{a<n\leqslant b} e^{2\pi i f(n)} = O\left(h^{2/K}(b-a)\lambda_k^{\frac{1}{2K-2}}\right) + O\left((b-a)^{1-\frac{2}{K}}\lambda_k^{-\frac{1}{2K-2}}\right), \qquad (3.4)$$

where $K = 2^{k-1}$, the O's implying constants which are independent of k.

As before, this can be applied to sums of the form $\sum_N^{N'} n^{-it}$, hence to the zeta-function of Riemann, to obtain the following result:

If l is a positive integer, $l \geqslant 3$, $L = 2^{l-1}$, $\sigma = 1 - \frac{l}{2L-2}$, then

$$\zeta(s) = O\left(t^{\frac{1}{2L-2}} \log t\right), \qquad s = \sigma + it.$$

If $l = 3$, $L = 4$, we obtain as a special case,

$$\zeta\left(\frac{1}{2} + it\right) = O(t^{1/6} \log t).$$

This pair (3, 4) can be made to yield another pair which gives a slight improvement of the order of $\zeta(s)$, and so on. E. C. Titchmarsh (Quart. J. Math., **2**, 1931, 161; ibid., 313) and Eric Phillips (Quart. J. Math., **4**, 1933, 209) have developed this method of «exponent pairs» a little further, but to use Phillips's own words, «the improvements are always extremely small, while the results become very unwieldy».

There is a two-dimensional extension of van der Corput's method which is of importance in its application to lattice-point problems (see *E. C. Titchmarsh*. Proc. Lond. Math. Soc., (2) **36**, 1934, 485; (2)**38**, 1935, 96—115; *I. M. Vinogradov*. Izv. Leningr. Polytechn. Univ., **33**, 1927, 31). There is also an n-dimensional extension (*B. R. Srinivasan*. Acta Arithm., 8, 1963, 173).

van der Corput used his method to obtain the following estimate (Math. Ann., 98, 1928, 697) in the divisor problem: *if $d(n)$ denotes the number of positive divisors of the positive integer n, then*

$$\sum_{1\leqslant n\leqslant x} d(n) = x \log x + (2\gamma - 1)x + E(x), \qquad E(x) = O\left(x^{\frac{27}{82}+\varepsilon}\right), \quad \varepsilon > 0,$$

where γ is Euler's constant. The conjectured estimate for $E(x)$ is $O(x^{1/4+\varepsilon})$, while the best so far claimed is by Yin, Wen-lin (Sci. Rec. N. S., **3**, 1959, 6), namely $O(x^{13/40+\varepsilon})$.

Another application is to the lattice-point problem for the circle. If $r(n)$ denotes the number of lattice-points on a circle $x^2+y^2=n$ in the (x, y)-plane, then van der Corput proved (1923) that

$$\sum_{0\leqslant n\leqslant x} r(n)=\pi x+P(x),\quad P(x)=O(x^{\theta+\varepsilon});\qquad \theta<\frac{1}{3},\quad \varepsilon>0.$$

The conjectured estimate for $P(x)$ is $O(x^{1/4+\varepsilon})$. L. K. Hua proved (Quart. J. Math., **13**, 1942, 18) that $P(x)=O(x^{13/40+\varepsilon})$, and Yin, Wen-Lin (Sci. Sinica, **11**, 1962, 10) and Chen, Jing-run (Sci. Sinica, **12**, 1963, 633) have claimed the exponent $12/37+\varepsilon$.

4. The method of I. M. Vinogradov

Estimates much stronger than Weyl's, and capable of a variety of applications, were obtained by I. M. Vinogradov in the years following 1935, and published in a book (1937). An improved version appeared in his second book (1947) on the same subject, which has been translated into English by K. F. Roth and Anne Davenport (The method of trigonometric sums in the theory of numbers, Interscience, 1954). Some further improvements have appeared in his recent book (May 1971), which it has not been possible for me fully to take into account in this lecture.

Let k be an integer, $k\geqslant 2$, and $f(n)$ a polynomial in n of degree k with real coefficients, say

$$f(n)=\alpha_k n^k+\ldots+\alpha_1 n+\alpha_0, \tag{4.1}$$

and let

$$S(q)=\sum_{a<n\leqslant a+q} e^{2\pi i f(n)}, \tag{4.2}$$

where a and q are integers.

We consider the mean-value of $|S(q)|^{2l}$, where l is a positive integer, over a unit-cube in a space of k dimensions, namely

$$J(q,\ l)=\int_0^1\ldots\int_0^1 |S(q)|^{2l}\,d\alpha_1\ldots d\alpha_k. \tag{4.3}$$

Trivially $|S(q)|\leqslant q$, and $|J(q,\ l)|\leqslant q^{2l}$. To obtain a non-trivial estimate, we observe that

$$|S(q)|^{2l}=\sum_{\substack{m_1,\ldots,m_l\\ a<m_j\leqslant a+q}}\ \sum_{\substack{n_1,\ldots,n_l\\ a<n_j\leqslant a+q}} \exp 2\pi i\alpha_k(m_1^k+\ldots$$

$$\ldots+m_l^k-n_1^k-\ \ldots\ -n_l^k)+\ldots,$$

and use the fact that for an integer h, we have

$$\int_0^1 e^{2\pi i h x}dx=\begin{cases}1, & \text{if } h=0,\\ 0, & \text{if } h\neq 0.\end{cases}$$

It follows that $J(q,\ l)$ equals the number of solutions in integers of the system of equations

$$m_1^h+\ldots+m_l^h=n_1^h+\ldots+n_l^h,\qquad h=1,\ 2,\ldots,k,$$

with $a < m_\nu \leqslant a + q$, $a < n_\nu \leqslant a + q$. We also see that $J(q, l)$ is a non-decreasing function of q, and independent of a, for if we put $M_\nu = m_\nu - a$, $N_\nu = n_\nu - a$, then

$$\sum_{\nu=1}^{l} (M_\nu + a)^h = \sum_{\nu=1}^{l} (N_\nu + a)^h, \qquad h = 1, 2, \ldots, k,$$

and, on expanding the h-th powers, we find that this system is equivalent to

$$\sum_{\nu=1}^{l} M_\nu^h = \sum_{\nu=1}^{l} N_\nu^h, \qquad h = 1, 2, \ldots, k, \quad 0 < M_\nu \leqslant q, \quad 0 < N_\nu \leqslant q.$$

To estimate $J(q, l)$, one obtains an inequality between $J(q, l)$ and $J(q_1, l_1)$ for $q_1 < q$, $l_1 < l$. This idea is Vinogradov's. The actual proof of this inequality has undergone changes and simplifications over the years. Instead of finding the number of solutions in integers of a system of equations as above, one can reduce the question, after Vinogradov, Linnik, and Karacuba, to finding the number of solutions of a system of congruences modulo prime powers (see, for instance, *K. Chandrasekharan*. Arithmetical Functions, 1970, p. 90). Thus if T denotes the number of solutions, in integers, of the following system of congruences:

$$x_1 + \ldots + x_k \equiv \mu_1 \pmod p,$$
$$\cdots\cdots\cdots\cdots$$
$$x_1^k + \ldots + x_k^k \equiv \mu_k \pmod{p^k},$$

where $0 \leqslant x_j \leqslant Mp^k - 1$, for $j = 1, 2, \ldots, k$, M an integer, $M \geqslant 1$, $\mu_1, \ldots, \mu_k$ are integers, p a prime, $p > k$, and $x_j \not\equiv x_{j'} \pmod p$ for $j \neq j'$, then

$$T \leqslant k!\, M^k p^{k(k-1)/2}. \tag{4.4}$$

With the help of this estimate, one can prove that for $k \geqslant 2$, $l \geqslant k^2 + k$, $q \geqslant (2k^2)^k$, we have

$$J(q, l) < 3 l^{2k} q_1^{2k} p_1^{2l - [k(k+1)/2]} J(q_1, l - k), \tag{4.5}$$

where p_1 is any prime, such that

$$\frac{1}{2} q_1^{1/k} \leqslant p_1 \leqslant q^{1/k}$$

and

$$q_1 = [q p_1^{-1}] + 1.$$

By repeated application of inequality (4. 5), we get the following form of Vinogradov's mean-value theorem:

Let r, k, l and q be positive integers,

$$r \geqslant 1, \quad k \geqslant 2, \quad l \geqslant k^2 + kr, \quad q \geqslant (2^k k^{2k})^{\left(1 + \frac{1}{k-1}\right)^{r-1}}.$$

Then

$$J(q, l) < c(l, k, r)\, q^{2l - \frac{1}{2}k(k+1) + \delta_r}, \tag{4.6}$$

where

$$\delta_r = \frac{1}{2} k(k+1)\left(1 - \frac{1}{k}\right)^r,$$

and

$$c(l, k, r) = (3l^{2k})^r \, 3^{4lr - \frac{1}{2}k(k+1)r}.$$

If we choose r appropriately, we can put (4. 6) in another form in which the restriction on q can be relaxed:

Let $k \geqslant 2$, $l \geqslant k^2 + 4k^2 \log k$, $q \geqslant 2$. *Then*

$$J(q, l) \leqslant e^{clk \log^2 k} q^{2l - \frac{1}{2}k(k+1) + \frac{1}{2}}, \tag{4.7}$$

where c is a positive, absolute constant.

Vinogradov's mean-value theorem can be used to obtain an inequality for sums of the form

$$C = \sum_{n=P+1}^{P+Q} \exp 2\pi i F(n),$$

where P and Q are integers, and F a real-valued function with a certain number of continuous derivatives in the interval $P+1 \leqslant x \leqslant P+Q$. The result is Vinogradov's inequality:

Let P, Q be positive integers, $Q \geqslant 2$, and let F be a real function of a real variable, with a continuous derivative of order $k+1$, $k \geqslant 7$, for $P+1 \leqslant x \leqslant P+Q$. Let λ be a real number such that

$$\lambda \leqslant \left| \frac{F^{(k+1)}(x)}{(k+1)!} \right| \leqslant 2\lambda, \quad \textit{for} \quad P+1 \leqslant x \leqslant P+Q,$$

and such that

$$\frac{1}{\lambda^{1/4}} \leqslant Q \leqslant \frac{1}{\lambda} \left(\textit{so that } 0 < \lambda \leqslant \frac{1}{2} \right).$$

Then

$$\left| \sum_{n=P+1}^{P+Q} \exp 2\pi i F(n) \right| < c_1 Q^{1-\rho} \exp(c_2 k \log^2 k), \quad \rho = \frac{1}{70k^2 \log k}, \tag{4.8}$$

where c_1 and c_2 are positive constants.

The idea of the proof consists in observing that

$$|C| \leqslant \frac{1}{q} \sum_{n=P+1}^{P+Q-q} |T(n)| + q, \tag{4.9}$$

where q is an integer, $1 \leqslant q < Q$, and

$$T(n) = \sum_{m=1}^{q} \exp 2\pi i (F(m+n) - F(n)),$$

where m is an integer, $P+1 \leqslant n \leqslant P+Q-q$. Then $T(n)$ can be approximated by the Weyl sum $S(q)$ with a polynomial exponent, in the sense that

$$|T(n)| \leqslant |S(q)| + 2\pi k \lambda q^{k+2}.$$

By Hölder's inequality, this implies that

$$|T(n)|^{2l} \leqslant 2^{2l} |S(q)|^{2l} + (4\pi k\lambda q^{k+2})^{2l}.$$

By integrating this in k dimensions, relative to $\alpha_1, \ldots, \alpha_k$, Vinogradov makes it possible to apply his mean-value theorem to the first term on the right-hand side, and by summing over n, and using (4. 9), one obtains (4. 8) (for the necessary details, see, for instance, the author's Arithmetical Functions, p. 102). Inequality (4. 8) can be put in a form in which n does not necessarily run through the whole range $P+1 \leqslant n \leqslant P+Q$.

Thus, if F is real-valued, with a continuous derivative of order $(k+1)$, $k \geqslant 7$, for $P+1 \leqslant x \leqslant P+N$, where $N \leqslant Q$, Q real, and

$$\lambda \leqslant \left|\frac{F^{(k+1)}(x)}{(k+1)!}\right| \leqslant 2\lambda, \qquad \frac{1}{\lambda^{1/3}} \leqslant Q \leqslant \frac{1}{\lambda},$$

then

$$\left|\sum_{n=P+1}^{P+N} \exp 2\pi i F(n)\right| < c_3 Q^{1-\rho} \exp(c_4 k \log^2 k), \tag{4.10}$$

where

$$\rho = \frac{1}{70k^2 \log k}.$$

This is a powerful inequality, especially for large k, compared with Weyl's result in (2. 8), namely

$$S = O(\mu^{1-\rho}), \qquad \rho = \frac{1}{2^{k-1}} - \eta, \quad \eta > 0, \tag{2.8}$$

where

$$S = \sum_{m=1}^{\mu} \exp 2\pi i P(m), \qquad P(m) = \alpha m^k + \alpha_{k-1} m^{k-1} + \ldots + \alpha_1 m,$$

$\alpha \neq 0$, $\alpha, \alpha_1, \ldots, \alpha_{k-1}$ being real. One cannot, however, replace ρ by any number greater than $1/k$, for as Vinogradov has shown, there exist infinitely many sums of the form

$$S = \sum_{m=1}^{q} \exp 2\pi i \frac{a}{q} m^k, \qquad q > 0, \ (a, q) = 1, \tag{4.11}$$

for which

$$S = q^{1-\frac{1}{k}}.$$

This also shows that the following result of L. K. Hua's (his book, 1947, loc. cit., p. 2) cannot be substantially improved:

Let

$$S = \sum_{m=1}^{q} \exp \frac{2\pi i}{q} (a_k m^k + \ldots + a_1 m),$$

where $q > 0$, $(a_k, \ldots, a_1, q) = 1$. Then

$$S = O\left(q^{1-\frac{1}{k}+\varepsilon}\right), \text{ for every } \varepsilon > 0. \tag{4.12}$$

The case $k=2$, and q a prime, had been considered earlier by L. J. Mordell in 1932 (Quart. J. Math., **3**, 1932, 161—167).

5. Applications to the zeta-function of Riemann, and to the prime number theorem

Vinogradov's method of estimating exponential sums gives a substantial improvement of Littlewood's theorem on the zero-free region of $\zeta(s)$, as was first made clear by Chudakov (1936): there exists a positive constant A, such that

$$\zeta(s)\neq 0, \quad \text{for} \quad \sigma\geqslant 1-\frac{A}{\log^{3/4}t\,(\log\log t)^{3/4}}, \tag{5.1}$$

for $t\geqslant 3$. By a standard argument, one can deduce that

$$\pi(x)-\mathrm{li}(x)=O(x\exp(-c(\log x)^{4/7}(\log\log x)^{-3/7})). \tag{5.2}$$

Analogous results hold for the L-functions of Dirichlet, and for the error-term in the prime-number theorem for arithmetical progressions. Vinogradov himself, and members of his school, have given (1958) improved estimates for $\zeta(s)$, namely

$$\zeta(1+it)=O(\log^{2/3}t), \qquad t\geqslant 3,$$

and

$$\zeta(s)\neq 0, \quad \text{for} \quad \sigma\geqslant 1-\frac{A}{(\log t)^{2/3}(\log\log t)^{1/3}},$$

for $t\geqslant 3$, and a positive constant A, which yields the result

$$\pi(x)-\mathrm{li}(x)=O(x\exp(-c(\log x)^{3/5}(\log\log x)^{-1/5}))$$

(see, for instance, *A. Walfisz.* Weylsche Exponentialsummen in der neueren Zahlentheorie, 1963).

6. Waring's problem

Given a fixed integer $k>1$, Waring's problem consists in proving the existence of an integer m, such that every integer $N\geqslant 1$ can be written as a sum of m perfect k-th powers of non-negative integers. The problem was solved by Hilbert in 1909. Using their powerful «circle method», Hardy and Littlewood proved the deeper result (1919—1920) that if $r_{k,s}(N)$ denotes the number *of positive integral* solutions (n_j), $j=1,\ldots,s$ of the equation $n_1^k+\ldots+n_s^k=N$, then for any *fixed* $s>(k-2)2^{k-1}+5$,

$$r(N)\equiv r_{k,s}(N)=\frac{\Gamma^s\left(1+\frac{1}{k}\right)}{\Gamma\left(\frac{s}{k}\right)}\sigma N^{\frac{s}{k}-1}+o\left(N^{\frac{s}{k}-1}\right), \tag{6.1}$$

as $N\to\infty$, where σ, «the singular series», is a function of N lying between finite, positive bounds. (If we consider only $k\geqslant 3$, then the condition $s>k2^{k-1}$ will suffice.) Actually

$$\sigma = \sum_{q=1}^{\infty} \frac{1}{q^s} \sum_{\substack{a=1 \\ (a, q)=1}}^{q} (S_{a,q})^s \exp\left(-\frac{2\pi i}{q} aN\right), \tag{6.2}$$

where

$$S_{a,q} = \sum_{h=0}^{q-1} \exp 2\pi i \frac{h^k}{q} a.$$

Associated with the validity of the asymptotic formula (6. 1) is the problem of finding $G(k)$, the least integer such that every integer *beyond a certain stage* is the sum of $G(k)$, perfect k-th powers of non-negative integers, and of finding $g(k)$, the least integer such that *every* positive integer is the sum of $g(k)$ k-th powers of non-negative integers. Clearly $G(k) \leqslant g(k)$. And the number $G(k)$ is more fundamental than $g(k)$ since a finite number of idiosyncratic integers can substantially affect the value of $g(k)$.

By Langrange's fundamental theorem, every positive integer can be represented as a sum of 4 squares, and since no integer of the form $8m+7$ can be the sum of three squares, $g(2)=G(2)=4$. It was proved by Wiefrich (1909) and Kempner (Math. Ann., **72**, 1912, 387) that $g(3)=9$. On the other hand, Landau proved (1909) that $G(3) \leqslant 8$, and L. E. Dickson showed (1939) that 23 and 239 are the only positive integers which require as many as 9 cubes to represent them. Linnik proved in 1942 that $G(3) \leqslant 7$. By elementary reasoning, one can show that $G(k) \geqslant k+1$ for $k \geqslant 2$, so that $G(3)$ is 4, 5, 6 or 7. One does not yet know exactly which.

Hardy and Littlewood were the first (1920—1928) to attack the problem of finding an upper bound for $G(k)$. Their first result was that $G(k) \leqslant (k-2)2^{k-1}+5$, as already mentioned. They improved this later into

$$G(k) \leqslant k 2^{k-2} h(k), \tag{6.3}$$

where $h(k) \to 1$ as $k \to \infty$. Their conjecture that

$$G(k) \leqslant 2k+1$$

except when $k=2^m$, $m>1$, in which case $G(k)=4k$, has been decided, one way or another, only for $k=2$ and $k=4$ (Davenport showed that $G(4)=16$, Ann. Math., **40**, 1939, 731). Thus the problem of finding an upper bound for $G(k)$ is of great interest.

Building on the foundations laid by Hardy and Littlewood, Vinogradov succeeded with his ingenious method (Ann. Math., **36**, 1935, 395) in proving that

$$G(k) \leqslant 6k \log k + (4 + \log 216) k,$$

which he later improved (cf. his book, 1947) into

$$G(k) < k(3 \log k + 11). \tag{6.4}$$

He further proved that if $k \geqslant 12$, and $s \geqslant [10k^2 \log k]$, then the asymptotic formula (6. 1) holds with the error-term

$$O\left(N^{\frac{s}{k}-1-\frac{1}{k^2}}\right). \tag{6.5}$$

Still more recently (cf. his book, 1971), he has improved (6. 4) into

$$G(k) < 2k \log k\,(1 + \delta_k),$$

where $\delta_k \to 0$ as $k \to \infty$, and also proved that (6. 1) holds for

$$s > 4k^2 \log k\,(1 + \delta'_k),$$

where $\delta'_k \to 0$ as $k \to \infty$.

We shall sketch the proofs of (6. 4) and (6. 5) in order to see how his method works. In addition to using his mean-value theorem and inequality mentioned earlier, he introduces in the proof a special kind of double sum in an ingenious way, and his handling of such sums seems to be a basic new idea in the theory of trigonometric sums, capable of yielding quite unexpected results in an «elementary» way.

Let $k \geqslant 3$, $s \geqslant 2k+1$, N a sufficiently large positive integer, and N_0 any positive integer not exceeding N.

Let $W(N_0)$ denote the number of representations of N_0 in the form

$$N_0 = n_1^k + \ldots + n_s^k$$

where $n_1, n_2, \ldots, n_s$ are *positive* integers.

Let $P = [N^{1/k}]$. Then $n_r \leqslant P$, for $r = 1, \ldots, s$.

Let $\tau = 2kP^{k-1}$, and β such that $\frac{1}{4} \leqslant \beta \leqslant 1 - \frac{1}{k}$.

Then

$$W(N_0) = \int_0^1 (L(\alpha))^s \exp(-2\pi i N_0 \alpha) d\alpha = \int_{-1/\tau}^{1-1/\tau} (L(\alpha))^s \exp(-2\pi i N_0 \alpha) d\alpha, \quad (6.6)$$

where

$$L(\alpha) = \sum_{m=1}^{P} \exp 2\pi i \alpha m^k. \quad (6.7)$$

We wish to consider the interval of integration in (6. 6), namely $[-1/\tau, 1 - -1/\tau]$ as the union of two classes of subintervals, one called the *basic intervals*, and the other the class of *supplementary intervals*. To each rational number a/q such that

$$(a, q) = 1, \quad 0 \leqslant a < q, \quad 0 < q \leqslant P^\beta,$$

we make correspond an interval $I(a, q)$ defined as the set of all real numbers α of the form

$$\alpha = \frac{a}{q} + y, \quad |y| \leqslant \frac{1}{q\tau}. \quad (6.8)$$

We call this a *basic interval*. If a/q and a_1/q_1 are two different rational numbers (of the special kind we started with), then

$$\left|\frac{a}{q} - \frac{a_1}{q_1}\right| \geqslant \frac{1}{qq_1} > \frac{1}{q\tau} + \frac{1}{q_1\tau},$$

since $\tau > P^{k-1} > P > q + q_1$. Therefore two basic intervals which correspond to two different rational numbers do *not* overlap. The intervals which remain after the removal of all the basic intervals from the interval $-\frac{1}{\tau} \leqslant \alpha \leqslant 1 - \frac{1}{\tau}$ are called the *supplementary intervals*. It can be shown that if α belongs to a supplementary interval, then

$$\alpha = \frac{a}{q} + y, \qquad |y| \leqslant \frac{1}{q\tau}, \qquad P^{\beta} \leqslant q \leqslant \tau = 2kP^{k-1}. \tag{6.9}$$

Let $W^*(N_0)$ denote that part of the integral in (6.6) which is contributed by *all* the basic intervals. Then

$$W^*(N_0) = \sum_{q \leqslant P^{\beta}} \sum_{a} \int_{-1/q\tau}^{1/q\tau} \left(L\left(\frac{a}{q}+y\right)\right)^s \exp\left(-2\pi i N_0 \left(\frac{a}{q}+y\right) dy\right)$$

$$= \sum_{q \leqslant P^{\beta}} \sum_{a} W^*(N_0, a, q), \text{ say.}$$

Vinogradov shows that if $N_0 \geqslant \frac{1}{2} N$, then

$$W^*(N_0) = \frac{\Gamma^s\left(1+\frac{1}{k}\right)}{\Gamma\left(\frac{s}{k}\right)} N_0^{\frac{s}{k}-1} \sum_{q \leqslant P^{\beta}} A(q, N_0, s) + O\left(P^{s-k-\frac{1}{k}}\right), \tag{6.10}$$

where

$$A(q, N, s) = \sum_{a} q^{-s} \{S(a, q)\}^s \exp\left(-2\pi i \frac{a}{q} N\right),$$

and

$$S(a, q) = \sum_{m=0}^{q-1} \exp 2\pi i \frac{a}{q} m^k.$$

To obtain an upper bound for $G(k)$, we specialize $\beta = \frac{1}{4}$, and $s = 4k$, and define a sum

$$S(\alpha) = \sum_{u} \exp 2\pi i \alpha u, \tag{6.11}$$

where u runs through U *distinct* integers, $\left(\frac{1}{5} P\right)^k < u < \left(\frac{1}{2} P\right)^k$, and each $u = \xi_1^k + \ldots + \xi_j^k$, where $\xi_r > 0$, and ξ_r is an integer for $1 \leqslant r \leqslant j$. They *can* be so chosen that

$$U \gg P^{k-\frac{1}{12}} \text{ (by suitable choice of j).} \tag{6.12}$$

Let $P_0 = [P^{1/2}]$, and determine a set U_0 of distinct integers u_0 in the same way as u, but with P_0 in place of P, and j_0 in place of j, so that

$$U_0 \gg P^{\frac{k}{2} - \frac{1}{12}\left(1-\frac{1}{k}\right)} \text{ (by suitable choice of } j_0). \tag{6.13}$$

This again *can* be done.

Finally let v run through the primes in the interval $\frac{1}{2}P_0 \leqslant v \leqslant P_0$ and V denote the number of such v's. Then

$$V \gg \frac{P_0}{\log P_0} \gg P^{\frac{1}{2}-\varepsilon}. \tag{6.14}$$

Now define the double sum

$$Q(\alpha) = \sum_v \sum_{u_0} \exp 2\pi i (\alpha v^k u_0). \tag{6.15}$$

Vinogradov proves that if α belongs to any *supplementary interval*,

$$Q(\alpha) \ll U_0^{1/2} P^{\frac{1}{4}k+\frac{3}{8}+\varepsilon} \tag{6.16}$$

in comparison with the trivial estimate $Q(\alpha) = O(VU_0)$. This estimate is crucial to the whole proof.

On the other hand, since the u's are distinct,

$$\int_0^1 |S(\alpha)|^2 d\alpha = U. \tag{6.17}$$

Now consider $I(N)$, which is the number of representations of N as

$$N = n_1^k + \ldots + n_s^k + u + u' + v^k u_0 \quad (n_r > 0,\ 1 \leqslant r \leqslant s),$$

where u' satisfies the same condition as u (defined above). N is a sum of $(s+2j+j_0)$ k-th powers of *positive* integers satisfying certain conditions. We shall prove that $I(N) > 0$ for *all large* N ,from which it will follow that

$$G(k) \leqslant s + 2j + j_0,$$

where $s = 4k$, and j, j_0 have been so chosen in (6.12) and (6.13) that

$$G(k) < k(3 \log k + 11).$$

Now

$$I(N) = \int_{-1/\tau}^{1-1/\tau} (L(\alpha))^s Q(\alpha) (S(\alpha))^2 \exp(-2\pi i N\alpha)\, d\alpha,$$

where $L(\alpha)$ is defined as in (6.7), and $Q(\alpha)$, $S(\alpha)$ as in (6.11), (6.15). We separate $I(N)$ into two parts:

$$I(N) = I^*(N) + I^{**}(N),$$

where I^* is the contribution from *all* the basic intervals, and I^{**} from the supplementary intervals.

Because of the estimate of Q given in (6.16), and the trivial estimate $|L(\alpha)| \leqslant P$, we have

$$I^{**}(N) = O\left(P^s U_0^{1/2} P^{\frac{1}{4}k+\frac{3}{8}+\varepsilon}\right) \int_0^1 |S(\alpha)|^2 d\alpha.$$

Since the U numbers u are *distinct*, $\int_0^1 |S(\alpha)|^2 d\alpha = U$. Hence

$$I^{**}(N) = O\left(UU_0^{1/2} P^{s + \frac{k}{4} + \frac{3}{8} + \varepsilon}\right).$$

By (6.12), (6.13), and (6.14), however, we have

$$VU_0^{1/2} U \gg P^{\frac{5}{4}k + \frac{3}{8} + \frac{1}{24k} - \varepsilon}.$$

Hence

$$I^{**}(N) = O\left(VU_0 U^2 P^{s-k-\frac{1}{25k}}\right). \tag{6.18}$$

The basic intervals, on the other hand, give the contribution

$$I^{*}(N) = \sum_{N_0} W^{*}(N_0), \tag{6.19}$$

where $N_0 = N - v^k u_0 - u - u'$, and the summation is over all permissible values of v, u_0, u, u'. By formula (6.10), with $\beta = 1/4$, $s = 4k$, we have

$$I^{*}(N) = C_k \sum_{N_0} N_0^{s/k-1} \sum_{0<q\leqslant P^{1/4}} A(q_0, N_0, s) + O\left(VU_0 U^2 P^{s-k-\frac{1}{k}}\right), \tag{6.20}$$

where C_k is a positive constant depending only on k. By elementary reasoning, $A(q) = O(q^{1-s/k}) = O(q^{-3})$, so that

$$\sum_{q>P^{1/4}} A(q, N_0, s) = O\left(\sum_{q>P^{1/4}} q^{-3}\right) = O(P^{-1/2}).$$

Hence

$$I^{*}(N) = C_k \sum_{N_0} N_0^{s/k-1} \sigma(N_0, s) + O\left(VU_0 U^2 P^{s-k-\frac{1}{k}}\right). \tag{6.21}$$

Here the «singular series» σ (see (6.2)) converges for $s = 4k$, and lies between two positive, finite bounds. The number of numbers N_0 is VU_0U^2, and each of them is

$$> N - P_0^k \left(\frac{1}{2} P_0\right)^k - \left(\frac{1}{2} P\right)^k - \left(\frac{1}{2} P\right)^k \gg P^k.$$

Hence the first term on the right-hand side of (6.21) is

$$\gg VU_0 U^2 P^{s-k}. \tag{6.22}$$

It follows that $I(N) > 0$ for all N large enough, hence

$$G(k) < k(3 \log k + 11),$$

as claimed in (6.4).

The proof of the asymptotic formula (6.1) with the error-term given in (6.5) runs along the same lines, but requires the use of the following two propositions:

A. Let $P > 1$ be an integer, $k \geqslant 12$, $h = k + 2$, $b = \left[\frac{5}{4}k + \frac{1}{2}\right]$,

$$\sigma = \left(1 - \frac{1}{k}\right)^k, \qquad f(x) = a_k x^k + \ldots + a_1 x,$$

where $a_k, \ldots, a_1$ are fixed, *positive* integers, and

$$T(\alpha) = \sum_{m=1}^{P} \exp 2\pi i \alpha f(m).$$

Then

$$\int_0^1 |T(\alpha)|^{2b(k+h)} d\alpha = O\left(P^{2b(k+h)-k+\frac{1}{2}k(k+1)\sigma}\right).$$

B. Let

$$\left|\alpha - \frac{a}{q}\right| \leqslant \frac{1}{q^2}, \ (a, q) = 1, \ P_1^{1-1/k} < q \leqslant 2kP_1^{k-1}, \ P_1 \text{ a positive integer,}$$

$$T_1(\alpha) = \sum_{m=1}^{P} \exp 2\pi i \alpha m^k.$$

Then

$$T_1 = O(P^{1-\rho}), \qquad \rho = \frac{1}{3k(k-1)\log(12k^2)}.$$

Vinogradov's upper bound for $G(k)$, combined with the elementary result that $g(k) \geqslant 2^k + [(3/2)^k] - 2$ for $k \geqslant 2$, can be used, in some cases, to determine $g(k)$. Thanks to the work of S. S. Pillai (J. Indian Math. Soc., (2) **2**, 1936, 16; Proc. Indian Acad. Sci., **(A)** 4, 1936, 261) and of L. E. Dickson (Amer. J. Math., 58, 1936, 521), supplemented by that of Pillai (Proc. Indian Acad. Sci., **(A) 12**, 1940, 30), R. K. Rubugunday (J. Indian Math. Soc., **(2) 6**, 1942, 192), I. Niven (Amer. J. Math., **66**, 1944, 137), and Chen, Jing-run (Chinese Math. Acta, **6**, 1965, 105), the value of $g(k)$ has been determined for all $k \geqslant 5$. The exact formula for $g(k)$, $k \geqslant 5$, depends on whether B is less than or greater than $2^k - A$, where

$$A = \left[\left(\frac{3}{2}\right)^k\right], \qquad B = 3^k - 2^k A.$$

In the former case, we have $g(k) = 2^k + A - 2$. K. Mahler has shown (Mathematika, 4, 1957, 122) that the latter case can occur for at most a finite number of values of k. The exact value of $g(4)$ is not yet known. Dickson has shown (Bull. Amer. Math. Soc., 39, 1933, 701—727) that $g(4) \leqslant 35$. In the case of $G(k)$, however, we know $G(2)$ and $G(4)$ but no others.

7. Waring's problem for algebraic number fields

Let $\mathbf{K}$ be an algebraic number field of degree n, and D the absolute value of the discriminant of $\mathbf{K}$. Let $\mathbf{K}^{(1)}, \ldots, \mathbf{K}^{(n)}$ be the n conjugate fields, $\mathbf{K}^{(l)}$, $l = 1, \ldots, n_1$ being real, and $\mathbf{K}^{(l+n_2)}$, $l = n_1 + 1, \ldots, n_1 + n_2$ conjugate complex, with $n_1 + 2n_2 = n$. A number ν of $\mathbf{K}$ is called *totally positive* if $\nu^{(l)} > 0$, $l = 1, \ldots, n_1$.

Siegel proved in 1921 the theorem, first stated by Hilbert, that every totally positive integer in $\mathbf{K}$ can be expressed as a sum of four squares of numbers from the same field (Math. Z., **11**, 1921, 246; Ges. Abh., (i), 47). For the proof he used the Hilbert—Furtwängler law of quadratic reciprocity. He went on to consider the problem of representing every totally positive integer in $\mathbf{K}$ as a sum of squares of *integers* from the same field, and proved the decisive result that: if the field $\mathbf{K}$ is totally real, and *every* totally positive integer is a sum of integral squares, then $\mathbf{K}$ is either the rational number field **Q**, or the real quadratic field **Q** $(\sqrt{5})$; and if the field **K** is not totally real, then *every* totally positive integer is a sum of *integral* squares if and only if the discriminant of **K** is odd (Ann. Math., **46**, 1945, 313; Ges. Abh., (iii), 12). In 1944, however, Siegel considered the analogue of Waring's problem (Amer. J. Math., **66**, 1944, 122; Ges. Abh., (ii), 406) using Vinogradov's idea of finite trigonometric sums instead of the Hardy—Littlewood idea of power series. Since the number of totally positive integral solutions $m_1, \ldots, m_s$ of the equation

$$\nu = m_1^k + \ldots + m_s^k$$

is not necessarily finite, if $n_2 > 0$, one restricts the solutions by the condition

$$|m_j^{(l)}|^k < |\nu^{(l)}|, \quad j = 1, \ldots, s; \qquad l = n_1 + 1, \ldots, n_1 + n_2,$$

and denotes the number of such solutions by $A(\nu)$.

Whereas in the field of rational integers, it is trivial that any positive integer ν is a sum of k-th powers of integers, namely ν times 1^k, the corresponding statement is not true without restriction for an arbitrary field. One therefore introduces a ring J_k generated by the k-th powers of all integers in **K**. It consists of all numbers $a_1\lambda_1^k + \ldots + a_h\lambda_h^k$, $h = 1, 2, \ldots$, where $\lambda_1, \ldots, \lambda_h$ are integers in **K**, and $a_1, \ldots, a_h$ are rational integers. Obviously $A(\nu) = 0$ if $\nu \notin J_k$. (J_k is an *order*). Siegel proved that *for any fixed* $s > (2^{k-1} + n)nk$, *and M tending to* ∞, *where M is the norm of the totally positive integer ν in* **K**, *we have*

$$A(\nu) = D^{\frac{1-s}{2}} \sigma_0 \sigma M^{s/k-1} + o(M^{s/k-1}),$$

where σ_0 is a positive number depending only on n_1, n_2, s, k, *and*

$$\sigma_0 = \left(\frac{\Gamma^s\left(1 + \frac{1}{k}\right)}{\Gamma(s/k)} \right)^n, \quad \text{if } n_2 = 0.$$

The «singular series» σ lies between finite, positive bounds, whenever $\nu \in J_k$, and $\sigma = 0$ otherwise.

In consequence, all totally positive numbers of J_k with sufficiently large norm are sums of a bounded number of k-th powers of *totally positive* integers in **K**.

Siegel raised the question whether the restriction $s > (2^{k-1}+n)nk$ could be replaced by the condition $s > c(k)$, where $c(k)$ is a function of k alone (that is, independent of n). C. P. Ramanujam (Mathematika, **10**, 1963, 137), and B. J. Birch, independently of each other, answered the question in the affirmative. Thus any totally positive integer of sufficiently large norm in any algebraic number field, belonging to the order generated by the k-th powers of integers of this field, is a sum of at most max $(2^k+1,\ 8k^5)$ k-th powers of totally positive integers of this field.

One would like to know the analogues of Vinogradov's results in the case of arbitrary algebraic number fields.

8. Goldbach's problem and Vinogradov's theorem

The problem dates from 1742, and consists in showing that every even integer > 2 is a sum of two primes. This is equivalent to proving that *every* integer > 5 is a sum of three primes. The first analytical attack on the problem was made by Hardy and Littlewood (Acta Math., 44, 1922, 1; Hardy's Collected Papers, (i), 561). On the basis of an unproved hypothesis concerning the location of the zeros of Dirichlet's L-functions, they proved that if $I(N)$ denotes the number of representations of an *odd*, positive integer N as a sum of 3 primes, then

$$I(N) \sim C \frac{N^2}{(\log N)^3} \prod_{p/N} \frac{(p-1)(p-2)}{p^2-3p+3},$$

where p runs through all (odd) prime factors of N, and

$$C = \prod_{\text{all odd } p} \left(1 + \frac{1}{(p-1)^3}\right).$$

The unproved hypothesis is that there exists a number $\theta < 3/4$, such that all zeros of all L-series $L(s, \chi)$, $s = \sigma + it$ formed with Dirichlet's characters χ lie in the half-plane $\sigma \leqslant \theta$.

The stimulus for Hardy and Littlewood's analytical attack came from Viggo Brun's result (1920), obtained by the «sieve» method, that every positive, even integer is representable as a sum of two integers, each of which is the product of not more than 9 primes. Many improvements of this result have since been made, culminating in A. A. Buhštab's result that every sufficiently large even integer is a sum of a prime, and an integer with not more than 3 prime factors. (Uspehi Math. Nauk, **22**, 1967, № 3 (135), 199; Russian Math. Surveys, **22**, 1967, 205).

Supplementing Viggo Brun's method with his own arguments about the density of sequences of integers, L. G. Schnirelman proved (1930) that *every* integer > 1 is representable as a sum of a *bounded* number of primes. Schnirelman's bound was 800.000. Yin, Wen-Lin has since proved that every *large* integer is a sum of at most 18 primes (Bull. Acad. Polon. Sci., cl. III, 4, 1956, 793).

Between 1935 and 1937, however, T. Estermann made the important observation that a result due to A. Page (Proc. Lond. Math. Soc., (2) **39**, 1935, 116) on the estimation of $\pi(x; k, l) = \sum_{p \leqslant x,\ p \equiv l \pmod k} 1$, (where k and l are integers, $k > 0$, $(k, l) = 1$ and p a prime) can be used to obtain an asymptotic formula for that part of the *integral for* $I(n)$ which is contributed by the «basic intervals». For, clearly

$$I(N) = \int_{-1/\tau}^{1-1/\tau} (S(\alpha))^3 \exp(-2\pi i \alpha N)\, d\alpha,$$

where

$$S(\alpha) = \sum_{p \leqslant N} \exp 2\pi i \alpha p.$$

This leads to interesting results in additive number theory, in particular to the proof that every sufficiently large integer is a sum of two primes and a square (Proc. Lond. Math. Soc., (2) **42**, 1937, 501), and that every sufficiently large odd integer is a sum of two primes and a third integer which is a product of two primes, and so on.

It was, however, Vinogradov who proved in 1937 (Rec. Math. Moscow, (2) **2**, 1937, 179) that every sufficiently large odd integer is a sum of 3 primes, as a particular case of an asymptotic formula for $I(N)$, where $N > 0$, and N is odd:

$$I(N) = \frac{N^2}{2(\log N)^3} \sum (N) + O\left(\frac{N^2}{(\log N)^{7/2-\varepsilon}}\right), \qquad \varepsilon > 0,$$

where

$$\sum(N) = \prod_{\text{all } p}\left(1 + \frac{1}{(p-1)^3}\right) \prod_{p/N}\left(1 - \frac{1}{p^2 - 3p + 3}\right),$$

and we have

$$\sum(N) > \frac{6}{\pi^2}, \quad \text{for all } N.$$

It follows that there exists an integer C_0, such that every odd integer $N \geqslant C_0$ is a sum of 3 primes.

Vinogradov's orginal proof has been simplified by T. Estermann (Introduction to modern prime number theory, Cambridge, 1952), in the sense that for treating the «basic intervals» he uses the theorem of Siegel (Acta Arithm., **1**, 1935, 83; Ges. Abh., (i), 21), which states that: *Given* $\varepsilon > 0$, *there exists a number* $A = A(\varepsilon) > 0$, *such that if* $k > A$, *and* χ *a real, non-principal character modulo* k, *then* $L(s, \chi)$ *has no zeros on the real axis between*

$$1 - \frac{1}{k^\varepsilon} \textit{ and } 1.$$

From this one can deduce, as Walfisz (Math. Z., **40**, 1936, 593) has done, that

$$\pi(x;\ k,\ l) - \frac{\operatorname{li}(x)}{\varphi(k)} = O(x \exp(-c\sqrt{\log x})),$$

as $x \to \infty$, uniformly for $3 \leqslant k \leqslant \log^\alpha x$, where α is any positive number however large, c a positive *absolute* constant, and φ denotes Euler's function.

This estimate suffices to take care of the «basic intervals». For the «supplementary intervals», however, one needs Vinogradov's ingenious method for estimating sums such as

$$S(\alpha, v) = \sum_{p \leqslant v} \exp 2\pi i \alpha p,$$

where $v \geqslant 0$, p a prime, α real. Vinogradov proves that if $n \geqslant n_0$, and $\frac{n}{\log^3 n} < v \leqslant n$, $\log^{15} n < q \leqslant \frac{n}{\log^{15} n}$, then

$$S\left(\frac{h}{q},\ v\right) \leqslant \frac{n}{\log^3 n},$$

where h, q are integers, $0 < h \leqslant q$, $(h, q)=1$. (Rec. Math., (2) 2, 1937, 179). U. V. Linnik has given a proof of the theorem of Vinogradov (Rec. Math. N. S., **19 (61)**, 1946, 3) by estimating the number of zeros of all L-functions of a given modulus in a particular region. Some improvement of Linnik's method has been made by N. Tchudakoff (Ann. Math., 48, 1947, 515). Linnik has also proved a conjecture of Hardy and Littlewood that *every* large integer is a sum of 2 squares and a prime. (The dispersion method for binary additive problems, Leningrad, 1961).

Combining Waring's problem with Goldbach's problem, and extending Vinogradov's method, L. K. Hua (loc. cit.) has obtained a number of results on the number of representations of an integer as a sum of prime powers.

Let k be positive integer, p a prime, $p^\theta \mid k$, $p^{\theta+1} \nmid k$, and let

$$\gamma = \begin{cases} \theta+2, & \text{if } p=2,\ 2 \mid k, \\ \theta+1, & \text{otherwise,} \end{cases} \qquad K = \prod_{(p-1)\mid k} p^\gamma.$$

Then Hua (loc. cit., p. 106) has proved that *every sufficiently large* integer $N \equiv s \pmod K$ *is a sum of s k-th powers of primes, if $s \geqslant s_0$; where*

$$s_0 \geqslant \begin{cases} 2^k+1, & \text{for } 1 \leqslant k \leqslant 10, \\ 2k^2\left(2\log k + \log\log k + \frac{5}{2}\right), & \text{for } k > 10. \end{cases}$$

The case $k=1$, $K=2$, $s=3$ corresponds to Vinogradov's theorem.

Vinogradov's estimate of sums of the type $\sum\limits_p \exp 2\pi i\alpha p$, where α is real, and p is a prime, depends on rational approximations to α. He has applied the method to more general sums of the type

$$\sum_p \exp 2\pi i f(p)$$

where $f(p)$ is a polynomial in p of degree > 1, in which case the estimates depend on rational approximations to some coefficient of a power of p greater than the first. He has also considered the case in which f is a function which approximates to a polynomial. This leads to a study of the distribution of the fractional parts of the value of $f(p)$ which has interesting arithmetical consequences. (cf. Vinogradov's book, English translation, loc. cit., 1954).

9. More general exponential sums

The scope for generalizing Vinogradov's method is far from exhausted. The order of magnitude of a relatively simple exponential sum like $\sum\limits_{n \leqslant T} a_n \exp 2\pi i n x$, where the coefficients are arithmetically significant, seems to be of some interest. For example, if a_n is the coefficient of the Dirichlet series representing a zeta-function of an ideal class in a quadratic number field, the order of magnitude of such a sum is related to the existence of an infinity of zeros of the corresponding zeta-function on the «critical line».

Specifically, for every irrational x, we have

$$\sum_{m \leqslant T} a_m \exp 2\pi i m x = o(T), \qquad \text{as } T \to \infty,$$

by Weyl's inequality, and this implies that $\zeta_K\left(\frac{1}{2}+it, \mathbf{C}\right)$ vanishes for an infinity of values of t. Here $\mathbf{K}$ is a quadratic number field, $\mathbf{C}$ an ideal class in $\mathbf{K}$. So also the function $\sum_{j=1}^{r} \alpha_j \zeta_K(s, \mathbf{C}_j)$, where α_j is real, and $\mathbf{C}_j$ an ideal class in a field $K = \mathbf{Q}(\sqrt{\pm d})$. This was proved in the case of the zeta-function of an ideal class in an imaginary quadratic field by E. Hecke by means of the properties of automorphic functions (München. Sitzungsbericht, II, 8, 1937, 73). The result in the case of a real quadratic field follows from the above estimate (*K. Chandrasekharan*, *R. Narasimhan*. Commentarii Math. Helvetici, 43, 1968, 18). The corresponding question in the case of fields of higher degree than 2 is open.

The estimate of Kloosterman's sums obtained by André Weil (Proc. Nat. Acad. Sci. USA, 34, 1948, 204) suggests a study of exponential sums over finite fields, à la Vinogradov, and Karacuba's proof of Vinogradov's mean-value theorem also suggests a similar generalization.

Kloosterman's sum is defined as (Acta Math., 49, 1926, 407)

$$S(u, v, q) = \sum_{h=1}^{q} \exp 2\pi i \left(\frac{uh + v\bar{h}}{q}\right),$$

where q is an integer > 0, $(h, q) = 1$, $h\bar{h} \equiv 1 \pmod q$, and Weil's estimate is that

$$|S(u, v, q)| = O(q^{1/2}), \quad q \text{ prime}.$$

This can be used, among other things, to obtain the estimate

$$\tau(n) = O\left(n^{6-\frac{1}{4}+\varepsilon}\right), \qquad \varepsilon > 0,$$

for Ramanujan's τ-function, which is defined by the relation

$$\sum_{n=1}^{\infty} \tau(n) x^n = x \left(\prod_{n=1}^{\infty} (1 - x^n)\right)^{24}, \qquad |x| < 1.$$

The conjectured estimate is that $\tau(n) = O(n^{11/2+\varepsilon})$ for every $\varepsilon > 0$.

Trudy Mat. Inst. Steklov.
132 (1973)

Proc. Steklov Inst. Math.
132 (1973)

RECENT WORKS OF I. M. VINOGRADOV

Ju. V. LINNIK

ABSTRACT. This article contains a brief survey of the results obtained by I. M. Vinogradov in the period 1951–1971.

Bibliography: 2 items.

The work of I. M. Vinogradov before 1951 is largely summarized in his *Selected Works* (Academy of Sciences of the USSR, Moscow, 1952). The recent book [1] contains in the most complete form the results obtained by the Vinogradov method in several areas of analytic number theory. However, this book presents far from all the important results obtained by the author in the interval between the two publications. In particular, it does not have his new investigations of the order of growth of the functions $G(n)$ in Waring's problem, of the zeros of the Riemann zeta-function, of the asymptotic theory of characters with increasing modulus, or other questions. The present essay gives a brief exposition both of the work in [1] and of other important papers.

First of all, in [1] we find a fundamental new result on Weil sums. We set

$$T = T(\alpha_n, \ldots, \alpha_1) = \sum_{0<x\leqslant P} \exp 2\pi i f(x),$$

where $f(x) = a_n x^n + \cdots + a_1 x$; $a_n, \cdots, a_1$ are real numbers, and x runs through integral values.

Let $n \geqslant 12$. We divide the points of n-dimensional space R^n into two classes. The first class contains all points of the form $(p_n/q_n + z_n, \cdots, p_1/q_1 + z_1)$, where the p_i/q_i are irreducible rational numbers with positive denominators whose least common multiple does not exceed P^ν $(\nu = 1/n)$, and $|z_s| < P^{-s+\nu}$ $(s = 1, \cdots, n)$. The second class is the complement of the first class in R^n.

Suppose that $T_m = T(ma_n, ma_{n-1}, \cdots, ma_1)$; $m > 0$ *is an integer. Then, setting*

$$\rho = \frac{1}{8n^2(\ln n + 0.5\ln\ln n + 1.3)},$$

we obtain the following estimate for the points of the second class for $m \leqslant P^{2\rho}$:

$$T_m \ll P^{1-\rho}.$$

AMS (MOS) subject classifications (1970). Primary 10–02, 10G05, 10H25; Secondary 10–03, 10H05, 01A65.

Further, setting $\delta_0 = \max(|z_n|P^n, \cdots, |z_1|P)$, *for* $m \leqslant P^{4\nu^2}$ *we obtain*

$$T_m \ll P(m, Q)^{\nu} Q^{-\nu+\varepsilon}$$

for any $\epsilon > 0$, *and also, if* $\delta_0 \geqslant 1$,

$$T_m \ll PQ^{-\nu+\varepsilon}\delta_0^{-\nu}.$$

The measure of the points of the first class in the unit cube of R_n *has the estimate* $P^{1+2\nu-n(n+1)/2}$.

This very general result is the deepest that we now know about Weil sums; it was the crowning achievement of Vinogradov's work in estimating Weil sums in the course of 36 years. It has applications for sums of the more general form

$$S = \sum_{n < u \leqslant n_1} \exp 2\pi i F(u),$$

where $F(u)$ is not even a polynomial. For example, we may take $F(u) = -t \ln u/2\pi$, which gives important facts concerning the zeros of the Riemann ζ-function. These estimates can be directly applied to the theory of the distribution of the fractional parts of the values of polynomials ([1], Chapter 5), where they also give the deepest information currently available on this question.

In Chapter 6 of [1], these estimates are used to prove an asymptotic formula for Waring's problem

$$N = x_1^n + \ldots + x_r^n, \qquad x_i > 0, \quad n > 12,$$

valid for

$$r \geqslant 2[n^2(2\ln n + \ln\ln n + 3)].$$

Chapter 7 of [1] examines the sum T', which differs from the sum T in that x only runs through prime values. An estimate is obtained for T' which differs only slightly from that for T. In Chapter 8 these estimates lead to remarkable results concerning the distribution of the fractional parts of a polynomial in the case when the argument runs through the prime numbers, and in Chapter 9 they lead to asymptotic information for the solution of Waring's problem with prime numbers

$$N = p_1^n + \ldots + p_r^n, \qquad n > 12,$$

with a lower limit for r which is almost the same as the earlier one, i.e.

$$r \geqslant [2n^2(2\ln n + \ln\ln n + 3)].$$

We now proceed to the work not presented in [1] and published by Vinogradov in the interval between the two publications mentioned above, for the most part in "Dokl. Akad. Nauk SSSR" and "Izv. Akad. Nauk SSSR Ser. Mat.".

In 1959 a new estimate for $G(n)$ was obtained in Waring's problem:

$$G(n) < 2n \ln n(1 + o(1)) \quad \text{as} \quad n \to \infty.$$

It should be noted that, ever since the first estimate of this order for $G(n)$ obtained by Vinogradov (1934), many authors have worked on improving it, but no one was able to. This new estimate of Vinogradov for $G(n)$ is unsurpassed.

In 1958 the new estimates of Weil sums and their generalizations were applied to give a new boundary for the region where $\zeta(s) = \zeta(\sigma + it)$ is known not to have zeros, namely

$$\sigma \geqslant 1 - \frac{A}{(\ln t)^{2/3}},$$

where A is a constant. This gives the following expression for the prime number theorem:

$$\pi(x) = \int_2^x \frac{du}{\ln u} + O(x \exp(-c_1 (\ln x)^{0.6}));$$

$c_1 > 0$ is a constant.

Vinogradov's investigations into the asymptotic behavior of Dirichlet characters are of great importance. As early as 1952, he obtained an estimate for the sum of the Dirichlet characters of shifted primes $\Sigma_{p \leqslant N} \chi(p + h)$, which gave a degree of decrease relative to N as soon as $N > q^{0.75+\epsilon}$, $\epsilon > 0$, where q is the modulus of the character. This has fundamental significance, since it is deeper than what can be obtained by direct application of the extended Riemann hypothesis, and in this direction seems to carry a deeper truth than that hypothesis (if the hypothesis is valid). Recently A. A. Karacuba has been able to improve this estimate.

In 1964 a deep estimate was obtained for the well-known summation formula for the class number $h(-t)$, which goes back to Gauss. Vinogradov's result is as follows:

$$\sum_{t \leqslant n} h(-t) = \frac{4\pi}{21\zeta(3)} n^{3/2} - \frac{2}{\pi} n + O(n^{2/3+\varepsilon}).$$

In the same year the best remainder term now known for the problem of integral points in the ball $x^2 + y^2 + z^2 \leqslant a^2$ was obtained: $O(a^{4/3}(\ln a)^6)$. In this same period several fundamental works were completed on elementary methods in analytic number theory. The article [2] was especially remarkable. Here, instead of trigonometric sums, the idea of estimating simple sums is applied to the elementary step-function: $\psi(z)$ is periodic with period 1; $\psi(z) = 1 - \sigma$ $(0 \leqslant z < \sigma)$; $\psi(z) = -\sigma$ $(\sigma < z < 1)$, so that its average value on $[0, 1]$ is zero. Using this simple function in many ways alters the trigonometric sums, and seems to open up a new approach in analytic number theory for obtaining strong analytic results without using imaginary numbers or analysis.

The recent results of Vinogradov have been used by many authors both here and in other countries to solve new problems, but they have very rarely managed to improve upon these results. The methods of Vinogradov have been applied outside number theory–in probability theory (expansion of probability laws, limit theorems) and in some problems in integral programming.

BIBLIOGRAPHY

1. I. M. Vinogradov, *The method of trigonometrical sums in the theory of numbers,* rev. ed., "Nauka", Moscow, 1971. (Russian)

2. ————, *An elementary proof of a theorem from the theory of prime numbers,* Izv. Akad. Nauk SSSR Ser. Mat. **17** (1953), 3–12. (Russian) MR **15**, 855.

Translated by N. KOBLITZ

Trudy Mat. Inst. Steklov.
132 (1973)

Proc. Steklov Inst. Math.
132 (1973)

SOME EXPONENTIAL SUMS*

L. J. MORDELL([1])

ABSTRACT. Precise expressions, and also estimates, are given for some exponential sums.

Let $f(x)$ be a polynomial in x, say,

$$f(x) = a_0x^n + a_1x^{n-1} + \dots + a_{n-1}x,$$

where the a are integers. We may suppose $a_0 \not\equiv 0 \pmod p$ where p is an odd prime. Write $e(X) = e\left(\frac{2\pi i}{p} X\right)$, and([2]) if $X \not\equiv 0 \pmod p$, define $1/X$ as $\bar{X}$ where $X\bar{X} \equiv 1$.

The study of the exponential sum

$$S = \sum_{x=0}^{p-1} e(f(x))$$

has led to results and developments of the greatest importance revealing new and unexpected fields of study and research whose possibilities have not yet been exhausted.

When $n=2$, S becomes a Gaussian sum for which there is the classical evaluation

$$S = \sqrt{p}\, i^{\left(\frac{p-1}{2}\right)^2} e\left(-\frac{a_1^2}{4a_0}\right).$$

When $n > 2$, simple closed expressions such as this do not, in general, exist for S. The problem now is to find estimates of the form

$$|S| = O(p^\alpha), \quad \text{or say} \quad |S| < kp^\alpha,$$

where k is independent of the coefficients of $f(x)$, $\alpha = \alpha(n)$ and usually n is a fixed number independent of p and p is large.

A general result was given in 1932 by Mordell who found by elementary means an estimate $\alpha = 1 - \frac{1}{r}$. Then Weil by a very deep argument found the best possible result,

$$|S| < (n-1)p^{1/2}.$$

A specially interesting case arises when

AMS (MOS) subject classifications (1970). Primary 10G05.

**Editor's note.* The body of this article is an unaltered reprinting of the original paper. Details of the bibliography have been checked and put in standard form, and MR citations have been added.

([1]) Professor Mordell died on 12 March, 1972.

([2]) We shall hereafter omit the mod p in congruence.

$$f(x) = a_0 x^{p-2} + a_{p-3} x.$$

Since $x^{p-1} \equiv 1$ if $x \not\equiv 0$, it suffices to consider the sum in the form

$$S_1 = \sum_{x=1}^{p-1} e\left(ax + \frac{b}{x}\right), \qquad ab \not\equiv 0.$$

This is the well known Kloosterman sum for which another difficult proof gives

$$|S_1| < 2\sqrt{p}.$$

It is fairly easy to find an estimate,

$$|S_1| < kp^{2/3},$$

but no method seems to be known for an improvement on $p^{2/3}$ other than $p^{1/2}$.

More general results for exponential sums have been found by Weil, Perelman, Bombieri, who give the best possible estimates $O(\sqrt{p})$ for sums([3]) of the form

$$S_2 = \sum_x{}' \chi(P)\, e(Q),$$

where P and Q are rational functions of x and χ is a multiplicative character mod p.

Closed expressions do not exist for sums such as S_2 except in some classical cases and a few special ones. Salié found a very simple instance related to the Kloosterman sum, namely,

$$S_3 = \sum_x{}' e\left(ax + \frac{b}{x}\right)\left(\frac{x}{p}\right),$$

where (x/p) is the Legendre symbol. We may suppose that a and b are not both congruent to zero.

I have found [1] a more perspicuous proof of his result, which now has many applications, and which I state as follows. Write

$$S_4 = \sum_h e(h),$$

where the sum in h is taken over the roots of the congruence

$$h^2 \equiv 4ab.$$

The sum is zero if $\left(\frac{ab}{p}\right) = -1$. Then we can write

$$pS_4 = \sum_{t,h} e(h + t(h^2 - 4ab)),$$

since this sum in t is zero unless $h^2 \equiv 4ab$ when it is p. The last sum in h is a Gaussian sum. The contribution given by $t = 0$ is zero. Hence

([3]) The summations for all variables hereafter will be 0 to $p-1$ unless otherwise specified. A Σ' denotes that those values of the variables for which the summand is undefined are excluded.

$$pS_4 = \sqrt{p}\, i^{\left(\frac{p-1}{2}\right)^2} \sum_t{}' e\left(-4abt - \frac{1}{4t}\right)\left(\frac{t}{p}\right).$$

If $a \not\equiv 0$, write now $t \to -\frac{1}{4at}$. Then

$$pS_4 = \sqrt{p}\, i^{\left(\frac{p-1}{2}\right)^2} \left(\frac{-a}{p}\right) S_3,$$

and so

$$S_3 = \sqrt{p}\, i^{\left(\frac{p-1}{2}\right)^2} \left(\frac{a}{p}\right).$$

If $b \not\equiv 0$, we put $t \to -\frac{t}{4b}$ and so we replace (a/p) by (b/p) in the last formula. Similarly [2] by considering a sum

$$S_5 = \sum_h e(Ah)\left(\frac{h}{p}\right),$$

where the summation is extended over the solutions of the congruence

$$Bh + \frac{c}{h} \equiv D$$

we are led to a simple closed expression for the sum of a series of the type

$$S_6 = \sum{}' e\left(\frac{ax+b}{x^2+c}\right)\left(\frac{x^2+c}{p}\right).$$

In general, best possible estimates for exponential sums in several variables do not exist, but my form of Salié's result proves useful [1] for some special sums and gives best possible results. An instance is the

Theorem. *Let $f(X)$ be a rational function of X. Then*

$$S_7 = \sum_{x,\,y}{}' e(ax + by + f(xy))\left(\frac{x}{p}\right) = O(p).$$

The sum is zero if both $a \equiv 0$, $b \equiv 0$. The contribution to the sum is zero when $x \equiv 0$. Write $y \equiv v/x$, $x \not\equiv 0$. Then

$$S_7 = \sum_{x,\,v}{}' e\left(ax + \frac{bv}{x} + f(v)\right)\left(\frac{x}{p}\right).$$

The contribution when $v \equiv 0$ is $O(\sqrt{p})$. The sum in x gives

$$S_7 = i^{\left(\frac{p-1}{2}\right)^2} \sqrt{p}\, e(h + f(v))\left(\frac{c}{p}\right),$$

where $c = a$ or bv according as $a \not\equiv 0$, or $a \equiv 0$ and $h^2 \equiv 4abv$. Hence S_7 is an exponential sum in h and so $S_7 = O(p)$.

A particular case of some interest is given by

$$S_8 = \sum_{x,\,y}{}' e\left(ax + by + \frac{c}{xy}\right)\left(\frac{x}{p}\right) = O(p).$$

It has been conjectured that

$$S_9 = \sum_{x,\, y}{}' e\left(ax + by + \frac{c}{xy}\right) = O(p).$$

This has not been proved but Davenport, Hooley and Carlitz have found the result

$$S_9 = O(p^{5/4}).$$

Another application [3] is to the series

$$S_{10} = \sum_{x,\, y} e\,(ax + by + cxy + dx^2y^2)\left(\frac{x}{p}\right).$$

If this is summed for y and the result compared with that obtained from the proof of the theorem, we are led to a result which suggests the interesting

Problem. *To find two rational functions of x, say $f(x)$ and $g(x)$, such that for every integer m, either $f(x) \equiv m$ or $g(x) \equiv m$ has a solution.*

There is the obvious generalization to more functions. A trivial instance is $f(x) = x^2$, $g(x) = nx^2$ where n is any quadratic nonresidue of p. A significant one is with $ac \not\equiv 0$

$$f(x) = ax^4 + bx^2 + cx, \qquad g(x) = \frac{c^2}{x} - \frac{1}{4a}\left(b + \frac{c^2x}{4}\right)^2.$$

This was communicated to me by Schinzel after I gave him the special case when $b \equiv 0$. His proof was an arithmetic one based on Skolem's result [4] on quartic congruences. I give now [5] one simpler than my original one. It suffices to take $a = 1$ as is obvious on dividing $f(x)$ by a, and replacing x in $g(x)$ by $\frac{x}{a}$.

Consider now the sum

$$S_{11} = \sum_{x} e\,(x^4 + bx^2 + cx), \qquad b \not\equiv 0.$$

Multiply both sides by $e(-y^2)$ and sum for y, but on the right hand side replace y by $y + x^2$. Then

$$S_{11} \sum_{y} e\,(-y^2) = \sum_{x,\, y} e\,(x^4 + bx^2 + cx - (y + x^2)^2) = \sum_{x,\, y} e\,(x^2\,(b - 2y) + cx - y^2).$$

The contribution to this sum when $b - 2y \equiv 0$ is zero since $c \not\equiv 0$. When $b - 2y \not\equiv 0$, we sum for x and so

$$S_{11} \sqrt{p}\, i^{\left(\frac{p-1}{2}\right)^2} \left(\frac{-1}{p}\right) = \sqrt{p}\, i^{\left(\frac{p-1}{2}\right)^2} \sum_{y}{}' e\left(-\frac{c^2}{4\,(h - 2y)} - y^2\right)\left(\frac{b - 2y}{p}\right).$$

Write $y \to \frac{4b - y}{8}$. Then

$$S_{11}\left(\frac{-1}{p}\right)=\sum_{y}{}' e\left(-\frac{c^2}{y}-\left(\frac{4b-y}{8}\right)^2\right)\left(\frac{y}{p}\right).$$

This identity is of the form

$$\sum_{r} d_r e(r)=0,$$

and since the cyclotomic equation is irreducible in the rational field, the identity must reduce to

$$\sum_{r} d_0 e(r)=0,$$

unless of course all the d are zero.

This cannot occur here since there are p terms on the left hand side and only $p-1$ on the right hand side. Hence every residue mod p must occur either on the left hand side or the right hand side. Schinzel's result follows on writing $\frac{1}{y}$ for y.

For another application of the idea just used, we prove that

$$S_{12}=\sum_{x,\,y}{}' e\left(ax+by+\frac{c}{x}+\frac{d}{y}+j(fx+gy)^2\right)\left(\frac{xy}{p}\right)=O(p).$$

Multiply both sides by $\sum e(-jz^2)$ and on the right hand side replace z by $z+j(fx+gy)$. Then

$$\sqrt{p}\, i^{\left(\frac{p-1}{2}\right)^2}\left(\frac{-j}{p}\right)S_{12}=\sum_{x,\,y,\,z}{}' e\left(ax+by+\frac{c}{x}+\frac{d}{y}-2j(fx+gy)z-jz^2\right)\left(\frac{xy}{p}\right).$$

The sums in x, y are Salié sums, and so the right hand sum is

$$p(-1)^{\left(\frac{p-1}{2}\right)^2}\sum_{z} e(2h+2k-jz^2)\left(\frac{cd}{p}\right) \text{ if } cd\not\equiv 0,$$

where

$$h^2\equiv(a-2tjz)c, \qquad k^2\equiv(b-2gjz)d.$$

Hence

$$gdh^2-fck^2=(ag-bf)(cd).$$

Hence unless $ag-bf\equiv 0$ and $\left(\frac{-fgcd}{p}\right)=-1$ when the only solution is $h=$ $=k=0$, we have a parametric solution in the form

$$h=\frac{A_1+B_1t+C_1t^2}{A+Bt+Ct^2}, \qquad k=\frac{A_2+B_2t+C_2t^2}{A+Bt+Ct^2}, \qquad z=\frac{A_3+B_3t+C_3t^2}{A+Bt+Ct^2},$$

where t is a parameter. The right hand summation becomes

$$\sum_{t}{}' e\left(\frac{F(t)}{(A+Bt+Ct^2)^2}\right)=O(p),$$

where $F(t)$ is a quartic in t.

BIBLIOGRAPHY

1. L. J. Mordell, *On some exponential sums related to Kloosterman sums*, Acta Arith. **21** (1972), 65–69. MR **46** #5260.

2. ———, *A finite evaluation of a special exponential sum*, Proc. Cambridge Philos. Soc. **71** (1972), 75–84.

3. ———, *On rational functions representing all residues* mod p, J. London Math. Soc. **5** (1972), 166–168.

4. P. A. Leonard, *On factoring quartics* (mod p), J. Number Theory **1** (1969), 113–115. MR **38** #5751.

5. L. J. Mordell, *Rational functions representing all residues* mod p. II, Proc. Amer. Math. Soc. **35** (1972), 411–412. MR **46** #8945.

Trudy Mat. Inst. Steklov.
132 (1973)

Proc. Steklov Inst. Math.
132 (1973)

ARTIN'S CONJECTURES AND THE LAW OF RECIPROCITY(1)

A. I. VINOGRADOV

ABSTRACT. A method of argument is presented for proving the connection between two Artin conjectures and the nonabelian reciprocity law.

Bibliography: 16 items.

There are two conjectures due to E. Artin. In the first he proposed that his L-series [3]–[7] with nonabelian characters are entire functions in all of the complex plane. In the second conjecture ([3], Introduction) he proposed that any rational integer a not equal to a_1^2 is a primitive root for infinitely many prime numbers. In this report we shall delineate a brief method of argument which implies that both of the Artin conjectures are closely related to the nonabelian reciprocity law.

After the fundamental papers by Artin himself [4]–[7] and then by R. Brauer [8], [9], it became clear that the method of induced characters puts Artin L-series in the class of meromorphic functions. An essential difficulty is encountered in narrowing down the class of functions they belong to further from meromorphic to entire functions. The trouble is essentially that Brauer's fundamental theorem [8] on representing any character as a sum of induced characters with rational integer coefficients becomes false in the general case if we suppose that all the coefficients are nonnegative.

Of course, we may attempt to avoid this difficulty by using the new subtle properties of induced characters. But there is another way—finding a new approach to this conjecture. Here we shall explain some ideas in this direction. Their general content is as follows.

After articles by Tate [10] and by Iwasawa [11] on the theory of L-series with abelian characters, we had a new adèle technique which uses the properties of the Euler product of L-series. The language of this technique explicitly clarifies the leading role of the Artin reciprocity law [7] in the proof that the Hecke L-series is entire in all of the complex plane. Hence the possibility arises of trying to find a nonabelian reciprocity law in a form which in the language of the adèle technique implies that the Artin L-series is entire.

Before presenting the basic idea of a nonabelian reciprocity law, we discuss some

---

AMS (MOS) subject classifications (1970). Primary 12A70; Secondary 10G05.

(1) See [1], [2].

suggestive considerations. Let K/Q be a normal extension of the rational numbers Q with nonabelian Galois group $G(K/Q)$, let A be an irreducible representation of the group G of degree > 1, and let A_p be the matrix of the representation in the p-class. The Artin L-series is defined as the Euler product

$$\prod_{p \nmid d} \det\left(E - \frac{A_p}{p^s}\right)^{-1}, \operatorname{Re} s > 1, \ d \text{ is the discriminant of } K/Q.$$

But the determinant definition of the Artin L-series is unsuitable for formulating its properties in the language of adèle groups, since the operation of integration does not commute with the determinant sign. In addition, in the determinant definition the Artin L-series actually depends not on the representation A but on its character χ. But a nonabelian character χ does not have the group property, i.e. $\chi(p_1)\chi(p_2) \neq \chi(p_1 p_2)$. This fact prevents us from connecting all the p-components into an idèle group. Hence the idea appears of introducing the matrix L-series

$$\prod_{p \nmid d} \left(E - \frac{A_p}{p^s}\right)^{-1}, \qquad \operatorname{Re} s > 1. \tag{1}$$

Both difficulties disappear here. In fact,

$$\left(E - \frac{A_p}{p^s}\right)^{-1} = \sum_{n=0}^{\infty} \frac{A_p^n}{p^{ns}} = \sum_{n=0}^{\infty} \frac{A_{p^n}}{p^{ns}} = \int_{O_p^*} A_{v_p} \| v_p \|^s \, dv_p,$$

where $A_{v_p} = A_{p^n}$ for $\| v_p \| = p^{-n}$.

The local integral is considered in the multiplicative topology; Q_p^* are the p-adic integers. In addition, the matrices A_{v_p} have a group structure, i.e. $A_{v_p} A_{v'_p} = A_{v_p v'_p}$, and we may introduce the idèle representation matrix for all $p \nmid d$. The problem consists in defining this matrix at points $p \mid d$ so that the principal idèle representation matrix becomes one. To do this we must obtain a nonabelian reciprocity law for the Galois group $G(K/Q)$. But the one-dimensional adèle group and all of its subgroups are commutative. Hence it is not possible to obtain a noncommutative reciprocity law in the language of one-dimensional adèle groups.

We have the idea of using matrix adèle groups. If the full linear group $G_p = GL_p(n, Q)$ is divided by its unipotent subgroup U_p, then the quotient group G_p/U_p has for its p-component a semisimple product of n conjugate cyclic subgroups. The generators of these cyclic subgroups are n conjugate integer matrices with determinant $\pm p$.

To connect class field theory with the matrix point of view, we must translate it into the language representation matrices. The essence of this translation is as follows. Let $K = Q(\sqrt[q]{1}, \sqrt[q]{a})$; i.e., the field K is obtained by two cyclic extensions:

$$Q \to Q(\sqrt[q]{1}) = k \to k(\sqrt[q]{a}) = K,$$

where $q \geqslant 3$ is a prime, and a is a rational integer which is not a qth power. The subgroups $G(k/Q)$ and $G(K/k)$ are cyclic, of degrees $q-1$ and q, respectively. The Galois group $G(K/Q)$ is nonabelian of degree $q(q-1)$.

The factorization laws and the reciprocity laws are well known for the cyclic extensions k/Q and K/k, so the problem consists in connecting them by single nonabelian factorization and reciprocity laws for the entire Galois group $G(K/Q)$. To do this we must learn how to descend from the field k with factorization laws for the extension $k \to K$ to the field Q while preserving these laws. It is precisely here that the matrix point of view helps us.

Suppose that $\alpha \in k$ and f is the conductor of class field theory for the extension $k \to K$; let $\sigma \in G(K/k)$ be the automorphism corresponding to α. Then, by the law of recirpocity, there exists a residue l_σ in the σ-class (mod f) such that

$$\alpha \equiv l_\sigma \pmod{f}.$$

Using all the conjugates of the field k, we obtain the vector congruence

$${}^t(\alpha^1, \alpha^2, \ldots, \alpha^{q-1}) \equiv {}^t(l_\sigma^1, l_\sigma^2, \ldots, l_\sigma^q) [\mathrm{mod}\ {}^t(f^1, f^2, \ldots, f^{q-1})].$$

Passing to representation matrices, we obtain

$$A_\alpha \equiv L_\sigma \pmod{F}, \tag{2}$$

where A_α, L_σ and F are rational integer representation matrices for α, l_σ and f relative to some fixed basis $\zeta_1, \zeta_2, \ldots, \zeta_{q-1}$ of the field k/Q.

If π is a prime ideal of the field k/Q and $\pi \mid \alpha$, then

$$A_\alpha = A_{\alpha/\pi} C_\pi,$$

where $A_{\alpha/\pi}$ is the representation matrix for α relative to π, and C_π is the representation matrix for π. In addition, α can always be chosen so that

$$A_{\alpha/\pi} \equiv E \pmod{F},$$

and, since F is diagonal, from (2) we obtain

$$C_\pi \equiv L_\sigma \pmod{F}. \tag{3}$$

Thus the factorization and reciprocity laws for the extension K/k have been translated into the language of representation matrices. The matrix modulus F that arises here explains the arithmetic nature of the Artin conductors [7] for nonabelian characters. They are the determinants of F, and their norm tells us the number of matrix residues mod F. In order to include these factorization laws in general factorization laws for the entire nonabelian Galois group $G(K/Q)$, we analyze the structure of this group in more detail.

As already noted, this group has order $q(q-1)$. The simplest model of this group consists of pairs of integers $[a, b]$, where a runs through the multiplicative group of residues mod q, and b runs through a complete system of residues mod q. We consider

$$[a_1, b_1] = [a_2, b_2]$$

only if we have both

$$a_1 \equiv a_2 \pmod q), \qquad b_1 \equiv b_2 \pmod q).$$

The law for composing two pairs $[a, b]$ and $[c, d]$ is given by the equation

$$[a, b] \cdot [c, d] = [ac, bc + d].$$

In the language of the pairs $[a, b]$, the cyclic subgroup $G(K/k)$ is isomorphic to the group of elements $[1, b]$, $0 \leqslant b \leqslant q - 1$, and the subgroup $G(k/Q)$ is isomorphic to the group of elements $[a, 0]$, $1 \leqslant a \leqslant q - 1$. The conjugacy classes of elements are given by the table

$[1, 0]$	$[1, 1]$	$[2, 0]$	$\cdots$	$[q-1, 0]$
	$[1, 2]$	$[2, 1]$	$\cdots$	$[q-1, 1]$
	$\vdots$	$\vdots$		$\vdots$
	$[1, q-1]$	$[2, q-2]$	$\cdots$	$[q-1, q-2]$
		$[2, q-1]$	$\cdots$	$[q-1, q-1]$

The columns are the classes.

If $p \equiv 2 \pmod q$ and $p \nmid d$,

$$p = \pi_1 \pi_2 \ldots \pi_{q-1},$$

where the π_i are the prime ideals of degree one in the field k/Q. In the adèle group generated by the quotient group G_p/U_p, we take for p-generators of the cyclic subgroups the representation matrices C_{π_i}, $1 \leqslant i \leqslant q - 1$. For the multiplicative character of the element C_{π_i} we take the representation matrix A_{π_i} of the automorphism $\sigma_{\pi_i} \in (K/Q)$, which has the form $[1, b_i]$. Here if at least one π_i lies in the principal class $[1, 0]$, then all of its conjugates also lie in the principal class. In this case $b_i = 0$ for all $i \in [1, q - 1]$. If at least one of the π_i corresponds to an element $[1, b_i]$, where $b_i \neq 0$, then its conjugates fill out the entire class of the element $[1, 1]$.

By the law of reciprocity, we can define the multiplicative character between the elements $[1, b_i]$ and the matrix residues (mod F) on the p-units mod F_p for $p \mid \det F$ so that the residues of the representation matrices C_π correspond to A_π^{-1}, i.e. the inverse of the matrix which was defined as the p-generator for C_π. Then, when C_π is imbedded in the group of p-units for $p \mid \det F$, we obtain the unit representation matrix for the principal idèle C_π.

If p does not split completely in the field k/Q, i.e. where $m = (q - 1)/\delta$, $\delta > 1$, then we apply the following procedure. We take a representation matrix C_π for which $\pi \mid p$. This matrix consists of δ identical blocks. Hence the matrix equation

$$p = \pi_1 \pi_2 \cdots \pi_m$$

is solvable in the ring of rational integer matrices. We take one of these solutions, denote it by C_p, and take the $q - 1$ conjugate matrices C_p^i, $1 \leqslant i \leqslant q - 1$, where for conjugacy elements we use the matrices of the Weil group, i.e. the $q - 1$ matrices Y_i with the condition

$$X^{\delta} = C_{\pi}.$$

In the two-dimensional case $Y_1 = \begin{pmatrix} 1 & \\ & 1 \end{pmatrix}$, $Y_2 = \begin{pmatrix} & 1 \\ 1 & \end{pmatrix}$. We use these $q - 1$ conjugate matrices with determinant $\pm p$ as p-generators of conjugate cyclic subgroups for the p-component of the adèle group. For multiplicative p-characters we let the elements C_p^i correspond to the representation matrices $A_{[a,i]}$, $1 \leqslant i \leqslant q - 1$, where $a^{\delta} \equiv 1 \pmod q$, and the choice of the residue a is given by the congruence $p \equiv a \pmod q$.

This choice of the multiplicative character is compatible with the operation of raising the matrix C_p^i (respectively $A_{[a,i]}$) to the δth power, because all the prime ideals π of degree $\delta > 1$ lie in the principal class of the extension K/k in the sense of class field theory. It remains to solve the problem of the law of reciprocity between the residues mod F for the matrices C_p^i and the representation matrices $A_{[a,i]}$ which correspond to them as multiplicative characters.

We may at this point expect that the nonabelian reciprocity law in this form is possible because of the noncommutative nature of the group of matrix residues. If it is obtained, then our adèle approach immediately leads to a functional equation for $L^{q-1}(s, \chi)$, where $\chi = \operatorname{tr} A$, and, in addition, we find that the $(q-1)$th power of the Artin L-series is entire in all of the complex s-plane. Combining this with Brauer's theorem [8] that $L(s, \chi)$ is meromorphic, we obtain the result that the Artin series itself is entire. Thus, from the point of view of the nonabelian reciprocity law and the adèle technique, it is most natural to consider not the Artin L-series itself but the power of it equal to the multiplicity of the simple character in the regular representation.

We now discuss Artin's conjecture on the primitive root and see how it is connected with the above material. In 1967 Hooley [12] proved that this conjecture of Artin holds under certain conditions. Namely, he assumed the Riemann hypothesis for zeta-functions of algebraic number fields and obtained the Artin conjecture in a strengthened version; that is, Hooley obtained not only the infinity of primes, but also an asymptotic estimate for the number of them. More precisely, his result has the following form. If p_a denotes a prime number for which a is a primitive root and $\pi(x, a) = \sum_{p_a \leqslant a} 1$, then it follows from [12] that

$$\pi(x, a) = \mathfrak{S}(a)\, \pi(x) + O\left(\frac{\pi(x)}{\ln x} \ln \ln x\right), \tag{4}$$

where $\mathfrak{S}(a)$ is a certain series (for details see [12]) and $\pi(x)$ is the total number of primes $\leqslant x$. Equation (4) shows that the numbers p_a have positive asymptotic density.

In this report we show that the conditional equation (4) can be replaced by an unconditional upper estimate

$$\pi(x, a) \leqslant \mathfrak{S}'(a)\, \pi(x) + c_a \frac{\pi}{\sqrt[4]{\ln x}} (\ln \ln x)^2 \tag{5}$$

and a conditional lower estimate

$$\pi(x, a) \geqslant \mathfrak{S}(a)\pi(x) - c_a \frac{\pi(x)}{\sqrt[6]{\ln x}}(\ln\ln x)^2. \tag{6}$$

The condition for the lower estimate consists in taking the average density conjecture for Artin L-series in the spirit of the theorems of Bombieri and Montgomery [13], [14].

We begin with the lower estimate (6), since for natural reasons it is more interesting and correspondingly more difficult. We introduce the function $\pi_q(x, a, z)$, where q is a prime number. This function equals the number of primes $p \leqslant x$ with the condition that

$$p \equiv 1 (\operatorname{mod} q), \quad a^{\frac{p-1}{q}} \equiv 1 (\operatorname{mod} p), \quad ((p-1, \operatorname{ind} a), P_z) = 1, \tag{7}$$

where P_z is the product of all primes $< z$ (strictly less); ind is the index function with respect to some primitive root (mod p).

Using these functions, we obtain the expansion

$$\pi(x, a) = \pi(x) - \sum_{q \leqslant x} \pi_q(x, a, q).$$

The part of the sum over q in the interval $\sqrt{x}\exp(-\sqrt{\ln x}) \leqslant q \leqslant x$ is estimated exactly as is done in [12] for the case of large moduli. As a result we obtain the estimate

$$O\left(\frac{\pi(x)}{\sqrt{\ln x}}\ln\ln x\right).$$

The part

$$\pi(x) - \sum_{q < \sqrt[6]{\ln x}} \pi_q(x, a, q)$$

is examined using the modern estimates on the zeros of zeta-functions, and for this part we obtain the lower estimate

$$\geqslant \mathfrak{S}(a)\pi(x) - c_a \frac{\pi(x)}{\sqrt[6]{\ln x}}(\ln\ln x)^2. \tag{8}$$

It remains to consider the sum over q in the interval

$$\sqrt[6]{\ln x} \leqslant q < \sqrt{x}\exp(-\sqrt{\ln x}).$$

We denote this interval by Ω. In it we apply the estimate

$$\pi_q(x, a, q) \leqslant \pi_q(x, a).$$

The function $\pi_q(x, a)$ on the right of the inequality means that the last of the three conditions in (7) defining the function $\pi_q(x, a, q)$ has been omitted.

As a result we obtain the lower estimate

$$\pi(x, a) \geqslant \mathfrak{S}(a)\pi(x) - \sum_{q \in \Omega} \pi_q(x, a) - r(x), \tag{9}$$

where $r(x)$ has the form of the remainder on the right side of (8).

For $\pi_q(x, a)$ we have the estimate from above

$$\pi_q(x, a) \leqslant \frac{1}{q(q-1)} \sum_{N\mathfrak{p} \leqslant x} 1,$$

where $\mathfrak{p}$ runs through all prime ideals of the field $K_q = Q(\sqrt[q]{1}, \sqrt[q]{a})$, and Q is the rational numbers. In addition, we have the estimate

$$\sum_{N\mathfrak{p} \leqslant x} 1 \leqslant \frac{2}{\ln x} \sum_{N\mathfrak{a} \leqslant x} \Lambda(\mathfrak{a}) + O(\sqrt{x}\, q^2),$$

where Λ is the Mangoldt function of the field K_q and $\mathfrak{a}$ runs through all integral ideals of the field K_q.

If we let $\psi_q(x) = \Sigma_{N_\mathfrak{a} \leqslant x} \Lambda(\mathfrak{a})$, then the estimate (9) is transformed to

$$\pi(x, a) \geqslant \mathfrak{S}(a)\, \pi(x) - \frac{2}{\ln x} \sum_{q \in \mathfrak{Q}} \frac{\psi_q(x)}{q(q-1)} - r(x). \tag{10}$$

We note that the $r(x)$ are different in (9) and (10), but they have the same order, equal to the size of the remainder in (8).

Using the logarithmic derivative of the zeta-function of the field K, we obtain

$$\psi_q(x) = x - \sum_{|\rho| < \sqrt{x}} \frac{x^\rho}{\rho} + O(q^2 \sqrt{x} \ln^2 x), \tag{11}$$

where ρ are the zeros of $\zeta_{K_q}(s)$ in the critical strip.

The Galois group $G(K_q/Q)$ has q conjugacy classes, so that the character of the representation of G has q simple characters in its expansion. It follows from the structure of G and its subgroups that among the q simple characters there is only one nonabelian character of "thickness" $q-1$, i.e. $\chi_q(1) = q-1$. The other $q-1$ simple characters are one-dimensional abelian characters of the subgroups of the cyclotomic extension $Q(\sqrt[q]{1})$. Hence, if θ_q is the character of the regular representation of the group $G(K_p/Q)$, then

$$\theta_q = \sum_{i=1}^{q-1} \chi_i + (q-1)\chi_q,$$

where χ_i, $1 \leqslant i \leqslant q-1$, are abelian cyclotomic characters, and χ_q is a nonabelian character.

In connection with this equation we obtain the Artin expansion

$$\zeta_{K_q}(s) = \zeta(s) \prod_{i=2}^{q-1} L(s, \chi_i)\, L^{q-1}(s, \chi_q),$$

where ζ is the Riemann zeta-function, the $L(s, \chi_i)$ are Dirichlet series (mod q), and

$L(s, \chi_q)$ is the Artin series with nonabelian character.

It can be shown that, for the Galois groups $G(K_q/Q)$ under consideration, the Artin series with nonabelian character χ_q is an entire function in all of the complex s-plane. Hence the sum over ρ in (11) can be written

$$\sum_{|\rho|\leqslant\sqrt{x}} \frac{x^\rho}{\rho} = \sum_{i=1}^{q-1} \sum_{|\rho_i|\leqslant\sqrt{x}} \frac{x^{\rho_i}}{\rho_i} + (q-1) \sum_{|\rho_q|\leqslant\sqrt{x}} \frac{x^{\rho_q}}{\rho_q},$$

where the ρ_i are the zeros of $L(s, \chi_i)$, and $L(s, \chi_1) = \zeta(s)$.

Substituting this equation in (11) and then in (10) and applying the Bombieri-Montgomery density theorem [13], [14] to the zeros of the Dirichlet L-series, we reduce the problem to the zeros of the Artin L-series with respect to moduli $q \in \Omega$:

$$\pi(x, a) \geqslant \mathfrak{S}(a)\, \pi(x) - \frac{2}{\ln x} \sum_{q\in\Omega} \frac{1}{q} \sum_{|\rho_q|\leqslant\sqrt{x}} \frac{x^{\rho_q}}{\rho_q} - r(x). \tag{12}$$

Estimate (12) is unconditional. It shows that Artin's conjecture on the primitive root is equivalent to the average density conjecture for Artin L-series.

In applying currently available density techniques to Artin L-series, we encounter an essential difficulty. It consists essentially in the large "thickness" of the nonabelian character χ_q. Roughly speaking, from the analytic point of view the Artin L-series behaves like the $(q-1)$th power of an ordinary Dirichlet L-series. Hence in the critical strip the length of the approximating intervals of the Dirichlet series for the Artin series has order $(c_a q|s|)^{(q-1)/2}$ (see [15]), and, if $q \in \Omega$, then the available density techniques are powerless to handle such a length of these intervals. The idea is then to find a way around this difficulty. Here there are two approaches. We first return again to the Artin matrix L-series which was introduced at the beginning:

$$\prod_{p\nmid d} \left(E - \frac{A_p}{p^s}\right)^{-1}, \quad \operatorname{Re} s > 1. \tag{13}$$

Unlike the classical Artin L-series, which is defined as the determinant of (13), the matrix (13) is not uniquely determined by the representation A. But we can handle this difficulty, and then try to extend the matrix (13) onto the entire complex s-plane or at least as far as the line $\operatorname{Re} s = ½$. If we could do this, then the proper values $L_i(s)$ of the matrix (13), $1 \leqslant i \leqslant q-1$, would also have the analogous property. But the $L_i(s)$ have the properties of one-dimensional Dirichlet series, whose coefficients are the proper values $\lambda_i(p)$ of the matrices A_p. The approximating intervals for $L_i(s)$ in the critical strip would have length $c_a q|s|$, and the available density techniques would be applicable to them in the spirit of Bombieri and Montgomery. But the Artin L-series is equal to the product of the $L_i(s)$, $1 \leqslant i \leqslant q-1$, so that the density techniques would be constructed for them as well.

For a concrete technique for extending the matrix (13) to the left half-plane it is natural to use the language of adèle groups and the nonabelian reciprocity law in the same spirit as described before for the first Artin conjecture. But, from the point of view of the adèle technique, there is a difference in principle between the power $L^{q-1}(s, \chi_q)$ and the Artin L-series $L(s, \chi_q)$ itself or the matrix (13). In the first part it was shown that, to construct the power $L^{q-1}(s, \chi_q)$ in the language of the full linear group $GL(q-1, Q)$, we must take the unipotent subgroup $U(q-1, Q)$ for the stationary subgroup. To obtain $L(s, \chi_q)$, or the matrix (13), in the language of $GL(q-1, Q)$, we must find a stationary subgroup $X(q-1, Q)$ which would give us one p-generator in the p-component G_p/X_p of the adèle group. In other words, the subgroup X_p must "be able to distinguish" a given matrix with determinant $\pm p$ from all of its conjugates. The existence of such a subgroup X seems very problematic. Most likely, it does not exist in general.

The second approach to obtaining the unconditional lower estimate (6) from (12) consists in trying to prove the Riemann hypothesis for Artin L-series, using Selberg's method [16] and its further development. This approach is encouraging because the power $L^{q-1}(s, \chi_q)$ can be expressed in the language of the quotient group GL/U. Of course, this is under the condition that the reciprocity law is obtained for the Galois group $G(K_q/Q)$.

BIBLIOGRAPHY

1. A. I. Vinogradov, *Artin L-series and the adèle group*, Trudy Mat. Inst. Steklov. **112** (1971), 105-122 = Proc. Steklov Inst. Math. **112** (1971), 105-122.

2. ———, *Artin L-series and his conjectures*, Trudy Mat. Inst. Steklov. **112** (1971), 123-140 = Proc. Steklov Inst. Math. **112** (1971), 124-142.

3. *The collected papers of Emil Artin*, Addison-Wesley, Reading, Mass., 1965. MR **31** #1159.

4. E. Artin, *Über eine neue Art von L-Reihen*, Abh. Math. Sem. Univ. Hamburg **3** (1923), 89-108.

5. ———, *Zur Theorie der L-Reihen mit allgemeine Gruppencharakteren*, Abh. Math. Sem. Univ. Hamburg **8** (1930), 292-306.

6. ———, *Die gruppentheoretische Struktur der Diskriminanten algebraischer Zahlköper*, J. Reine Angew Math. 164 (1931), 1-11.

7. ———, *Beweis des allgemeinen Reziprozitätsgesetzes*, Abh. Math. Sem. Univ. Hamburg **5** (1927), 353-363.

8. R. Brauer, *On Artin's L-series with general group characters*, Ann. of Math. (2) **48** (1947), 502-514. MR **8**, 503.

9. ———, *On the zeta-functions of algebraic number fields*, Amer. J. Math. **69** (1947), 243-250. MR **8**, 567.

10. J. Tate, *Fourier analysis in number fields, and Hecke's zeta-functions*, Dissertation, Princeton Univ., Princeton, N.J., 1950; reprinted in *Algebraic number theory* (Proc. Instructional Conf., Brighton, 1965), Thompson, Washington, D.C., 1967, pp. 305-347. MR **36** #121.

11. K. Iwasawa, *A note on functions*, Proc. Internat. Congress Math. (Cambridge, 1950), vol. I, Amer. Math. Soc., Providence, R. I., 1952, p. 322.

12. C. Hooley, *On Artin's conjecture*, J. Reine Angew. Math. **225** (1967), 209-220. MR **34** #7445.

13. E. Bombieri, *On the large sieve,* Mathematika **12** (1965), 201-225. MR **33** #5590.

14. H. L. Montgomery, *Zeros of L-functions*, Invent. Math. **8** (1965), 346-354.

15. A. F. Lavrik, *Approximate functional equations for Dirichlet functions*, Izv. Akad. Nauk SSSR Ser. Mat. **32** (1968), 134-185 = Math. USSR Izv. **2** (1968), 129-180. MR **36** #6361.

16. A. Selberg, *Harmonic analysis and discontinuous groups in weakly symmetric Riemannian spaces with applications to Dirichlet series*, J. Indian Math. Soc. **20** (1956), 47-87. MR **19**, 531.

Translated by N. KOBLITZ

Trudy Mat. Inst. Steklov.
132 (1973)

Proc. Steklov Inst. Math.
132 (1973)

THE ARITHMETIC OF K3 SURFACES

I. I. PJATECKIĬ-ŠAPIRO AND I. R. ŠAFAREVIČ*

ABSTRACT. In this paper we study some properties of K3 surfaces and give arithmetic applications.

Bibliography: 8 items.

§1. The main results

A complete algebraic surface X is called a K3 surface if $H^1(X, O_X) = 0$ and if the canonical class K_X of X equals zero.

It is known (see [1], Chapter VIII) that, in addition to K3 surfaces, two-dimensional abelian varieties satisfy the condition $K_X = 0$. Over the field of complex numbers $\mathbf{C}$ a two-dimensional abelian variety is uniquely determined by the periods of its holomorphic one-forms. From this it follows that it is also determined by the periods of its holomorphic two-form. On a K3 surface there are no holomorphic one-forms, but there is a holomorphic 2-form, which is unique up to a constant multiple. The question naturally arises of "to what extent do its periods determine the surface?" To answer this we need to introduce some more concepts.

One sees that for all K3 surfaces X the homology groups $H_2(X, \mathbf{Z})$ together with the scalar product defined by the intersection index are all isomorphic. We fix one such Euclidean lattice L and a vector $l \in L$. A marked K3 surface is a triple $\widetilde{X} \approx (X, \xi, \varphi)$, where X is a K3 surface, $\xi \in H_2(X, \mathbf{Z})$ is the homology class of a hyperplane section under some projective imbedding of X and $\varphi: H_2(X, \mathbf{Z}) \to L$ is an isomorphism of Euclidean lattices such that $\varphi(\xi) = l$.

In the space $\widetilde{\Omega} = \mathrm{Hom}\,(L, \mathbf{C})$ we define a bilinear scalar product by extending the one given in L. The conditions

$$\omega^2 = 0, \qquad \omega\bar{\omega} > 0$$

AMS (MOS) subject classification (1970). Primary 14G13, 14J25; Secondary 14G25, 14J20.

**Editor's note.* The present translation incorporates suggestions made by the authors.

define a domain Ω in projective space corresponding to the vector space $\widetilde{\Omega}$. We let $\Omega(l)$ denote the connected component of a section, defined by the condition $\omega l = 0$. This is a homogeneous symmetric domain on which the group $G(R)$ acts, where G is the isotropy subgroup of the vector l in the orthogonal group of the lattice L. For a form $\eta \in \Gamma(X, \Omega_X^2)$ and $\gamma \in H_2(X, \mathbf{Z})$ we put

$$f(\gamma) = \int_\gamma \eta.$$

Obviously $f \in \operatorname{Hom}(H_2(X, \mathbf{Z}), \mathbf{C})$, and we see that φf defines a point in $\Omega(l)$. We denote this point by $\tau(\widetilde{X})$.

THEOREM 1 (*Torelli theorem for* K3 *surfaces*). *A marked* K3 *surface* $\widetilde{X}$ *is uniquely determined by a point* $\tau(\widetilde{X}) \in \Omega(l)$.

This theorem is proved in [2]. Here we illustrate some applications of it. We are basically interested in arithmetic applications, but since our method uses the analytic-geometric Theorem 1, it is impossible to consider these applications divorced from at least some geometric results.

K3 surfaces possess an integral invariant which is analogous to the genus of a curve; this is the smallest value of the number $\frac{1}{2}(D^2)$, where D is an irreducible curve on the surface with $(D^2) > 0$. We shall call it the class of the surface. It was proved in [1], Chapter VIII, that the divisor $2D$ is always very ample. Therefore all the surfaces with the same class define an open set S' in some Hilbert scheme.

In [2] it was proved that the scheme S' is smooth. On it we have a universal family $X' \to S'$ of K3 surfaces of a given class. Also in [2] we constructed a family $X'' \to S''$ of marked K3 surfaces of given class. Its base S'' is a complex manifold and an unramified covering of S', where the Galois group of the covering $f\colon S'' \to S'$ is isomorphic to the discrete group $G(\mathbf{Z})$, and a point of the fiber $f^{-1}(S')$ corresponds to all the ways of introducing the structure of a marked surface on the polarized K3 surface $X'_{S'}$. The period mapping defines a holomorphic mapping $\tau\colon S'' \to \Omega(l)$. In [2] we showed that the group $Pl(n)$ of projective transformations of the projective space $\mathbf{P}^n$ in which the surfaces $X'_{s'}$, $s' \in S'$, are imbedded acts in a natural way on S' and on S'', where τ is factored through the projection $p\colon S'' \to S''/Pl(n)$, and we have $\tau = hp$, and now Theorem 1 follows from the fact that $h\colon S''/Pl(n) \to \Omega(l)$ is an open immersion.

THEOREM 2. S'' *is connected.*

Consider a normal subgroup $\Gamma \subset G(\mathbf{Z})$ which is a congruence subgroup and contains no elements of finite order different from the identity. The variety $S = S''/\Gamma$ is a finitely sheeted covering of S' with Galois group $G(\mathbf{Z})/\Gamma$, and by a general theorem S is an algebraic variety. The period mapping defines a holomorphic map

$\pi: S \to \Omega(l)/\Gamma$. Since Γ acts without fixed points in $\Omega(l)$, we apply to π a theorem proved independently by Borel and Kobayshi [3], which says that π is a morphism of algebraic varieties. Together with Theorem 1 this yields that $\Omega(l) - \tau(S'')$ is contained in a proper complex submanifold, and hence $\tau(S'')$ is connected. From this it follows that the manifold $S''/Pl(n)$, homeomorphic to $\tau(S'')$, is connected, and hence S'' is also.

This result has two applications.

COROLLARY 1. *The manifold* S' *of all* K3 *surfaces of a given class is connected.*

This is obvious since S' is the image of S'' under f.

COROLLARY 2. *The image of the fundamental group* $\pi(S', s'_0)$ *in* $\mathrm{Aut}\, H^2(X'_{s'_0}, \mathbf{Z})$, *defined by the action on the cohomology of a fiber, coincides with* $G(\mathbf{Z})$.

In fact, if $H \subset G(\mathbf{Z})$ is the image of $\pi(S', s'_0)$ and s''_0 is a point lying over s'_0, then, covering all the mapping of the unit interval $f: I \to S'$, $f(0) = s'_0$, for covering mappings g with $g(0) = s''_0$ we obtain as $g(1)$ the points of some component of the manifold S'' in which only the points $H \cdot s''_0$ among those lying over s'_0 will be contained. Since the fiber over s'_0 has the form $G(\mathbf{Z})s'_0$, and S'' is connected, we have $G(\mathbf{Z}) = H$.

For example, arbitrarily deforming a nonsingular surface of degree 4 in $\mathbf{P}^3(\mathbf{C})$, as an automorphism of its cohomology we can take any automorphism of the corresponding integral quadratic form of signature (2.19).

We now proceed to a statement of the arithmetic applications.

THEOREM 3. *For a* K3 *surface over a finite field the Riemann hypothesis is true.*

The method, based on an application of Theorem 1, uses a lifting of a surface X defined over a finite field to a surface defined over a field of characteristic zero. This means that X is contained in the scheme $Y \to B$ as closed fiber, where the structure morphism is smooth, and the general fiber is a K3 surface over a field of characteristic 0. We shall assume that such a lifting exists, although it follows from the definition of a K3 surface that $H^2(X, \Theta) = 0$, and hence all the obstructions to lifting vanish. Let X be a smooth projective variety over $\mathbf{C}$, and let X^n denote the n-fold product of X with itself. The Hodge group of X will be an algebraic subgroup of the group of all automorphisms of $H^*(X, \mathbf{Q})$, distinguished by the condition of having trivial action on all the subspaces $H^{p,p} \cup H^{2p}(X^n, \mathbf{Q})$ (n and p are arbitrary). For abelian varieties the Hodge group was introduced and studied in [4]. We shall say that a smooth projective variety is a variety of CM-type if its Hodge group is commutative.

CONJECTURE. *Any variety of CM-type is defined over an algebraic number field, and its zeta-function is expressed by means of L-functions of one-dimensional characters of a finite extension of the field of definition of the variety.*

The excellent book of Shimura and Taniyama [5] contains a proof of this conjecture for abelian varieties. In this paper we prove it for K3 surfaces.

THEOREM 4. *A K3 surface of CM-type is defined over a number field, and its zeta-function is expressed by means of L-functions of one-dimensional characters of a finite extension of the field of definition.*

Theorem 3 was proved independently by P. Deligne, who kindly sent us a manuscript.** Since our proof is somewhat different from his (the use of the Torelli theorem introduces significant simplications), we have resolved to introduce it here.

§2. Cohomology of fibers as Galois modules

Let X and S be irreducible algebraic varieties defined over a field k which is a finite extension of the field of rational numbers $\mathbf{Q}$, and let $f\colon X \to S$ be a smooth proper morphism over k. The Galois group $\operatorname{Gal}(\overline{k(S)}/k(S))$ of the algebraic closure $\overline{k(S)}$ of the field $k(S)$ acts on the l-adic étale cohomology $H^*(F, \mathbf{Q}_l)$ of the general fiber of the bundle $f\colon X \to S$ and so defines a homomorphism

$$\operatorname{Gal}(\overline{k(S)}/k(S)) \to \operatorname{Aut} H^*(F, \mathbf{Q}_l)$$

into the group of automorphisms of the cohomology algebra. We let $k(S)_l$ denote the subfield of $\overline{k(S)}$ corresponding to the kernel of this homomorphism. If we must emphasize its dependence on the bundle f, we will denote it by $k(S)_{f,l}$. Let k' be the extension of k obtained by adjoining all the l^nth roots of 1 to k. By the existence of a Verdier multiplication in the cohomology it follows that $k(S)_l$ contains a subfield over which k' has finite degree.

A bundle $f\colon X \to S$ is said to be *strictly nonconstant* if the algebraic closure of k in $k(S)_l$ has finite degree over its intersection with k'. Then we can find a finite extension of k such that k' will be algebraically closed in $(k(S))_l$.

LEMMA 1. *The bundle $f\colon X \to S$ is strictly nonconstant if and only if the group* $\operatorname{Gal}(K(S)_l/K'(S))$ *has finite index in* $\operatorname{Gal}(k(S)_l/k'(S))$ *for any extension K of the field k.*

Analogously to the field $k(S)_l$ we consider fields $k(S)_{l,n}$, for which the groups $\operatorname{Gal}(k(S)_{l,n}/k(S))$ are contained in $\operatorname{Aut} H^*(F, \mathbf{Z}/l^n\mathbf{Z})$. We have a commutative diagram

---

***Translator's note.* Deligne's paper has since appeared; see Invent. Math. **15** (1972), 206–226. Other important arithmetic results on K3 surfaces have been obtained by M. Artin and H. P. F. Swinnerton-Dyer; see Invent. Math. **20** (1973), 249–265.

$$\begin{array}{ccc} \mathrm{Gal}\,(K'(S)_{l,n}/K'(S)) & \xrightarrow{\varphi_n} & \mathrm{Gal}\,(k'(S)_{l,n}/k'(S)) \\ & \searrow \quad \swarrow & \\ & \mathrm{Aut}\, H^*(F, \mathbf{Z}/(l^n)), & \end{array}$$

in which the slanting arrows are inclusions, and therefore the homomorphism φ_n will also be an inclusion. Its kernel corresponds to the subfield $k'(S)_{l,n} \cdot K'(S)$, and hence

$$K'(S)_{l,n} = k'(S)_{l,n} \cdot K'. \tag{1}$$

Suppose that the bundle $f: X \to S$ is strictly nonconstant and that k' is algebraically closed in $k'(S)_l$. Then $k'(S)_{l,n}$ is a regular extension of k'. Therefore, by the general properties of regular extensions,

$$[k'(S')_{l,n} \cdot K' : K'(S)] = [k'(S)_{l,n} : k'(S)].$$

Together with (1) this proves that φ_n is an isomorphism, and hence φ is also one. The converse is obtained by reversing these arguments.

In the applications we shall encounter, the field K will be the field of complex numbers $\mathbf{C}$. We consider an embedding of the field k over which S, X and f are defined into $\mathbf{C}$, and we shall assume that $k \subset \mathbf{C}$. We let $S_{\mathbf{C}}$ denote the variety of complex points of S. The fundamental group $\pi(S_{\mathbf{C}}, s_0)$ acts on the complex cohomology $H^*(F_{s_0}, \mathbf{Z})$ of a fiber of f. Here, of course, the image of elements of $\pi(S_{\mathbf{C}}, s_0)$ is contained in the group $\mathrm{Aut}\, H^*(F_{s_0}, \mathbf{Z})$.

§3. Kuga-Satake varieties

A construction of certain abelian varieties corresponding to K3 surfaces was made by M. Kuga and I. Satake. This construction goes as follows. Let X be a K3 surface and let $l \in H^2(X, \mathbf{Z})$ be the cohomology class corresponding to a hyperplane section.

We let $H^2(X, \mathbf{Z})^l = V \subset H^2(X, \mathbf{Z})$ denote the orthogonal complement to l. The signature of the intersection index form on V is (2, 19).

Let $C^+ = C^+(V)$ be the even Clifford algebra associated with V. Now in $\mathrm{End}\,(C^+ \otimes \mathbf{R})$ we define a complex structure operator j.

Let $V_+ = (H^{2,0} + H^{0,2}) \cap V \otimes \mathbf{R}$ and let e_1, e_2 be any orthonormal basis of V. Put $e_+ = e_1 e_2$ and define a complex structure in V by

$$jx = e_+ x, \qquad x \in C^+. \tag{2}$$

As is shown in [6], on $C^+ \otimes \mathbf{R}$ there exists a skew symmetric form $\alpha(x, y)$ which is integral on C^+ such that the form $\alpha(jx, y)$ is symmetric and positive definite. Hence $(C^+ \otimes \mathbf{R})/C^+$ is an abelian variety. Denote it by J_X. In [4] Mumford gave a construction for Hodge families of abelian varieties. From the results of [4], [7] it follows that on $\Omega(l)/\Gamma$ there is a Hodge family of abelian varieties corresponding to the spinor group; here Γ is any congruence subgroup of the spinor group. The canonical homomorphism of $\pi_1(\Omega(l)/\Gamma)$ into $\operatorname{Aut} H^1(J_X, \mathbf{Z})$ coincides with the natural inclusion of Γ into the spinor group. We know [4] that $\Omega(l)/\Gamma$ is defined over a finite extension of the field $\mathbf{Q}$, and the rational functions on $\Omega(l)/\Gamma$ possess the following property: at all the points corresponding to abelian varieties of CM-type their values belong to $\overline{\mathbf{Q}}$.

LEMMA 2. *The family of abelian varieties corresponding to* $\Omega(l)/\Gamma$ *is strictly nonconstant.*

PROOF. In [8] it was proved that for any abelian variety A a subgroup of finite index in the image of the Galois group in $\operatorname{Aut} H^1(A, \mathbf{Q}_l)$ is contained in the Hodge group of A. For the family defined above the Hodge group coincides with the spinor group. On the other hand, by construction of this family the fundamental group of the base is a congruence subgroup of the spinor group.

LEMMA 3. *If the Hodge group of the* (K3) *surface* X *is commutative, the corresponding abelian variety* J_X *is of CM-type.*

PROOF. From the construction of J_X we get the existence of an inclusion

$$V = H^2(X, \mathbf{Q})^l \to H^2(J_X, \mathbf{Q}) \tag{3}$$

which preserves the Hodge structure. From the definition of the Hodge group of any abelian variety A it follows that this group takes any subspace $V \subset H^k(A, \mathbf{Q})$ which satisfies the following property into itself:

$$V \otimes \mathbf{C} = \sum_{p+q=k} (V \otimes \mathbf{C}) \cap H^{p,q}. \tag{4}$$

From this and from (3) we get the existence of a homomorphism:

$$\tilde{\rho}: \operatorname{Hg}(J_X) \to \operatorname{Aut} H^2(X, \mathbf{Q})^l. \tag{5}$$

At a generic point this homomorphism coincides with the homomorphism of the spinor group into the orthogonal group. Hence its kernel is the group of order two. But this occurs for any X. Now suppose that the Hodge group of X is commutative. Hence the Hodge group of J_X is an extension of a commutative group by a group of order two. We know that the Hodge group of any abelian variety is reductive, from which it follows that it must be commutative if its quotient by a subgroup of order two is commutative. As is known from [4], the commutativity of the Hodge group of an abelian variety is equivalent to the fact that this variety is a variety of CM-type.

COROLLARY. *If* X *is a singular* K3 *surface, the abelian variety* J_X *is of CM-type.*

It suffices to verify that the Hodge group of X is commutative. This follows from the fact that $\mathrm{Hg}\, X$ must act trivially on $H^{1,1}(X, \mathbf{Q}) = S_X$. Hence it takes the orthogonal complement T_X of S_X into itself and the intersection index preserves the form. It remains to take into account that the group of automorphisms of a two-dimensional module equipped with a scalar product is commutative. We now give an example of a K3 surface of CM-type.

First we will prove the following property of a Hodge group.

LEMMA 4. *Let X be any smooth projective variety. The algebra $\mathfrak{A}_X$ of all endomorphisms of the additive group of $H^*(X, \mathbf{Q})$ which preserve the Hodge bigrading commutes with* $\mathrm{Hg}\, X$.

PROOF. All such endomorphisms can be interpreted in a standard way as rational cohomology classes of type (n, n) on $X \times X$, where n is the dimension of X. By definition, the Hodge group preserves such classes, which gives the assertion of the lemma.

COROLLARY. *Let X be a* K3 *surface. If the algebra of endomorphisms of $H^2(X, \mathbf{Q})$ which preserve the Hodge bigrading has dimension $\geqslant 22$ and is semisimple, then X is a surface of CM-type.*

We give an example of K3 surfaces of CM-type. Let X be a CM-field, i.e. a field of the form $F(\sqrt{-\delta})$, where F is totally real and δ is totally positive, i.e. positive under any imbedding $F \to \mathbf{R}$. Now consider the triple (K, φ_1, λ), where φ_1 is an imbedding $F \to \mathbf{R}$ such that $\varphi_1(\lambda) > 0$ for $\lambda \in F$, and $\varphi(\lambda) < 0$ for the remaining imbeddings. For $\alpha, \beta \in K$ we define a scalar product in the following way:

$$(\alpha, \beta) = \mathrm{tr}\, (\alpha\lambda\beta^\sigma),$$

where σ is the automorphism of complex conjugation. We shall show that $(\alpha, \beta) = (\beta, \alpha)$ and that the quadratic form (α, α) has signature $(2, 2n - 2)$, where $2n = [K : \mathbf{Q}]$. We have $\mathrm{tr}\, x = \mathrm{tr}\, x^\sigma$ for any $x \in K$. Hence $(\alpha, \beta) = \mathrm{tr}\, \alpha\lambda\beta^\sigma = \mathrm{tr}\, \beta\lambda\alpha^\sigma = (\beta, \alpha)$ ($\lambda \in F$ and hence $\lambda^\sigma = \lambda$).

To compute the signature we introduce suitable coordinates on $K \otimes \mathbf{R}$. Let $\psi_1, \cdots, \psi_{2n}$ denote a complete system of imbeddings $K \to \mathbf{C}$, numbered so that ψ_{2i-1} and ψ_{2i} will be conjugate. Put

$$x_{2i-1} = {}^1/_2\, (\psi_{2i-1}(\alpha) + \psi_{2i}(\alpha)), \quad x_{2i} = \frac{1}{2i} \cdot (\psi_{2i-1}(\alpha) - \psi_{2i}(\alpha)).$$

We obviously have

$$(\alpha, \alpha) = 2 \sum \lambda_k (x_{2k-1}^2 + x_{2k}^2),$$

where $\lambda_k = \psi_{2k-1}(\lambda) = \psi_{2k}(\lambda)$. It remains to show that this form can be represented by a standard form–the intersection index on $H^2(X, \mathbf{Q})$. By Hasse's theorem [7] this is possible in every case if $2n \leqslant 16$, since the dimension of the standard module is 22.

We define a linear functional ω on the additive group of the field K as an imbedding of K in $\mathbf{C}$ that extends φ_1, and on the orthogonal complement of K in $H^2(X, \mathbf{Q})$ we set it equal to zero. It is obvious that in this way we determine a point in the domain $\Omega(l)$ for suitable l. Points thus obtained are dense in this domain. Therefore one of them corresponds to a surface of type K3 that will be a surface of CM-type.

§4. Zeta-functions of K3 surfaces

Our method of investigation is based on the construction of two bundles over the same base such that these bundles satisfy the conditions of Lemma 6 (below). One of these is the universal bundle $f\colon X \to S$ for K3 surfaces of a given class. The base was defined earlier. The other will be constructed beginning from the fibering into abelian varieties over the base $\Omega(l)/\Gamma$ introduced in the preceding section. An application of Lemma 6 allows us to carry the classical result about abelian varieties over to K3 surfaces.

On the variety $S = S''/\Gamma$ introduced in §1, we consider the family $f\colon X \to S$ obtained by pulling back the family $X' \to S'$ relative to the covering $S \to S'$. By Weil's criteiron the family f is defined over a finite extension of $\mathbf{Q}$.

LEMMA 5. *The bundle $f\colon X \to S$ is strictly nonconstant.*

We consider a homomorphism $\varphi\colon \pi(S_{\mathbf{C}}, s_0) \to \operatorname{Aut} H^2(F_{s_0}, \mathbf{Z})$, which is given by the action of the group $\pi(S_{\mathbf{C}}, s_0)$ on the cohomology of a fiber. The basis of the proof is the relation

$$\varphi \cdot \pi(S_{\mathbf{C}}, s_0) = \Gamma,$$

which immediately follows from Corollary 2 of §1.

We note that $H^*(F, \mathbf{Z}) = H^0(F, \mathbf{Z}) \oplus H^2(F, \mathbf{Z}) \oplus H^4(F, \mathbf{Z})$, and therefore the group $\operatorname{Aut} H^*(F, \mathbf{Z})$ contains $G(\mathbf{Z})$, where G is the orthogonal group of the quadratic form defined by multiplication of cohomology in the lattice $H^2(F, \mathbf{Z})^l$. We denote by G_l the completion of the group $\varphi \cdot \pi(S_{\mathbf{C}}, s_0)$ in the topology defined by congruence subgroups modulo l^n in the group $\operatorname{Aut} H^*(F_{s_0}, \mathbf{Z}) \supset G(\mathbf{Z})$. Since the approximation theorem holds for $G(\mathbf{Z})$, the completion of $G(\mathbf{Z})$ coincides with $G(\mathbf{Z}_l)$, and G_l is the closure of Γ in $G(\mathbf{Z}_l)$. By the hypothesis $(G(\mathbf{Z}_l)\colon G_l) < \infty$.

By the comparison theorem for étale cohomology the rings $H^*(F_{s_0}, \mathbf{Z}_l)$ and $H^*(F, \mathbf{Z}_l)$ are isomorphic, and the group G_l coincides with the image of the group $\operatorname{Gal}(\mathbf{C}(S)_l/\mathbf{C}(S))$. Since the image of the group $\operatorname{Gal} k(S)_l$ is obviously contained in $G(\mathbf{Z}_l)$, and since the image of $\operatorname{Gal}(\mathbf{C}(S)_l/\mathbf{C}(S))$ already has finite index in $G(\mathbf{Z}_l)$, we have

$$(\operatorname{Gal}(k(S)_l/k(S)) : \operatorname{Gal}(\mathbf{C}(S)_l/\mathbf{C}(S))) < \infty.$$

Thus, a necessary condition for f to be strictly nonconstant has been obtained for $K = \mathbf{C}$. Since $\mathbf{C}$ can be taken as a universal domain in the sense of A. Weil, this condition holds for arbitrary K.

We now consider two bundles $f: X \to S$ and $f': X' \to S$ over the same base. Having considered them over $\mathbf{C}$, we associate to each of them a "system of local coefficients" $H^*(F_S, \mathbf{Q})$, or, what amounts to the same, the sheaf $R^*f_*\mathbf{Q}$ on S. Suppose we are given a locally trivial subsheaf A of the sheaf $\operatorname{Aut} R^*f_*\mathbf{Q}$ and a homomorphism

$$\rho : A \to \operatorname{Aut} R^*f'_*\mathbf{Q}$$

of these sheaves, i.e. a system of homomorphisms

$$\rho_S : A_S \to \operatorname{Aut} H^*(F'_S, \mathbf{Q}),$$

compatible with the action of the fundamental group $\pi(S_\mathbf{C})$. By the same token, as we already saw, we have a homomorphism of groups

$$\rho_{\text{ét}} : A_\xi \to \operatorname{Aut} H^*(F', \mathbf{Q}_l),$$

where ξ is a generic point of the base S, and F' is the generic fiber of f'.

LEMMA 6. *If F and F' are strictly nonconstant bundles, then the homomorphism $\rho_{\text{ét}}$ commutes with the action of subgroups of finite index in the groups* $\operatorname{Gal}(k(S)_{f,l}/k'(S))$ *and* $\operatorname{Gal}(k(S)_{f',l}/k'(S))$.

In fact, by construction $\rho_{\text{ét}}$ commutes with the action of the group G_l, i.e. the group $\operatorname{Gal}(\mathbf{C}(S)_l/\mathbf{C}(S))$ (this group depends only on the base, and hence it is the same for both bundles f and f'). Since both bundles are strictly nonconstant, the group $\operatorname{Gal}(\mathbf{C}(S)_l/\mathbf{C}(S))$ is contained in the groups $\operatorname{Gal}(k(S)_{f,l}/k'(S))$ and $\operatorname{Gal}(k(S)_{f',l}/k'(S))$ as a subgroup of finite index. The assertion of the lemma now follows.

To construct the second bundle we make use of the period mapping. It is easy to see that this defines a holomorphic map $\pi: S \to \Omega/\Gamma$. On the other hand, recall that S and Ω/Γ have the structure of an algebraic variety defined over a finite extension of the field of rational numbers.

LEMMA 7. *The holomorphic mapping π coincides with the corresponding morphism of algebraic varieties, defined over a finite extension of the field of rational numbers.*

The fact that π is a morphism of varieties is true of any holomorphic mapping of the algebraic variety S into Ω/Γ. This is the content of the theorem of Borel and Kobayshi [3].

It remains to show that π is defined over a finite extension of $\mathbf{Q}$, or, what is the same, over $\overline{\mathbf{Q}}$. For this we recall that the group G of projective transformations of the space in which all the fibers of family f are imbedded acts on S. We consider an open set $U \subset S$ such that the quotient U/G exists. We must prove that the morphism $\pi: U \to \Omega/\Gamma$ is defined over $\overline{\mathbf{Q}}$, and that it can be represented as $\pi = h \cdot p$, where $p: U \to U/G$ is the natural projection, and $h: U/G \to \Omega/\Gamma$ is a morphism. It suffices to prove that h is defined over $\overline{\mathbf{Q}}$.

By the Torelli theorem proved in [2], h is an open immersion. We shall prove that $h^*(r) \in \overline{\mathbf{Q}}(U/G)$ for a function $r \in \overline{\mathbf{Q}}(\Omega/\Gamma)$. For this we use the fact that there are many points $x \in \Omega/\Gamma$ for which it is known that $r(x) \in \overline{\mathbf{Q}}$ for all $r \in \overline{\mathbf{Q}}(\Omega/\Gamma)$. These are the points corresponding to singular surfaces of CM-type (see the Corollary to Lemma 4 in §3). In particular, they are points corresponding to singular K3 surfaces. The point $h^{-1}(x)$ is defined over $\overline{\mathbf{Q}}$ if the point x corresponds to a singular K3 surface. This will be proved by a lemma. Thus the points $h^{-1}(x)$ corresponding to singular surfaces form an everywhere dense set in U/G, and for $r \in \overline{\mathbf{Q}}(\Omega/\Gamma)$ the function $h^*(r)$ takes algebraic values on this set. Then $h^*(r) \in \overline{\mathbf{Q}}(U/G)$ by Lemma 9.

Lemma 8. *If a point $x \in \Omega/\Gamma$ corresponds to a singular* K3 *surface, then the point $h^{-1}(x) \in U/G$ is defined over a finite extension of the field* $\mathbf{Q}$.

It suffices to prove that the point $h^{-1}(x)$ is invariant relative to a subgroup of finite index in the group of all automorphisms of $\mathbf{C}$. Since $h^{-1}(x)$ is the image of a G-orbit on U, our assertion will be proved if we verify that this orbit is left stable by a subgroup of finite index of Aut $\mathbf{C}$. This assertion is equivalent to the fact that if X is the polarized K3 surface corresponding to the point x (i.e. X is singular), then among the polarized surfaces X^σ, $\sigma \in \mathrm{Aut}\, \mathbf{C}$, there are only a finite number of nonisomorphic ones.

The last assertion follows from the classification of singular surfaces obtained in [2]. In fact, the Néron-Severi lattices S_X and S_{X^σ} of the surfaces X and X^σ are isomorphic. Therefore the orthogonal complements T_X and T_{X^σ} of these lattices in the homology group have the same discriminant, equal to the discriminant of S_X. But there exist only a finite number of nonisomorphic lattices with given discriminant, and the lattice T_X, as proved in [2] (the theorem in §8), uniquely determines the surface X up to isomorphism.

Lemma 9. *Let M be an irreducible algebraic variety defined over a finite extension k of the field* $\mathbf{Q}$. *If $f \in \mathbf{C}(M)$ is such that for some everywhere dense set $Z \subset M_{\overline{\mathbf{Q}}}$ the values of f belong to $\overline{\mathbf{Q}}$, then $f \in \overline{\mathbf{Q}}(M)$.*

Proof. Since M is defined over a finite extension k of $\mathbf{Q}$, it follows that there exist functions $f_0, f_1, \cdots, f_n$, $n = \dim M$, in $k(M)$ and an integer $m \geqslant 0$ such that any $f \in \mathbf{C}(M)$ can be represented uniquely in the form

$$f = \frac{1}{q} \sum_{k=0}^{m} p_k(f_1, \ldots, f_n) f_0^m,$$

where $p_k(f_1, \cdots, f_n)$, $k = 0, \cdots, m$, and $q(f_1, \cdots, f_n)$ are polynomials of $f_1, \cdots, f_n$ with complex coefficients; $(p_0, \cdots, p_m, q) = 1$.

Obviously, if $f = 0$ on an everywhere dense set of points, then $f \equiv 0$. By hypothesis, the infinite system of linear equations relative to the coefficients of q and p_k, $k = 0, \cdots, m$, of the form

$$q(f_1(x), \ldots, f_n(x)) f(x) = \sum p_k(f_1(x), \ldots, f_n(x)) f_0^k(x), \quad x \in M,$$

has a nonzero solution. This system is linear, and hence, if it has a nonzero solution in $\mathbf{C}$, it must have a nonzero solution in the field generated by the coefficients. From this it follows that it must have a solution in some finite extension of $\mathbf{Q}$. Thus there exists a function

$$\tilde{f} = \frac{1}{\tilde{q}} \sum \tilde{p}_k(f_1, \ldots, f_k) f_0^m$$

such that $\widetilde{q}, \widetilde{p}_k \in \overline{\mathbf{Q}}[T_1, \cdots, T_n]$ and $\widetilde{f}(x) = f(x)$ for $x \in \mathbf{Z}$. Then $\widetilde{f} = f$, which proves the lemma.

We now consider the family of abelian varieties A defined over $\Omega(l)/\Gamma$ in the introduction to the preceding section. We let X denote its pullback relative to the morphism $f\colon S \to \Omega(l)/\Gamma$. From what was proved earlier it follows that X' is defined over a finite extension of $\mathbf{Q}$ and it is strictly constant. The homomorphism $\widetilde{\rho}$ introduced in the preceding section defines a homomorphism $\rho\colon \mathrm{Hg}\,(A_\xi)(\mathbf{Q}) \to \mathrm{Aut}\, H^*(F_\xi, \mathbf{Q})$, where A_ξ and F_ξ are fibers of the bundles A and F.

We verify that ρ satisfies the conditions of Lemma 2, i.e. that it commutes with the action of the group $\pi(S_{\mathbf{C}}, s_0)$. The morphism $S \to \Omega(l)/\Gamma$ defines an epimorphism $\pi(S, s_0) \to \Gamma$, and the group $\pi(S, s_0)$ acts on the cohomology of the fibers of the bundle X via this homomorphism. For the bundle X' this is true by definition. That the homomorphism ρ commutes with the action of the group Γ is obvious from the invariant character of its definition.

Thus we have verified that the bundles X and X' constructed in this way satisfy the conditions of Lemma 2, and we can apply this lemma. We obtain a subgroup G of finite index in $\mathrm{Gal}\,(k(S)_l/k'(S))$, whose actions on $\mathrm{Aut}\, H^*(F_S, \mathbf{Q}_l)$ and $\mathrm{Hg}\,(A_S)(\mathbf{Q}_l)$ are connected by the homomorphism ρ.

We shall prove that the group $\mathrm{Gal}\,(k(S)_l/k(S))$ contains a subgroup G of finite index possessing the same property.

The diagram

$$\begin{array}{ccc} & \overset{\varphi}{\nearrow} & \mathrm{Hg}\,(A_S)(\mathbf{Q}_l) \\ \mathrm{Gal}\,(k(s)_l/k(s)) & \Psi & \downarrow \rho \\ & \searrow & \mathrm{Aut}\, H^*(F_S, \mathbf{Q}_l), \end{array}$$

where s is a generic point of the variety S, is commutative under a restriction on the subgroup Γ. On the other hand, passing to determinants, we obtain an analogous commutative diagram for the group $\mathrm{Gal}\,(k'(s)/k(s))$. From this it follows that the elements $\Psi(g)^{-1}(\rho\varphi)(g)$ commute with $\Psi(G)$. Since by hypothesis $\Psi(G)$ has finite index in the group of unimodular automorphisms of the algebra $H^*(F_S, \mathbf{Q}_l)$, and the center of the latter group is finite, it follows that the function $\Psi(g)^{-1}(\rho\varphi)(g)$ is equal to 1 on some neighborhood of 1, i.e. on the group $\widetilde{G}$.

Let $s_0 \in S$ be a point defined over a finite extension of the field $\mathbf{Q}$. In any finite Galois extension $k(s)_l \supset L \supset k'(s)$, $[L: k'(S)] < \infty$, we consider the normalization S_L of the variety S. The inverse images of the point s_0 in S_L are represented by the Galois group of $L/k'(s)$ and define a decomposition group in the residue class field. Considering the union of all such fields J', we obtain a residue class field $k(s_0)_l$ and a Galois group isomorphic to the decomposition group

$$\mathrm{Gal}\,(k\,(s_0)_l/k'\,(s_0)) \subset \mathrm{Gal}\,(k\,(s)_l/k'\,(s)). \tag{6}$$

From the functorial character of the cohomology groups and from the inclusion (6) we get the existence of a commutative diagram

$$\begin{array}{ccc} & \nearrow & \mathrm{Hg}\,(A_{s_0})(\mathbf{Q}_l) \\ \mathrm{Gal}\,(k(s_0)_l/k'(s_0)) & & \Big\downarrow \rho \\ & \searrow & \mathrm{Aut}\,H^*(F_{s_0}, \mathbf{Q}_l). \end{array}$$

In our case

$$\mathrm{Hg}\,(A_{s_0})\,(\mathbf{Q}_l) \subset \mathrm{Aut}\,\Lambda^2 H^1\,(F'_{s_0},\ \mathbf{Q}_l)$$

and ρ preserves the eigenvalues of the Frobenius automorphism.

Since A_{s_0} is an abelian variety, Theorem 3 will follow from this in view of the well-known results of Weil.

If F_{s_0} is a surface of CM-type, then, by Lemma 3, A_{s_0} is an abelian variety of CM-type. This implies Theorem 4 in view of the results of Shimura and Taniyama [5].

BIBLIOGRAPHY

1. I. R. Šafarevič, et al., *Algebraic surfaces*, Trudy Mat. Inst. Steklov. **75** (1965) = Proc. Steklov Inst. Math. **75** (1965). MR **32** #7557; **35** #6685.

2. I. I. Pjateckiĭ-Šapiro and I. R. Šafarevič, *A Torelli theorem for algebraic surfaces of type* K_3, Izv. Akad. Nauk SSSR Ser. Mat. **35** (1971), 530–572 = Math. USSR Izv. **5** (1971), 547–588. MR **44** #1666.

3. S. Kobayashi, Preprint.

4. D. Mumford, *A note on Shimura's paper "Discontinuous groups and Abelian varieties"*, Math. Ann. **181** (1969), 345–351. MR **40** #1400.

5. G. Shimura and Y. Taniyama, *Complex multiplication of abelian varieties and its applications to number theory,* Publ. Math. Soc. Japan, 6, Math. Soc. Japan, Tokyo, 1961. MR **23** #A2419.

6. M. Kuga and I. Satake, *Abelian varieties attached to polarized K_3-surfaces,* Math. Ann. **169** (1967), 239–242. MR **35** #1603.

7. Z. I. Borevič and I. R. Šafarevič, *Number theory,* "Nauka", Moscow, 1964; English transl., Pure and Appl. Math., vol. 20, Academic Press, New York, 1966. MR **30** #1080; **33** #4001.

8. I. I. Pjateckiĭ-Šapiro, *Interrelations between the Tate and Hodge conjectures for abelian varieties,* Mat. Sb. **85 (127)** (1971), 610–620 = Math. USSR Sb. **14** (1971), 615–625. MR **45** #3422.

Translated by J. S. JOEL

Trudy Mat. Inst. Steklov.
132 (1973)

Proc. Steklov Inst. Math.
132 (1973)

THE HILBERT MODULAR GROUP AND SOME ALGEBRAIC SURFACES *

F. HIRZEBRUCH

ABSTRACT. A detailed study is made of properties of modular groups for quadratic fields $Q(\sqrt{p})$, where p is a prime, $p \equiv 1 \pmod 4$.

Bibliography: 21 items.

Dedicated to D. C. Spencer on his 60-th birthday

In my Moscow lecture I reported about the resolution of the cusp singularities and gave some applications. My Bourbaki talk on these resolutions has appeared [12]. It sketches the applications only very briefly. Therefore I shall concentrate here on the applications. This preliminary report was written after the Moscow symposium. To keep it to reasonable size only the modular groups for the quadratic fields $\mathbf{Q}(\sqrt{p})$ where p is a prime $\equiv 1 \bmod 4$ will be discussed in detail. In my Moscow lecture and also already in the Bourbaki talk I mentioned the calculation of arithmetic genera. The fact that certain surfaces with arithmetic genus 1 are actually rational was not known to me at the time of the Moscow meeting and was added later.

Detailed proofs and discussions of arbitrary real quadratic fields have to appear elsewhere.

§ 1. The Euler number

Let k be a real quadratic field over $\mathbf{Q}$ and o the ring of algebraic integers in k. The group $SL_2(o)$ acts on $\mathfrak{H}\times\mathfrak{H}$ where $\mathfrak{H}$ is the upper half plane of $\mathbf{C}$. The Hilbert modular group $G = SL_2(o)/\{1, -1\}$ acts effectively.

If $z_1 = x_1 + iy_1$, $z_2 = x_2 + iy_2$ and $\mathfrak{H}\times\mathfrak{H}$ is given by $y_1 > 0$, $y_2 > 0$, then the Chern—Euler form (second Chern class) for an invariant Riemannian metric of the homogeneous domain $\mathfrak{H}\times\mathfrak{H}$ equals

$$c_2 = \frac{1}{4\pi^2}\frac{dx_1 \wedge dy_1}{y_1^2} \wedge \frac{dx_2 \wedge dy_2}{y_2^2}. \tag{1}$$

AMS (MOS) subject classifications (1970). Primary 10D05, 14G05; Secondary 12A25.

**Editor's note.* This reprinting of the original paper incorporates suggestions made by the author.

The c_2 — volume of $\mathfrak{H}\times\mathfrak{H}/G$ is[1] according to Siegel [19] (formula (19)) given by

$$c_2(\mathfrak{H}\times\mathfrak{H}/G)=\frac{1}{2}\pi^{-4}d^{3/2}\zeta_k(2), \tag{2}$$

where d is the discriminant of k and ζ_k the ζ-function of k which satisfies the well-known functional equation, namely, the expression

$$d^{s/2}\pi^{-s}\Gamma\left(\frac{s}{2}\right)^2\zeta_k(s)$$

remains unaltered under $s\to 1-s$. (Compare [13], p. 254). Therefore,

$$c_2(\mathfrak{H}\times\mathfrak{H}/G)=2\zeta_k(-1). \tag{3}$$

I know this formula from G. Harder and J. P. Serre (see [17], p. 159). The number $2\zeta_k(-1)$ is the Euler number of the discrete group G in the sense of Wall [21] and therefore a rational number.

There exist beautiful formulas for $\zeta_k(-1)$. They can be taken from Siegel [20] (a little bit reformulated). We give an example.

Let p be a positive integer $\equiv 1 \bmod 4$ and $k=\mathbf{Q}(\sqrt{p})$. Then

$$2\zeta_k(-1)=\frac{1}{15}\sum_{1\leqslant b<\sqrt{p},\ b\text{ odd}}\sigma_1\left(\frac{p-b^2}{4}\right), \tag{4}$$

where $\sigma_1(n)$ is the sum of the divisors of n.

We wish to determine the ordinary Euler number of the space $\mathfrak{H}\times\mathfrak{H}/G$. This is from the topological point of view a four-dimensional non-compact V-manifold in the sense of Satake [16] with finitely many quotient singularities. Such a singularity is locally the quotient of $\mathbf{C}^2$ by a linear action of the group of r-th roots of unity acting freely outside the origin. We call r the order of the singularity. Prestel [14] has determined the number a_r of quotient singularities of order r.

Theorem. *The Euler number* of $\mathfrak{H}\times\mathfrak{H}/G$ *is given by the formula*

$$e(\mathfrak{H}\times\mathfrak{H}/G)=2\zeta_k(-1)+\sum_r\frac{r-1}{r}a_r, \tag{5}$$

where a_r is the number of quotient singularities of order r.

Proof. We compactify $\mathfrak{H}\times\mathfrak{H}/G$ by the finitely many cusps and get an algebraic surface $\overline{\mathfrak{H}\times\mathfrak{H}/G}$. We remove «nice» open neighborhoods of the cusp singularities and the quotient singularities and obtain a manifold Y with boundary.

Its Euler number is

$$e(Y)=e(\mathfrak{H}\times\mathfrak{H}/G)-\sum_r a_r. \tag{6}$$

By the Gauss—Bonnet theorem of Allendoerfer—Weil—Chern [2],

[1] We will consistently write $\mathfrak{H}\times\mathfrak{H}/G$ to denote $(\mathfrak{H}\times\mathfrak{H})/G$.

$$e(Y)=\int_Y c_2+\int_{\partial Y}\Pi, \tag{7}$$

where Π is the 3-from defined on ∂Y induced by the field of unit outward normal vectors of ∂Y from a 3-from defined on the bundle of unit tangent vectors of Y. If we make the neighborhoods of the singularities smaller and smaller and change Y accordingly, then $\int_Y c_2$ converges to $2\zeta_k(-1)$. The boundary integral $\int_{\partial Y}\Pi$ is a sum of integrals over the various boundary components. For a quotient singularity the summand converges to $-1/r$ (see [2] formula (16)). For a cusp the corresponding summand converges to 0. This can be seen as follows. Up to an isometry of $\mathfrak{H}\times\mathfrak{H}$ we can assume that the boundary component is given by $y_1y_2=K$, where K is a sufficiently large positive constant. We denote this 3-dimensional compact manifold by B_K. It is locally homogeneous under the group of isometries of $\mathfrak{H}\times\mathfrak{H}$, and also any $\widetilde{B}_K$ can be carried in any other $\widetilde{B}_{K'}$ by an isometry of $\mathfrak{H}\times\mathfrak{H}$. (Here $\widetilde{B}_K$ denotes the subset $y_1y_2=K$ of $\mathfrak{H}\times\mathfrak{H}$.) Therefore there exists a number b independent of K such that

$$\int_{B_K}\Pi=b\,\text{volume}\,(B_K).$$

An invariant metric of $\mathfrak{H}\times\mathfrak{H}$ is

$$\frac{(dx_1)^2}{y_1^2}+\frac{(dx_2)^2}{y_2^2}+(d\,(\log y_1))^2+(d\,(\log y_2))^2.$$

For this metric we have

$$\text{volume}\,(B_K)=\frac{\sqrt{2}\,\log\varepsilon}{K}\int_T dx_1dx_2,$$

where ε is greater 1 and generates U_M^+ and $T=\mathbf{R}^2/M$, the cusp being given by the module $M\subset k$ and the group U_M^+ of totally positive units acting on M (compare [12]). Thus

$$\lim_{K\to\infty}\int_{B_K}\Pi=\lim_{K\to\infty}\text{volume}\,(B_K)=0. \tag{8}$$

Passing to the limit and applying formulas (6) and (7) yields (5).

R e m a r k. I know the above argument to prove (8) from Harder and Serre. There is a much more general result in [7]. The above theorem and its proof remain valid for the Hilbert modular group of any totally real algebraic number field of degree n over $\mathbf{Q}$. The Hilbert modular group acts on $\mathfrak{H}\times\ldots\times\mathfrak{H}$ (n factors) in this case.

§ 2. Euler number and arithmetic genus

We consider the field $k=\mathbf{Q}(\sqrt{p})$ where p is a prime $\equiv 1 \bmod 4$. There are also results (e. g. formulas for the arithmetic genus) for all real quadratic fields. Here we restrict ourselves for simplicity to some special cases with the hope to present a more complete treatment on some other occasion.

Theorem. *Let p be a prime $\equiv 1 \bmod 4$ and G the Hilbert modular group for the field $k=\mathbf{Q}(\sqrt{p})$. The Euler number of $\mathfrak{H}\times\mathfrak{H}/G$ is given by the formula*

$$e(\mathfrak{H}\times\mathfrak{H}/G)=2\zeta_k(-1)+\frac{1}{2}a_2+\frac{2}{3}a_3+\frac{4}{5}a_5. \tag{9}$$

We have $a_r=0$ for $r\neq 2, 3, 5$ and any p.

For $p=5$: $a_2=a_3=a_5=2$.

For $p>5$: $a_5=0$.

For any p: $a_2=h(-p)$, $a_3=h(-3p)$,

where $h(-b)$ denotes the class number of the field $\mathbf{Q}(\sqrt{-b})$.

This is just a special case of the theorem in § 1 using the results of Prestel [14].

Examples. For $p=5$ formula (4) gives

$$2\zeta_k(-1)=\frac{1}{15}\sigma_1(1)=\frac{1}{15},\qquad e(\mathfrak{H}\times\mathfrak{H}/G)=\frac{1}{15}+1+\frac{4}{3}+\frac{8}{5}=4.$$

For $p=29$ we get

$$2\zeta_k(-1)=\frac{1}{15}(\sigma_1(7)+\sigma_1(5)+\sigma_1(1))=1,\qquad e(\mathfrak{H}\times\mathfrak{H}/G)=1+\frac{6}{2}+\frac{2}{3}\cdot 6=8.$$

Theorem. *As in the preceding theorem we assume $k=\mathbf{Q}(\sqrt{p})$ where p is a prime $\equiv 1 \bmod 4$. The Euler number $e(\mathfrak{H}\times\mathfrak{H}/G)$ is always divisible by 4. Let χ_p be the arithmetic genus of the field of modular functions, i. e. the arithmetic genus of an arbitrary non-singular model of the compactification of $\mathfrak{H}\times\mathfrak{H}/G$. Then*

$$\chi_p=\frac{1}{4}e(\mathfrak{H}\times\mathfrak{H}/G). \tag{10}$$

Before proving (10) we give a table for χ_p calculated by using (4), (9), (10). Don Zagier using the IBM 7090 computer of Bonn University calculated χ_p for $p<2000$.

p	χ_p	p	χ_p	p	χ_p
5	1	101	5	229	10
13	1	109	5	233	12
17	1	113	5	241	14
29	2	137	6	257	13
37	2	149	7	269	12
41	2	157	7	277	14
53	3	173	8	281	16
61	3	181	8	293	13
73	3	193	10	313	19
89	4	197	9	317	14
97	4				

Proof of the theorem. For a non-singular compact connected algebraic surface X the arithmetic genus is defined as

$$\chi(X) = 1 - g_1 + g_2,$$

where g_j is the dimension of the space of holomorphic differential forms of degree j. The g_j are birational invariants. Therefore we can speak of the arithmetic genus of the function field of a (possibly singular) surface. We mean the arithmetic genus of some non-singular model. In terms of Chern numbers

$$\chi(X) = \frac{1}{12}(c_1^2 + c_2)[X] = \frac{1}{4}\left(c_2[X] + \frac{1}{3}(c_1^2 - 2c_2)[X]\right).$$

Thus

$$\chi(X) = \frac{1}{4}(e(X) + \operatorname{sign}(X)), \tag{11}$$

where $e(X)$ is the Euler number and $\operatorname{sign}(X)$ the signature of X (see [9], where $\operatorname{sign}(X)$ is denoted by $\tau(X)$ and called index). Let $\overline{\mathfrak{H} \times \mathfrak{H}/G}$ be the compactification. A non-singular model was constructed in [12]. Each cusp singularity has a cyclic resolution. See diagram

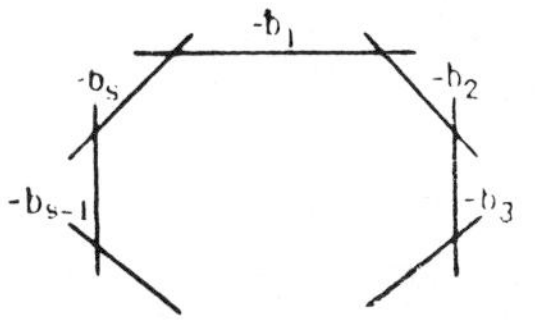

The lines indicate non-singular rational curves intersecting each other transversally. The $-b_j$ are the selfintersection numbers. Each quotient singularity has a linear resolution ([8], 3.4). For each singularity we consider a tubular neighborhood of the system of curves occuring in the resolution. This is a manifold with boundary. For such a manifold the Euler number and the signature are well defined. For a cusp the Euler number (as alternating sum of the Betti numbers) is $1-1+q=q$ where q is the number of curves in the resolution. The signature is $-q$ because the matrix of intersection numbers is negative definite for a singularity.

Thus

$$\text{Euler number} + \text{signature} = 0 \textit{ for a cusp.} \tag{12}$$

Clearly,

$$\text{Euler number} + \text{signature} = 1 \textit{ for a quotient singularity.} \tag{13}$$

For $\mathfrak{H}\times\mathfrak{H}/G$ the signature is well-defined since $\mathfrak{H}\times\mathfrak{H}/G$ is a (non-compact) rational homology manifold. If we remove from $\mathfrak{H}\times\mathfrak{H}/G$ «nice» open neighborhoods of the quotient singularities we obtain a manifold Y with boundary. Y and $\mathfrak{H}\times\mathfrak{H}/G$ have the same signature whereas the Euler number of Y equals the Euler number of $\mathfrak{H}\times\mathfrak{H}/G$ minus the number of quotient singularities. Our non-singular model X is obtained by pasting together Y and the tubular neighborhoods of the systems of curves belonging to the cusps and the quotient singularities. Since the Euler number and the signature behave additively under such pastings (compare [1], Prop. 7.1, and [10]) we have, by (11), (12) and (13),

$$\chi(X)=\frac{1}{4}(e(\mathfrak{H}\times\mathfrak{H}/G)+\operatorname{sign}(\mathfrak{H}\times\mathfrak{H}/G)). \tag{14}$$

We did not use yet our assumption that p is a prime $\equiv 1 \bmod 4$. We shall use it now.

There is a positive unit e with $N(e)=ee'=-1$. The bijection $(z_1, z_2)\to$ $\to(ez_1, e'\bar{z}_2)$ of $\mathfrak{H}\times\mathfrak{H}$ induces an orientation-reversing map of $\mathfrak{H}\times\mathfrak{H}/G$ onto itself. Therefore $\operatorname{sign}(\mathfrak{H}\times\mathfrak{H}/G)=0$, which proves (10).

R e m a r k. For our non-singular model X the number g_1 equals 0; thus the first Betti number of X is zero ([3], Satz 8). We have

$$\chi_p-1=g_2\geqslant 0.$$

g_2 equals the dimension of the space of cusp forms of weight 1 ([3a]).

We wish to determine the primes p with $\chi_p=1$. According to (2), (3), (9) and (10),

$$\chi_p>\frac{\pi^{-4}}{8}p^{3/2}\zeta_k(2).$$

Since $\zeta_k(2)\geqslant\zeta(4)$ where ζ is the ordinary ζ-function, we have

$$\chi_p>\frac{\pi^{-4}}{8}p^{3/2}\frac{\pi^4}{90}=\frac{p^{3/2}}{720}.$$

Thus $\chi_p>1$ for $p>81$. The above table then shows

$$\chi_p=1 \text{ if and only if } p=5,\ 13,\ 17. \tag{15}$$

Thus the non-singular model X of $\mathfrak{H}\times\mathfrak{H}/G$ can be a rational surface only if $p=5$, 13 or 17. We shall show later that it is indeed rational in these cases.

§ 3. The symmetric Hilbert modular group

Let X be a compact connected non-singular algebraic surface. The fixed point set of a holomorphic involution T of X (different from the identity) consists of finitely many points $p_1, \ldots, p_r$ and a disjoint union of connected non-singular curves $D_1, \ldots, D_s$. To get a non-singular model for X/T we blow up $p_1, \ldots, p_r$ into rational curves $E_1, \ldots, E_r$ of self-intersection number -1. Then we have an involution $\tilde{T}$ on the new model $\tilde{X}$ which was obtained from X by blowing up. The fixed-point set of $\tilde{T}$ equals

$$\bigcup_{i=1}^{s} D_i \cup \bigcup_{j=1}^{r} E_j.$$

The space $\tilde{X}/\tilde{T}$ is non-singular. The Lefschetz fixed-point theorem in the form given in [11] (formula (11)) and the equivariant signature theorem of [1] (see also [10], (7) and (9)) imply, in virtue of (11),

$$2\chi(\tilde{X}/\tilde{T}) = \chi(X) + \frac{1}{4}\sum_{i=1}^{s}(e(D_i) + D_i \cdot D_i) + \frac{1}{4}\sum_{j=1}^{r}(e(E_j) + E_j \cdot E_j).$$

Here e denotes the Euler number. Clearly, $e(E_j) + E_j \cdot E_j = 2 - 1 = 1$. By the adjunction formula,

$$e(D_i) + D_i \cdot D_i = c_1[D_i],$$

where c_1 is the first Chern class of X. Therefore, the arithmetic genus of any non-singular model of X/T (which will also be denoted by $\chi(X/T)$) is given by the formula

$$2\chi(X/T) = \chi(X) + \frac{r}{4} + \frac{1}{4}\sum_{i=1}^{s} c_1[D_i], \tag{16}$$

where r is the number of isolated fixed points of T and $D_1, \ldots, D_s$ are the components of the one-dimensional fixed-point set.

As in § 1 we consider the Hilbert modular group G for a real quadratic field k. The involution $(z_1, z_2) \rightarrow (z_2, z_1)$ of $\mathfrak{H}\times\mathfrak{H}$ induces an involution T of $\mathfrak{H}\times\mathfrak{H}/G$ and also of the compactification $\overline{\mathfrak{H}\times\mathfrak{H}/G}$. It can be lifted to an involution of our non-singular model because this was obtained by the canonical minimal resolutions of all occurring singularities. The lifted involution will also be called T. In the beginning of this section we have seen how to construct a non-singular model of $\overline{(\mathfrak{H}\times\mathfrak{H}/G)/T}$ from the non-singular model of $\overline{(\mathfrak{H}\times\mathfrak{H}/G)}$. We wish to calculate the arithmetic genus of the field of symmetric Hilbert modular functions, i. e. the arithmetic genus of a non-singular model of $\overline{(\mathfrak{H}\times\mathfrak{H}/G)/T}$. So far we succeeded only for $k = \mathbf{Q}(\sqrt{p})$, p a prime $\equiv 1 \bmod 4$. *We make this assumption in the following discussion.*

In $\mathfrak{H}\times\mathfrak{H}/G$ there are exactly two curves D_1, D_2 which are fixed pointwise under T. The first curve D_1 corresponds to the diagonal $z_1 = z_2$ of $\mathfrak{H}\times\mathfrak{H}$, whereas D_2 comes from $z_1=e_0^2z_2$. (Here e_0 is a fundamental unit with $N(e_0)=e_0e_0'=-1$.) The curve D_1 is isomorphic to $\mathfrak{H}$ divided by the ordinary modular group $SL_2(\mathbf{Z})$. The curve D_2 is isomorphic to $\mathfrak{H}$ divided by $\Gamma^*(p)$. This group was introduced by Fricke ([4], p. 357). It contains the matrix

$$\begin{pmatrix} 0 & -p^{1/2} \\ p^{-1/2} & 0 \end{pmatrix}$$

and is an extension of index 2 of $\Gamma^0(p)$:

$$\Gamma^0(p)=\left\{\begin{pmatrix} a & b \\ c & d \end{pmatrix}\in SL_2(\mathbf{Z}), \quad b\equiv 0 \bmod p\right\}.$$

In $\mathfrak{H}\times\mathfrak{H}$ there $h(-p)$ quotient singularities of order 2 and $h(-3p)$ of order 3. Half of those of order 3 have rotation numbers (ζ, ζ) for $\zeta^3=1$ (first kind) and half of them have rotation numbers (ζ, ζ^{-1}) for $\zeta^3=1$ (second kind), (see [14]). For $p=5$ there are in addition two quotient singularities of order 5. Of course, T acts on the set of quotient singularities. The description of this action is in Prestel [15]. It depends on the residue class of p mod 24. Therefore we define

$$\begin{aligned} \varepsilon&=1 \text{ for } p\equiv 1 \bmod 3, & \varepsilon&=0 \text{ for } p\equiv 2 \bmod 3,\\ \delta&=1 \text{ for } p\equiv 1 \bmod 8, & \delta&=0 \text{ for } p\equiv 5 \bmod 8. \end{aligned} \tag{17}$$

Theorem (Prestel [15]). *Let p be a prime*$\equiv 1$ mod 4. *Of the $h(-p)$ singularities of order 2, half of them lie on D_2 and not on D_1, and one of them (namely (i, i)) lies on D_1 and D_2 and is the only intersection point of D_1 and D_2 in $\mathfrak{H}\times\mathfrak{H}/G$. There are in addition δ quotient singularities of order 2 which are fixed under T. «They» lie neither on D_1 nor D_2. The remaining order 2 singularities are interchanged pairwise under T. The $h(-3p)/2$ singularities of order 3 and of second kind all lie on D_2. There is one singularity of order 3 and of first kind which lies on D_1 whereas ε such singularities lie on D_2. The remaining singularities of order 3 and of first kind are interchanged pairwise under T. For $p=5$, the two singularities of order 5 are interchanged under T. The involution T acts freely outside D_1, D_2, and the quotient singularities.*

We now look at our non-singular model X of $\overline{\mathfrak{H}\times\mathfrak{H}/G}$. The involution T and the curves D_1, D_2 lift to X. How do D_1 and D_2 pass through the resolved cusps? The following can be proved: D_1, D_2 only pass through the cusp at ∞. The remaining $h(p)-1$ cusps are interchanged pairwise under T. The cusp at ∞ is symmetric. It has a period (see [12])

$$\{2a+1, \; b_1, \ldots, b_s, \; b_s, \ldots, b_1\},$$

where $2a+1$ is the smallest odd number greater then $\sqrt{p}$. The curve D_1 intersects the curve with self-intersection number $-(2a+1)$ transversally. It does not intersect the other curves of the resolution. The curve D_2 passes transversally through the intersection of two curves of self-intersection number $-b_s$ (see (18)).

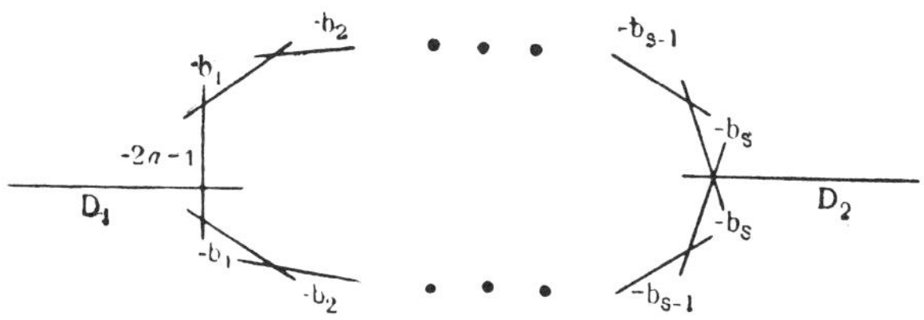

(18)

The involution T has one isolated fixed point on the curve with self-intersection number $-(2a+1)$. The curve D_1 passes through the resolution of one quotient singularity of order 3 and of first kind. On the single curve F_1 of self-intersection number -3 in the resolution there is one isolated fixed point of T. The same holds for D_2 when passing through the ε singularities of order 3 and of first kind (see (19)).

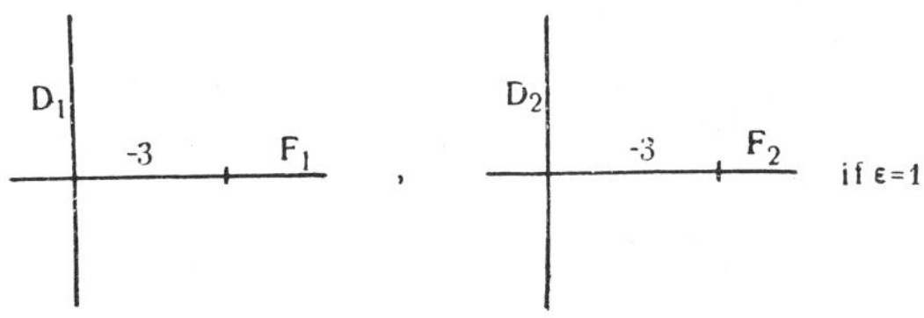

(19)

If we resolve the δ singularities of order 2, it can be shown that T has two isolated fixed points on the resulting curve with self-intersection number -2. Thus in our non-singular model X of $\overline{(\mathfrak{H}\times\mathfrak{H}/G)}/T$ the involution T has $2+2\delta+\varepsilon$ isolated fixed points and two fixed curves D_1 and D_2 which pass through the resolved singularity of order 2 at (i, i) in the way shown in the picture below.

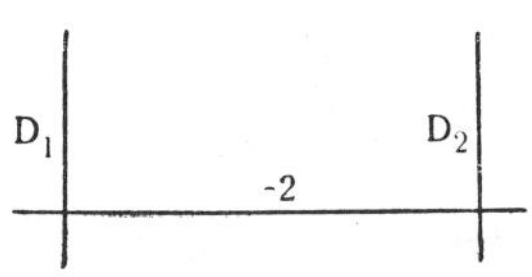

If c_1 is the first Chern class of X we have, according to (16),

$$2\hat{\chi}_p = \chi_p + \frac{1}{4}(2 + 2\delta + \varepsilon) + \frac{1}{4}(c_1[D_1] + c_1[D_2]). \tag{20}$$

Here $\hat{\chi}_p$ denotes the arithmetic genus of the function field of symmetric Hilbert modular functions ($p \equiv 1 \bmod 4$); in other words, it is the arithmetic genus of any non-singular model of $\overline{\mathfrak{H}\times\mathfrak{H}/G}/T$.

It remains to determine $c_1\,[D_j]$. We can only sketch this. The first Chern class c_1 (*with rational coefficients*) of our non-singular model X is a direct sum (à la Mayer—Vietoris) of the first Chern class of a neighborhood of the resolved cusps (whose Poincaré dual is simply *the sum of the curves in the resolution of each cusp*) *plus* the first Chern class of a neighborhood of the resolved quotient singularities *plus* the first Chern c_1 of $\mathfrak{H}\times\mathfrak{H}/G$ (with all quotient singularities removed). The adjunction formula shows that the first Chern class of a quotient singularity equals zero if only curves with self-intersection number -2 occur in its resolution. This happens for quotient singularities of order 2 and of order 3 (second kind). For order 3 and first kind the resolution is

$$\underline{D \quad -3} \quad (D \text{ non-singular rational curve}),$$

and by the adjunction formula the first Chern class is the Poincaré dual of $(1/3)D$.

The value of c_1 on a such a «diagonal» curve D_j equals twice the «volume» of D_j (normalized in the Euler sense, i. e. the volume of the fundamental domain of $SL_2(\mathbf{Z})$ on $\mathfrak{H}$ equals $-1/6$).

Therefore by (18) and (19)

$$c_1[D_1] = -\frac{1}{3} + \frac{1}{3} + 1 = 1, \quad c_1[D_2] = -\frac{p+1}{6} + \frac{\varepsilon}{3} + 2. \tag{21}$$

Here we used that $\Gamma^0(p)$ has index $p+1$ in $SL_2(\mathbf{Z})$.

Therefore the «volume» of $\mathfrak{H}/\Gamma^*(p)$ equals $-(p+1)/12$. From (20) and (21) we finally get the theorem.

Theorem. *Let p be a prime $\equiv 1 \bmod 4$. By χ_p we denote the arithmetic genus of the field of Hilbert modular functions, by $\hat{\chi}_p$ the arithmetic genus of the field of symmetric Hilbert modular functions for $k = \mathbf{Q}(\sqrt{p})$. Then*

$$\hat{\chi}_p = \frac{1}{2}\chi_p + \frac{1}{8}\left(-\frac{p+1}{6} + \frac{4}{3}\varepsilon + 2\delta + 5\right) = \frac{1}{2}\left(\chi_p - \left[\frac{p-5}{24}\right] + 1\right). \tag{22}$$

Here $[\cdot]$ denotes the greatest integer function.

Since

$$\hat{\chi}_p > \frac{p^{3/2}}{1440} - \frac{p+1}{48}$$

(see the remark in § 2) there are only finitely many p for which $\hat{\chi}_p = 1$. The tables mentioned in § 2 contain also the values for $\hat{\chi}_p$ for $p < 2000$. Above that bound certainly $\hat{\chi}_p > 1$. It turns out that there are exactly 24 primes $\equiv 1 \bmod 4$ for which the arithmetic genus equals 1, namely all primes ($\equiv 1 \bmod 4$) smaller than the prime 193, and 197, 229, 269, 293, 317. This was already mentioned in [12]. The reader may check that in the tables in § 2

$$\chi_p - 1 = \left[\frac{p-5}{24}\right] \quad \text{for} \quad p < 193.$$

For later use we give here information on the self-intersection numbers of D_1, D_2 in our non-singular model (obtained by minimal resolutions of all singularities of $\overline{\mathfrak{H} \times \mathfrak{H}/G}$).

By the adjunction formula,

$$c_1[D_j] - D_j \cdot D_j = e(D_j),$$

where e is the Euler number. D_1 is rational, i. e. $e(D_1) = 2$. Thus we have by (21)

$$D_1 \cdot D_1 = -1. \tag{23}$$

Moreover,

$$\textit{If } D_2 \textit{ is rational, then } D_2 \cdot D_2 = -\frac{p+1}{6} + \frac{\varepsilon}{3}. \tag{24}$$

According to Fricke ([4], p. 367) the curve D_2 is rational for $p=5$, 13, 17, 29, 41.

§ 4. Questions of rationality

We shall show that the field of Hilbert modular functions is rational, i. e. a purely transcendental extension of **C** of transcendence degree 2, for $k = \mathbf{Q}(\sqrt{p})$ and $p=5$, 13, 17. (See the remark at the end of § 2.) For this purpose we study a configuration of non-singular curves on our non-singular model X of $\overline{\mathfrak{H}\times\mathfrak{H}/G}$ and try to blow down some curves in order to show that X is a rational surface. In the following diagrams of curves on X which can be drawn for any prime $p \equiv 1 \bmod 4$ we give sometimes names to the curves and indicate the self-intersection numbers. On the nonsingular model X, the curves D_1 and D_2 intersect the curve E of self-intersection number -2 coming from the quotient singularity of order 2 located at $(i, i) \in \mathfrak{H}\times\mathfrak{H}$.

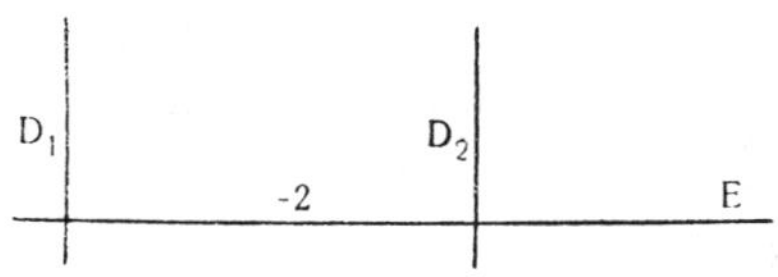

(25)

The other $h(-p)/2$ quotient singularities of order 2 on D_2 give rise to a configuration

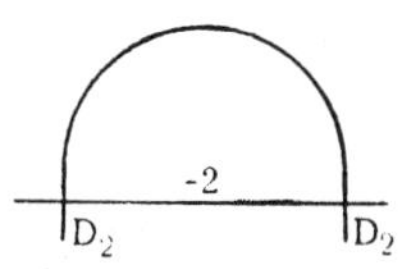

(26)

But we shall not need these $h(-p)/2$ curves. Important is however that we have a quotient singularity of order 3 and first kind on D_1 and sometimes on D_2 (see (19)). It can be shown that the $h(-3p)/2$ quotient singularities of order 3 and second kind on D_2 each give rise to a diagram

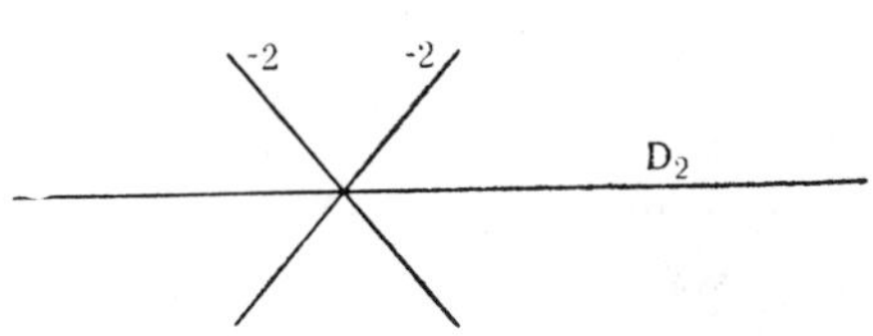

(27)

For $p=5$, 13, 17 there is only one cusp. The cycles in the sense of [12] are

$$\{3\} \text{ for } p=5,$$
$$\{5,\ 2,\ 2\} \text{ for } p=13,$$
$$\{5,\ 3,\ 2,\ 2,\ 3\} \text{ for } p=17.$$

In the following diagrams (fig. 1—fig. 3) we only draw the cusp curves and the curves needed for blowing down.

The diagrams show that many curves can be blown down. Van de Ven told me the following rationality criterion and proposed to apply it to these diagrams.

T h e o r e m. *Let X be a non-singular connected algebraic surface with vanishing first Betti number. If X has an irreducible non-singular rational curve D with non-negative self-intersection number, then X is rational.*

It suffices to prove this for $D \cdot D = 0$. It is not difficult to show in this case that $\dim H^0(X,\ \Omega(D)) = 2$. Then the rationality is a consequence of a theorem of M. Nöther (see for example [18] Proposition 3).

T h e o r e m. *Let p be a prime* $\equiv 1 \bmod 4$ *and* $k = \mathbf{Q}(\sqrt{p})$. *Let G be the Hilbert modular group. The non-singular model for* $\overline{\mathfrak{H} \times \mathfrak{H}/G}$ *is a rational surface if and only if* $p = 5,\ 13,\ 17$.

P r o o f. By (15) we only have to show the «if»-part and prove rationality in these cases. The assumption on the first Betti number is satisfied (§ 2). Consider the preceding diagrams.

For $p=5$, blow down D_1 and D_2 to get a nonsingular model in which E

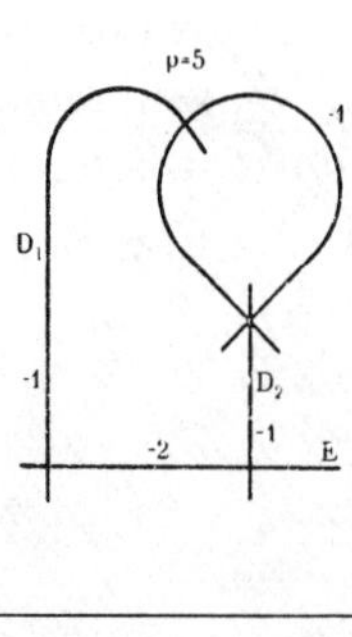

Fig. 1

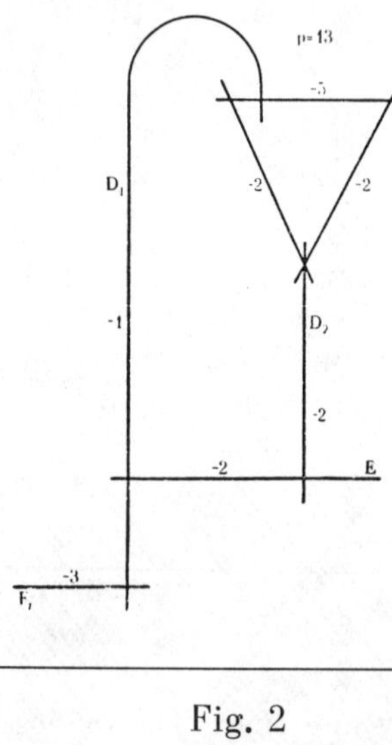

Fig. 2

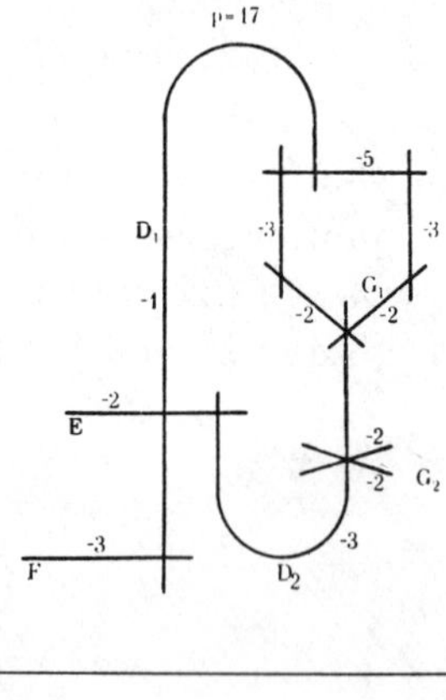

Fig. 3

is a rational non-singular curve with $E \cdot E = 0$.

For $p=13$, blow down consecutively D_1, E, F_1 to get a non-singular model in which D_2 is a rational non-singular curve with $D_2 \cdot D_2=0$.

For $p=17$, blow down consecutively D_1, E, F, D_2, G_1 to get a non-singular model in which G_2 is a rational curve with $G_2 \cdot G_2=0$.

Using the above rationality criterion it was possible to prove rationality also in some other cases. We state the following theorems without proof.

Theorem. *Let p be an arbitrary prime and G the Hilbert modular group for* $k=\mathbf{Q}(\sqrt{p})$. *Then the field of Hilbert modular functions is rational if and only if* $p=2, 3, 5, 7, 13, 17$.

Theorem. *Let p be an arbitrary prime and G the Hilbert modular group as before. Then the field of symmetric Hilbert modular functions is rational if* $p=2, 3, 5, 7, 11, 13, 17, 19, 23, 29, 37$.

For $p\equiv 1 \bmod 4$, according to § 3, only 24 primes have $\hat{\chi}_p=1$. For 19 of them it remains unsettled whether the field of symmetric Hilbert modular functions is rational or not.

Rationality was known up to now, as far as I know, only for $p=2, 3, 5$ (H. Cohn, E. Freitag, Gundlach [5, 6], W. F. Hammond).

BIBLIOGRAPHY

1. M. F. Atiyah and I. M. Singer, *The index of elliptic operators.* III, Ann. of Math. (2) **87** (1968), 546–604. MR **38** #5245.

2. S. S. Chern, *On the curvature integral in a Riemannian manifold*, Ann. of Math. (2) **46** (1945), 674–684. MR **7**, 328.

3. E. Freitag, *Über die Struktur der Funktionenkörper zu hyperabelschen Gruppen.* I, J. Reine Angew. Math. **247** (1971), 97–117. MR **43** #7404.

3a. ———, *Lokale und globale Invarianten der Hilbertschen Modulgruppe*, Invent. Math. **17** (1972), 106–134.

4. R. Fricke, *Die elliptischen Funktionen und ihre Anwendungen.* Vol. II, Teubner, Leipzig, 1922.

5. K.-B. Gundlach, *Die Bestimmung der Funktionen zur Hilbertschen Modulgruppe des Zahlkörpers* $Q(\sqrt{5})$, Math. Ann. **152** (1963), 226–256. MR **29** #1186.

6. ———, *Die Bestimmung der Funktionen zur einigen Hilbertschen Modulgruppen*, J. Reine Angew. Math. **220** (1965), 109–153. MR **33** #1290.

7. G. Harder, *A Gauss-Bonnet formula for discrete arithmetically defined groups*, Ann. Sci. École Norm. Sup. (4) **4** (1971), 409–455. MR **46** #8255.

8. F. Hirzebruch, *Über vierdimensionale Riemannsche Flächen mehrdeutiger analytischer Funktionen von zwei komplexen Veranderlichen,* Math. Ann. **126** (1953), 1–22. MR **16**, 26.

9. ———, *Topological methods in algebraic geometry,* 3rd ed., Die Grundlehren der math. Wissenschaften, Band 131, Springer-Verlag, New York, 1966. MR **34** #2573.

10. ———, *Involutionen auf Mannigfaltigkeiten,* Proc. Conf. on Transformation Groups (New Orleans, La., 1967), Springer-Verlag, Berlin and New York, 1968, pp. 148–166. MR **40** #3578.

11. ———, *Elliptische Differentialoperatoren auf Mannigfaltigkeiten*, Arbeitsgemeinschaft für Forschung des Landes Nordrhein-Westfalen, Natur-, Ingenieur- und Gesellschaftswissenschaften, Heft 157, Westdeutscher Verlag, Cologne, 1966, pp. 33–60. MR **36** #5961.

12. ———, *The Hilbert modular group, resolution of the singularities at the cusps and related problems,* Séminaire Bourbaki. 1970/71, Exposé 396, Lecture Notes in Math., vol. 244, Springer-Verlag, Berlin and New York, 1971.

13. S. Lang, *Algebraic number theory*, Addison-Wesley, Reading, Mass., 1970. MR **44** #181.

14. A. Prestel, *Die elliptischen Fixpunkte der Hilbertschen Modulgruppen*, Math. Ann. **177** (1968), 181–209. MR **37** #4019.

15. ———, *Die Fixpunkte der symmetrischen Hilbertschen Modulgruppe zu einem reell-quadratischen Zahlkörper mit Primzahldiskriminante*, Math. Ann. **200** (1973), 123–140.

16. I. Satake, *The Gauss-Bonnet theorem for V-manifolds*, J. Math. Soc. Japan **9** (1957), 464–492. MR **20** #2022.

17. J.-P. Serre, *Cohomologie des groupes discrets*, Prospects in Mathematics, Ann. of Math. Studies, vol. 70, Princeton Univ. Press, Princeton, N. J., 1971, pp. 77–169.

18. ———, *Critère de rationalité pour les surfaces algébriques*, Séminaire Bourbaki 1956/57, Exposé 146, Benjamin, New York, 1966. MR **33** #5420e.

19. C. L. Siegel, *The volume of the fundamental domain for some infinite groups*, Trans. Amer. Math. Soc. **39** (1936), 209–218; reprinted in *Gesammelte Abhandlungen*. Band I, Springer-Verlag, Berlin and New York, 1966, pp. 459–468. MR **33** #5441.

20. ———, *Berechnung von Zetafunktionen an ganzzahligen Stellen*, Nachr. Akad. Wiss. Göttingen Math.-Phys. Kl. II **1969**, 87–102. MR **40** #5570.

21. C. T. C. Wall, *Rational Euler characteristics*, Proc. Cambridge Philos. Soc. **57** (1961), 182–184. MR **23** #A185.

Trudy Mat. Inst. Steklov. 132 (1973)
Proc. Stekov Inst. Math. 132 (1973)

RECENT ADVANCES IN TRANSCENDENCE THEORY*

A. BAKER

ABSTRACT. Results relating to estimates for linear forms in logarithms are presented.

1. Originating from the well-known work of Gelfond and subsequent to the discoveries of 1966, there have appeared a succession of refinements giving positive lower bounds for linear forms in logarithms and I shall begin with an account of the latest of these. Suppose that $\alpha_1, \ldots, \alpha_n$ are non-zero algebraic numbers with degrees at most d and suppose further that $\alpha_1, \ldots, \alpha_{n-1}$ and α_n have heights at most A' and A respectively. Then it has recently been shown that for some $C = C(n, d, A')$, the inequalities

$$0 < |b_1 \log \alpha_1 + \ldots + b_n \log \alpha_n| < C^{-\log A \log B}$$

have no solution in rational integers $b_1, \ldots, b_n$ with $|b_j| \leqslant B$ [2]. The theorem is best possible with respect to B when A is fixed and with respect to A when B is fixed. Moreover the constant C is effectively computable. The proof incorporates many of the newer developments in this field; in particular it utilizes some fundamental studies of Feldman, and it also involves certain techniques introduced recently by Stark and myself [3]. We give now a brief sketch of the arguments.

Assuming that the assertion is false, one can prove, by means of Dirichlet's principle, that there exist integers $p(\lambda_{-1}, \ldots, \lambda_n)$, not all 0, such that

$$\sum_{\lambda_{-1}=0}^{L_{-1}} \cdots \sum_{\lambda_n=0}^{L_n} p(\lambda_{-1}, \ldots, \lambda_n) \Lambda(l) \alpha_1^{\lambda_1 l} \ldots \alpha_n^{\lambda_n l} = 0,$$

where

$$\Lambda(z) = \Delta(z + \lambda_{-1}; h, \lambda_0 + 1, m_0) \prod_{r=1}^{n-1} \Delta(b_n\lambda_r - b_r\lambda_n; m_r)$$

and

$$\Delta(z; k) = \frac{(z+1)\ldots(z+k)}{k!}, \quad \Delta(z; k, l, m) = (m!)^{-1}\left(\frac{d}{dz}\right)^m (\Delta(z; k))^l.$$

Here l runs through the range $1 \leqslant l \leqslant h$, where $h = [\log B]$, $m_0, \ldots, m_{n-1}$ run through all non-negative integers with $m_0 + \ldots + m_{n-1} \leqslant k \log A$ and

$$L_{-1} = h - 1, \quad L = L_0 = \ldots = L_{n-1} = \left[k^{1-\frac{1}{4n}} \log A\right], \quad L_n = [k^{1/2}],$$

k being a large parameter. Moreover, the integers $p(\lambda_1, \ldots, \lambda_n)$ can be chosen to have absolute values at most $A^{c_1 nk}$, where c_1, like $c_2, c_3, \ldots$

AMS (MOS) subject classifications (1970). Primary 10C15, 10F35; Secondary 12A25.

**Editor's note.* The body of this article is an unaltered reprinting of the original paper. Details of the bibliography have been checked and put in standard form, and MR citations have been added.

below, depend only on n, d and A'. The main arguments rest on an extrapolation procedure of now familiar kind which enables one to show that the same equations hold over the ranges

$$1 \leqslant l \leqslant hk^{n+1}, \qquad m_0 + \ldots + m_{n-1} \leqslant c_2 k \log A.$$

From this we further conclude, by another application of the extrapolation algorithm, that the equations hold with l replaced by l/q and with

$$1 \leqslant l \leqslant hk, \quad 0 \leqslant m_r \leqslant L \quad (0 \leqslant r < n), \quad 2 < q < 2L_n.$$

We can take q to be a prime p between L_n and $2L_n$ exclusive, and it then follows that the degree of the field generated by $\alpha_n^{1/p}$ over $Q(\alpha_1^{1/p}, \ldots, \alpha_{n-1}^{1/p})$ is less than p. Hence, by simple deductions relating to Kummer theory, we conclude that

$$\alpha_n = \alpha_1^{j_1} \ldots \alpha_{n-1}^{j_{n-1}} \alpha_n'^{p}$$

for some α_n' in the field generated by the α's over the rationals. But now α_n' has height at most $c_3 A'^{1/2}$, and on taking logarithms and substituting for $\log \alpha_n$ in the original inequality we obtain a similar expression but with α_n replaced by α_n' and $b_1, \ldots, b_n$ replaced by integers with absolute values at most $c_4 B$. An inductive argument then completes the proof of the theorem. It will be noticed that there has been no reference to an auxiliary function, the familiar starting point for work in this context. A function of this kind is, however, present in our exposition though it is here slightly obscured; it is, in fact, given by

$$\Phi(z_0, \ldots, z_{n-1}) = \sum_{\lambda_{-1}=0}^{L_{-1}} \cdots \sum_{\lambda_n=0}^{L_n} p(\lambda_{-1}, \ldots, \lambda_n) \Lambda(z_0) \alpha_1^{\gamma_1 z_1} \ldots \alpha_{n-1}^{\gamma_{n-1} z_{n-1}},$$

where

$$\gamma_r = \lambda_r - \frac{b_r \lambda_n}{b_n} \quad (1 \leqslant r < n)$$

and $\Lambda(z)$ is defined as above.

2. We mention now a few applications of the theorem. First it leads to a strengthening of several of the estimates obtained earlier in connexion with the effective resolution of Diophantine equations; in particular, it yields a further effective improvement on Liouville's inequality. In the course of this Colloquium I learnt that Feldman [4] has succeeded recently in obtaining a still sharper result; indeed he has proved that $|\alpha - p/q| > cq^{-\varkappa}$ for all rationals p/q $(q > 0)$, where α is any algebraic number with degree $n \geqslant 3$ and c, $\varkappa$, with $\varkappa < n$, are positive effectively computable numbers depending only on α. Rather curiously, in certain special cases, our theorem gives substantially more; we have, in fact, for any algebraic number α and positive integer m,

$$\left| \alpha^{1/m} - \frac{p}{q} \right| > cq^{-\varkappa \log m}$$

and this is better than the Thue—Siegel inequality when m is large. Further, the theorem gives another improvement on Ridout's generalization of Roth's theorem; we have namely

$$\left|\alpha - \frac{pp'}{qq'}\right| > Q^{-\varkappa \log\log Q'}$$

where p', q' are comprised solely of powers of fixed sets of primes and Q and Q' are the maxima of the absolute values of p, q and p', q' respectively.

The theorem, or rather a slight generalization of it, also enables one to determine effectively all the imaginary quadratic fields with class number 2; the analogous deduction for class number 1 was, of course, described by Gelfond and Linnik in 1949. The generalization in question relates to the solubility of the given inequality in algebraic numbers $b_1, \ldots, b_n$ and not merely rational integers; it is, in fact, easy to modify the arguments to show that, for any $\varepsilon > 0$, there exists $C = C(n, d, A', \varepsilon)$ such that the inequalities have no solution in algebraic numbers $b_1, \ldots, b_n$ with degrees at most d and heights at most B if log A is replaced by $(\log A)^{1+\varepsilon}$. For a full discussion of applications to class number problems we refer to the survey [1]. It would naturally be of interest to eliminate ε from the theorem cited above and it would also be of much value to derive inequalities of the same kind with A signifying the heights of all the α' s and not just α_n.

3. Finally we mention briefly some other results that have been obtained recently with the aid of the theory of linear forms in logarithms. Some twenty years ago Chowla considered the problem whether there exists a rational-valued function $f(n)$, periodic with prime period p, such that

$$\sum_{n=1}^{\infty} \frac{f(n)}{n} = 0;$$

he showed that this certainly could not hold for odd functions f if $^1/_2(p-1)$ is a prime, a condition subsequently removed by Siegel, and recently he showed that the same is true for even functions f if $f(0)=0$. In a paper by Birch, Wirsing and myself to appear in the Journal of Number Theory, it is shown that there is in fact no function f with the asserted properties. Actually we prove somewhat more than this, treating functions f that take algebraic values and are periodic with any modulus q. For instance we show that if $f(r)=0$ when $1 < (r, q) < q$ and if the q-th cyclotomic polynomial Φ_q is irreducible over $Q(f(1), \ldots, f(q))$, then $\sum \frac{f(n)}{n} \neq 0$. This implies, in particular, that, subject to the condition $(q, \varphi(q))=1$, the numbers $L(1, \chi)$ as χ runs through all nonprincipal characters mod q are linearly independent over the rationals. It is an open question whether this holds when $(q, \varphi(q)) > 1$.

Apart from the advances just discussed, a number of other developments in the subject have occurred recently; in particular several new results have been obtained in connexion with elliptic functions. In summary, it may perhaps be said that the work of the pioneers, and especially Gelfond, in the study of transcendental numbers has now grown into an extensive theory enriching diverse branches of mathematics, and one would hope for considerable progress in the future.

BIBLIOGRAPHY

1. Alan Baker, *On the class number of imaginary quadratic fields,* Bull. Amer. Math. Soc. 77 (1971), 678–684. MR 44 #3984.

2. ———, *A sharpening of the bounds for linear forms in logarithms,* Acta Arith. 21 (1972), 117–129. MR 46 #1717.

3. Alan Baker and H. M. Stark, *On a fundamental inequality in number theory,* Ann. of Math. (2) 94 (1971), 190–199. MR 46 #1716.

4. N. I. Fel'dman, *An effective refinement of the exponent in Liouville's theorem,* Izv. Akad. Nauk SSSR Ser. Mat. 35 (1971), 973-990 = Math. USSR Izv. 5 (1971), 985–1002. MR 44 #6609.

Trudy Mat. Inst. Steklov. 132 (1973)

Proc. Steklov Inst. Math. 132 (1973)

THE PRINCIPLE OF THE THEORY OF NONSTANDARD FUNCTIONAL EQUATIONS FOR DIRICHLET FUNCTIONS, CONSEQUENCES AND APPLICATIONS OF IT

A. F. LAVRIK

ABSTRACT. We state the general principle in the theory of nonstandard functional relations for Dirichlet functions possessing an ordinary functional equation of Riemann type. On the basis of this principle, we investigate periodic Dirichlet functions of the class of ζ-functions for function fields belonging to a class W.

Bibliography: 20 items.

The presence of functional equations analogous to Riemann's equation for $\zeta(s)$ is an important property of many Dirichlet functions used in number theory. In the general case it is stated in the following form:

The functions $\varphi(s)$ *and* $\psi(s)$, *representable for* $\mathrm{Re}\, s > \delta > 0$ *as absolutely convergent Dirichlet series*

$$\varphi(s) = \sum_{n=0}^{\infty} a_n e^{-\lambda_n s}, \qquad \psi(s) = \sum_{n=0}^{\infty} b_n e^{-\xi_n s}, \tag{1}$$

are meromorphic functions of finite order and satisfy the functional equation

$$A^s \Gamma_k(s)\varphi(s) = A^{\delta - s}\Gamma_k(\delta - s)\psi(\delta - s), \tag{2}$$

where $A > 0$, k *is an integer* $\geqslant 1$,

$$\Gamma_k(z) = \prod_{\nu=1}^{k} \Gamma(\alpha_\nu z + \beta_\nu), \tag{3}$$

$\Gamma(\omega)$ *is the gamma-function, the* α_ν *are nonnegative real numbers, the* β_ν *are complex numbers, and* $\beta_r \neq -m$ *for* $m = 0, 1, \cdots$ *if* $\alpha_r = 0$.

The class of Dirichlet functions $\varphi(s)$ thus defined is denoted by W. Functional equation (2), which describes this class, is the standard form for a functional equation of Riemann type.

In addition to this equation, we can find other forms of the functional equations for the same functions $\varphi(s)$; in particular, those which are in some sense more

---

AMS (MOS) subject classifications (1970). Primary 30A16, 39A15; Secondary 10H05, 30A08.

general or, at least, more directly useful in the theory of these functions and their applications.

EXAMPLE. *For the Riemann zeta-function, along with equation* (2) *in the form*

$$\pi^{-\frac{s}{2}}\Gamma\left(\frac{s}{2}\right)\zeta(s)=\pi^{-\frac{1-s}{2}}\Gamma\left(\frac{1-s}{2}\right)\zeta(1-s), \tag{4}$$

for any τ *with* $|\arg\tau|<\pi/2$ *we have*

$$\pi^{-\frac{s}{2}}\Gamma\left(\frac{s}{2}\right)\zeta(s)=\pi^{-\frac{s}{2}}\sum_{n=1}^{\infty}\frac{\Gamma\left(\frac{s}{2},\pi n^2\tau\right)}{n^s}$$
$$+\pi^{-\frac{1-s}{2}}\sum_{n=1}^{\infty}\frac{\Gamma\left(\frac{1-s}{2},\frac{\pi n^2}{\tau}\right)}{n^{1-s}}-\frac{\tau^{\frac{s-1}{2}}}{1-s}-\frac{\tau^{\frac{s}{2}}}{s}, \tag{5}$$

where $\Gamma(z,x)$ *is an ordinary incomplete gamma-function.*

Since for $|x|\longrightarrow\infty$ and $|\arg x|<3\pi/2$ we have

$$\Gamma(z,x)\sim e^{-x}x^{z-1},$$

the series in (5) are absolutely convergent for any complex s. The last relation for $\zeta(s)$ is more immediately interesting than its standard equation (3), since (4) follows from it by means of the simple replacement of s by $1-s$ and τ by $1/\tau$. Differentiating with respect to the parameter τ yields the well-known modular relation corresponding to $\zeta(s)$ [1]. For

$$\tau=\Delta\exp\left[i\left(\frac{\pi}{2}-\frac{1}{|\operatorname{Im}s|}\right)\operatorname{sgn}\operatorname{Im}s\right]$$

we obtain the approximate functional equation for $\zeta(s)$. For $\tau=1$ we have the differently written representation of Riemann [1] for $\zeta(s)$, by means of which he considered the properties and stated his hypotheses concerning the zeros of this function.

We can obtain the same kind of functional relations for other Dirichlet functions $\varphi(s)$ of the class W defined above. However, in reality a significantly more general proposition holds: the general principle in the theory of nonstandard functional relations for Dirichlet functions possessing an ordinary functional equation of Riemann type. Indeed, the following theorem is valid:

THEOREM A. *For* $0\leqslant m\leqslant k$ *and* $1\leqslant\nu\leqslant k$ *let* $\Gamma_m(z)$ *denote the product of any* m *factors* $\Gamma(\alpha_\nu z+\beta_\nu)$ *from* (3),

$$\Gamma_0(z)\equiv 1,\qquad \Gamma_{k-m}(z)\,\Gamma_m(z)=\Gamma_k(z),$$

$\beta_\nu^*=\operatorname{Re}\beta_\nu/\alpha_\nu$ *if* $\alpha_\nu\neq 0$, *and* $\beta_\nu^*=0$ *if* $\alpha_\nu=0$. *For* $\Delta>0$ *let the numbers* γ *and* σ *satisfy the inequalities*

$$\gamma < -\Delta, \qquad \sigma > \max_{\nu} (\delta + \Delta; \beta_\nu^*),$$

and let $H_{\gamma\sigma}$ *be the strip of the complex z-plane with* $\gamma \leqslant \operatorname{Re} z \leqslant \sigma$. *Then for* $\varphi(s) \in W$ *and* s *such that* $-\Delta \leqslant \operatorname{Re} s \leqslant \Delta$ *we have the relation*

$$A^s \sum_{n=0}^{\infty} a_n e^{-\lambda_n s} \int_{(\Delta')} \frac{\Gamma_r(z+s)}{\Gamma_m(-z+\delta-s)} \left(\frac{A}{e^{\lambda_n}}\right)^z \Phi(z)\, dz$$

$$-A^{\delta-s} \sum_{n=0}^{\infty} b_n e^{-\xi_n(\delta-s)} \int_{(\Delta'')} \frac{\Gamma_{k-m}(z+\delta-s)}{\Gamma_{k-r}(-z+s)} \left(\frac{A}{e^{\xi_n}}\right)^z \Phi(-z)\, dz$$

$$= 2\pi i \sum_{z \in H_{\gamma\sigma}} \operatorname{res} \left\{ \frac{\Gamma_r(z)}{\Gamma_m(\delta - z)} A^z \varphi(z) \Phi(z-s) \right\}, \tag{6}$$

which is valid for any meromorphic function $\Phi(z-s)$ *in the region* $H_{\gamma\sigma}$ *such that in* (6) *we can change the order of summation and integration, performed in the positive direction along the vertical lines* $\operatorname{Re} z = \Delta' = \sigma - \Delta$ *and* $\operatorname{Re} z = \Delta'' = \Delta - \gamma$, *and*

$$\left(\int_{\gamma+iT_j}^{\sigma+iT_j} - \int_{\gamma-iT_j}^{\sigma-iT_j} \right) \frac{\Gamma_r(z)}{\Gamma_m(\delta-z)} A^z \varphi(z) \Phi(z-s)\, dz \to 0$$

for some sequence of positive numbers $T_j \to \infty$. *The right side of* (6) *denotes the sum of the residues in strips of the region* $H_{\gamma\sigma}$, *or zero if there are none here.*

In the case $r = k$, $\alpha = \alpha_1 + \cdots + \alpha_k > 0$, and $\Phi(z) = F(z)/z$, identity (6) holds for any entire function $F(z)$ up to order 1 of type less than $\pi\alpha/2$. Moreover, we have $\Delta' = \Delta'' = \Delta^*$, where $\Delta^* = \operatorname{Re} z$ is any straight line that leaves the singularities of the integrand on its left. It is clear that an analogous identity can be written down for the functions $\psi(s) \in W$.

Functional relation (6) contains equation (2) but differs from it essentially in the fact that in a nontrivial way it includes a function Φ of a sufficiently arbitrary nature. Hence there is a big variety of functional relations between φ and ψ that preserve the symmetry property $s \leftrightarrow \delta - s$ and differ in the properties introduced by the nature of this function. As easy corollaries of Theorem A we obtain many well-known functional relations (f.r.). For example:

the f.r. of Siegel, from which follow Hamburger's theorem on the uniqueness of a Dirichlet series that satisfies Rièmann's equation for $\zeta(s)$ [1] and generalizations of it for similar purposes by Bochner and Chandrasekharan [2] and Chandrasekharan and Mandelbrojt [3];

the f.r. of Hardy and Littlewood [4] and Chandrasekharan and Narasimhan [5], which form the basis for their methods of obtaining approximate functional equations for Dirichlet functions;

the f.r. of Heilbronn [6], on which is based the derivation of Siegel's theorem on the real zero of $L(s, \chi)$ and the derivation of the Siegel-Brauer [7] asymptotic law for parameters of algebraic number fields;

the f.r. of Linnik [8], [9], which play an important role in the solution of additive problems by his dispersion method;

the f.r. of Schmidt [10], which were used by Hasse and Weil in their proof of the analogs of the Riemann hypothesis for ζ- and L-functions for algebraic function fields over a finite constant field, etc.

The significance of the general functional relation (6) is not only the fact that it joins, according to their form, methods of proof and applications, already known functional relations for Dirichlet functions. From it one can also obtain a number of useful new results. We shall be able to go into two such results briefly. The first is related to the problem of approximate functional equations for various ζ- and L-functions with uniform estimates in all their fundamental parameters. The second concerns arbitrary periodic Dirichlet functions defined by equations (2).

It is well known that in the theory and applications of, let us say, the zeta-function for algebraic number fields over the field of rational numbers, what is important is their estimates not only with respect to s, but also, sometimes even more so than with respect to s, with respect to the modulus of the discriminant of the field, the number of ideal classes, the regulator of the field, and for L-functions even the character modulus. The nature of the dependence on these parameters is different from the nature of the dependence on s, and the approximate equations obtained by known methods either in general do not concern such aspects of Dirichlet functions (Chandrasekharan and Narasimhan [5], et al.) or yield, with respect to them, estimates that differ significantly from the expected ones (Gel′fond [11], Tatuzawa [12], et al.). In 1960 Ju. V. Linnik [10] found a f.r. for $L(s, \chi)$ which, together with the f.r. of R. O. Kuz′min, reduced $L(s, \chi)$ to good uniform estimates both with respect to s and with respect to the modulus of the character d. This result was the starting point in our work. In the case

$$\Phi(z) = \frac{\tau^{\alpha z}}{z}, \qquad \arg \tau = \left(\frac{\pi}{2} - \frac{1}{|\operatorname{Im} s|}\right) \operatorname{sgn} \operatorname{Im} s$$

Theorem A allows us to solve, in general form, the problem concerning the approximate equations for Dirichlet functions of the class W with uniform estimates of the proper accuracy with respect to all their fundamental aspects.

Corresponding results were obtained by us in [13]–[18], and we shall not cite them here.

Furthermore, periodic (of arbitrary period) Dirichlet functions of the class W have been studied with the help of Theorem A. The interest in such functions is explained by the fact that among them are ζ_Ω- and L_Ω-functions for Ω-algebraic function fields over a finite constant field, for which Weil proved that the analog of the Riemann hypothesis was valid.

In the general case the following theorem holds:

THEOREM B. I. *The periodic Dirichlet functions* $\varphi, \psi \in W$ *have*:

a) *period* $2\pi i/\ln q$ *with* $q > 1$;

b) *Dirichlet series in the form*

$$\varphi(s) = \sum_{n=0}^{\infty} a_n q^{-ns}, \qquad \psi(s) = \sum_{n=0}^{\infty} b_n q^{-ns};$$

c) *functional equations in the form*

$$q^{\frac{ks}{2}} \varphi(s) = q^{\frac{k(\delta - s)}{2}} \psi(\delta - s)$$

with integral k;

d) *truncated functional equations in the form*

$$\varphi(s) = \sum_{n=0}^{m}{}' a_n q^{-ns} + q^{k\left(\frac{\delta}{2} - s\right)} \sum_{n=0}^{k-m}{}' b_n q^{-n(\delta - s)} + \sum_{z \neq s} \operatorname{res}\left\{ q^{m(z-s)} \frac{\varphi(z)}{s - z} \right\},$$

where m *is an arbitrary real number, the empty sum is assumed to be zero, and the prime on the summation sign means that in the case of integral* m *the last term is taken with weight* $1/2$.

II. a) *For given values* q, k *and* δ, *the entire set* $W(q, k, \delta; \omega_1, \cdots, \omega_p)$ *of functions* $\varphi, \psi \in W$ *having simple poles at the points*

$$z \equiv \omega_j \ (\operatorname{mod} 2\pi i/\ln q), \qquad \omega_i \not\equiv \omega_j \ (\operatorname{mod} 2\pi i/\ln q) \qquad \text{for } i \neq j,$$

consists of functions described by the relations

$$a_\nu^* = b_{k-\nu}^* q^{\delta\left(\nu - \frac{k}{2}\right)} + \ln q \sum_{j=1}^{p} q^{\nu\omega_j} A_j,$$

where for integral $\nu \geqslant 0$ *the numbers* $a_\nu^* = a_\nu$ *and* $b_\nu^* = b_\nu$ *are the coefficients of the Dirichlet series for the functions* φ *and* ψ *respectively, and* $a_\nu^* = b_\nu^* = 0$ *for other values of* ν; $A_j = D_j/D$,

$$D = \begin{vmatrix} q^{(k+1)\omega_1} & \cdots & q^{(k+j)\omega_1} & \cdots & q^{(k+p)\omega_1} \\ \cdots & \cdots & \cdots & \cdots & \cdots \\ q^{(k+1)\omega_j} & \cdots & q^{(k+j)\omega_j} & \cdots & q^{(k+p)\omega_j} \\ \cdots & \cdots & \cdots & \cdots & \cdots \\ q^{(k+1)\omega_p} & \cdots & q^{(k+j)\omega_p} & \cdots & q^{(k+p)\omega_p} \end{vmatrix},$$

and D_j *is the determinant obtained from* D *by replacing the* jth *column by the column of numbers*

$$a_{k+1}/\ln q, \ \ldots, \ a_{k+j}/\ln q, \ \ldots, \ a_{k+p}/\ln q.$$

b) *The functions* $\varphi \in W(q, k, \delta; \omega_1, \cdots, \omega_p)$ *have the form*

$$\varphi(s) = \sum_{n=0}^{m}{}' a_n q^{-ns} + q^{k\left(\frac{\delta}{2}-s\right)} \sum_{n=0}^{k-m}{}' b_n q^{-n(\delta-s)} + \ln q \sum_{j=1}^{p} q^{m(\omega_j - s)} \left(\frac{1}{q^{s-\omega_j}-1} + \frac{1}{2}\right) A_j,$$

where m *and* A *are the same as above.*

With the help of this theorem we can obtain substantial information concerning periodic Dirichlet functions of the class of ζ-functions for function fields belonging to the class W.

Let Ω *be an algebraic function field of one variable over a finite field of* q *constants, and let* ρ *be the genus of the field. Then* $\zeta_\Omega(s)$*—the zeta-function for* Ω*—is defined for* $\operatorname{Re} s > 1$ *in the form*

$$\zeta_\Omega(s) = \sum_{n=0}^{\infty} c_n q^{-ns},$$

where c_n *is the number of integral divisors of order* n *in* Ω, *is a meromorphic periodic function of period* $2\pi i/\ln q$, *has simple poles at the points*

$$s \equiv 1 \text{ and } 0 \pmod{2\pi i/\ln q}$$

and satisfies the functional equation

$$q^{(\rho-1)s}\zeta_\Omega(s) = q^{(\rho-1)(1-s)}\zeta_\Omega(1-s).$$

The question is, to what extent is the function $\zeta_\Omega(s)$, *i.e. the coefficients of its Dirichlet series, determined by these functional properties?*

Thanks to Hamburger's theorem, we know that only $\zeta(s)$ itself satisfies the functional Riemann equation for $\zeta(s)$ for sufficiently general assumptions on the regularity of the functions. Thus $\zeta(s)$ is completely determined by its functional equation. In the case of $\zeta_\Omega(s)$ the situation is different. Whatever the parameters q and ρ are, in addition to $\zeta_\Omega(s)$ an unbounded set of other Dirichlet functions satisfies the functional equation for $\zeta_\Omega(s)$ with other analytic properties identical with it. Among them are arbitrary functions for which, like $\zeta_\Omega(s)$, the analog of the Riemann hypothesis concerning zeros holds and arbitrary functions for which it does not hold.

Indeed, let $\zeta_{q\rho}$ *denote the class of meromorphic function* $Z(s)$ *representable for* $\operatorname{Re} s > 1$ *as absolutely convergent Dirichlet series*

$$Z(s) = \sum_{n=0}^{\infty} a_n q^{-ns}$$

with $q > 1$ *and* $a_0 = 1$ *having simple poles at the points* $s = 1$ *and* $0 \pmod{2\pi i/\ln q}$ *and functional equation*

$$q^{\rho s} Z(s) = q^{\rho(1-s)} Z(1-s)$$

with a nonnegative integer ρ. *Then the following theorem is valid:*

RIEMANN-ROCH THEOREM. *The function* $Z(s)$ *belongs to the class* $\zeta_{q\rho}$ *if and only if the coefficients* a_n *of its Dirichlet series are connected by the relation*

$$a_n = a_{2\rho-n} + (a_{\rho+1} - qa_{\rho-1}) \frac{q^{n-\rho} - 1}{q - 1},$$

where $a_\nu = 0$ *if* $\nu < 0$ *and, in particular, for* $\gamma \geqslant 1$ *we have*

$$a_{2\rho+\gamma} = (a_{\rho+1} - qa_{\rho-1}) \frac{q^{\rho+\gamma} - 1}{q - 1}.$$

COROLLARY.

$$q^{\rho s} Z(s) = \sum_{n=0}^{\rho}{}' a_n [q^{(\rho-n)s} + q^{(\rho-n)(1-s)}] - \frac{a_{\rho+1} - qa_{\rho-1}}{(q^s - 1)(q^{1-s} - 1)}.$$

Thus the functional properties of the functions $Z(s)$ completely determine the coefficients a_n of their Dirichlet series for all $n > \rho + 1$. As to the first $\rho + 1$ coefficients $a_1, \cdots, a_{\rho+1}$, they do not depend on these properties and can assume any values. Furthermore, obviously, the zeros of the Dirichlet functions $Z(s)$ coincide with the zeros of the expressions

$$(q + 1 - q^s - q^{1-s}) \sum_{n=0}^{\rho}{}' a_n [q^{(\rho-n)s} + q^{(\rho-n)(1-s)}] + qa_{\rho-1} - a_{\rho+1},$$

which are polynomials of degree $\rho + 1$ with respect to $Q = q^s + q^{1+s}$. Therefore the location of the zeros of the functions $Z(s) \in \zeta_{q\rho}$ depends only on q, ρ, and the first coefficients $a_1, \cdots, a_{\rho+1}$ of their Dirichlet series

$$Z(s) = \sum_{n=0}^{\infty} a_n q^{-ns}.$$

Moreover, for any values q and ρ we can find conditions on $a_1, \cdots, a_{\rho+1}$ under which all the zeros of $Z(s)$ will lie on the one vertical line $\operatorname{Re} s = 1/2$, and we can also construct any unbounded set of $Z(s) \in \zeta_{q\rho}$ for which this does not hold. Here we restrict ourselves to statements of these conditions for the simplest classes ζ_{q0} and ζ_{q1}.

THEOREM C. I. *All the zeros of* $Z(s) \in \zeta_{q0}$ *lie on the straight line* $\operatorname{Re} s = 1/2$ *if and only if the coefficient* a_1 *of the Dirichlet series for* $Z(s)$ *satisfies the inequality*

$$|a_1 - q - 1| \leqslant 2\sqrt{q}, \tag{$*$}$$

where the ordinates t *of the zeros are computed by the formula*

$$\cos(t \ln q) = (a_1 - q - 1)/2\sqrt{q}.$$

But if the inequality $(*)$ *does not hold, then the corresponding* $Z(s)$ *has a real zero* *or a zero with ordinate* $t = \pi/\ln q$ *lying in the half-plane* $\operatorname{Re} s > 1/2$.

II. *The analog of the Riemann hypothesis concerning zeros holds for those and only those functions* $Z(s) \in \zeta_{q1}$ *for which the coefficients* a_1 *and* a_2 *of the Dirichlet series for* $Z(s)$ *and the parameter* q *satisfy the inequalities*

$$|A - B| \leqslant 4\sqrt{q}, \quad |A + B| \leqslant 4\sqrt{q}, \quad C \geqslant 0,$$

where

$$A = q + 1 - a_1, \quad C = A^2 - 4(a_2 - q - a_1(q+1)), \quad B = \sqrt{|C|}.$$

Moreover, the ordinates t_1 *and* t_2 *of the zeros are computed by the formula*

$$\cos(t_\nu \ln q) = [A + (-1)^\nu B]/4\sqrt{q}.$$

A more detailed picture of the location of the zeros of the functions $Z(s) \in \zeta_{q1}$ is painted in [19], [20].

BIBLIOGRAPHY

1. E. C. Titchmarsh, *Theory of the Riemann zeta-function*, Cambridge Univ. Press, Cambridge, 1930; Russian transl., IL, Moscow, 1953.

2. S. Bochner and K. Chandrasekharan, *On Riemann's functional equation*, Ann. of Math. (2) **63** (1956), 336–360. MR **18**, 19.

3. K. Chandrasekharan and S. Mandelbrojt, *On Riemann's functional equation*, Ann. of Math. (2) **66** (1957), 285–296. MR **19**, 635.

4. G. H. Hardy and J. E. Littlewood, *The zeros of Riemann's zeta-function on the critical line*, Math. Z. **10** (1921), 238.

5. K. Chandrasekharan and R. Narasimhan, *Functional equations with multiple gamma factors and the average order of arithmetical functions*, Ann. of Math. (2) **76** (1962), 93–126. MR **25** #3911.

6. H. Heilbronn, *On Dirichlet series which satisfy a certain functional equation*, Quart. J. Math. Oxford Ser. **9** (1938), 194–195.

7. R. Brauer, *On the zeta-functions of algebraic number fields*. II, Amer. J. Math. **72** (1950), 739–746. MR **12**, 482.

8. Ju. V. Linnik, *An asymptotic formula in an additive problem of Hardy and Littlewood*, Izv. Akad. Nauk SSSR Ser. Mat. **24** (1960), 629–706; English transl., Amer. Math. Soc. Transl. (2) **46** (1965), 65–148. MR **23** #A130.

9. ————, *All large numbers are sums of a prime and two squares* (*A problem of Hardy and Littlewood*). II, Mat. Sb. **53** (**95**) (1961), 3–38; English transl., Amer. Math. Soc. Transl. (2) **37** (1964), 197–240. MR **22** #10964.

10. F. K. Schmidt, *Analytische Zahlentheorie in Körpern der Charakteristik p*, Math. Z. **33** (1931), 1–32.

11. A. O. Gel'fond, *Some functional equations implied by equations of Riemann type*, Izv. Akad. Nauk SSSR Ser. Mat. **24** (1960), 469–474. (Russian) MR **22** #7981.

12. T. Tatuzawa, *The approximate functional equation for Dirichlet's L-series*, Japan. J. Math. **22** (1952), 19–25. MR **16**, 16.

13. A. F. Lavrik, *Functional equations of Dirichlet functions*, Dokl. Akad. Nauk SSSR **171** (1966), 278–280 = Soviet Math. Dokl. **7** (1966), 1471–1473. MR **34** #4464.

14. ———, *A functional equation for Dirichlet L-series and the problem of divisors in arithmetic progressions*, Izv. Akad. Nauk SSSR Ser. Mat. **30** (1966), 433–448; English transl., Amer. Math. Soc. Transl. (2) **82** (1969), 47–65. MR **34** #4223.

15. ———, *On functional equations of Dirichlet functions*, Izv. Akad. Nauk SSSR Ser. Mat. **31** (1967), 431–442 = Math. USSR Izv. **1** (1967), 421–432. MR **35** #4174.

16. ———, *An approximate functional equation for the zeta-function of Hecke in an imaginary quadratic field*, Mat. Zametki **2** (1967), 475–482 = Math. Notes **2** (1967), 779–783. MR **36** #2571.

17. ———, *An approximate functional equation for the Dirichlet L-function*, Trudy Moskov. Mat. Obšč. **18** (1969), 91–104 = Trans. Moscow Math. Soc. **18** (1969), 101–116. MR **38** #4424.

18. ———, *Approximate functional equations for Dirichlet functions*, Izv. Akad. Nauk SSSR Ser. Mat. **32** (1968), 134–185 = Math. USSR Izv. **2** (1968), 129–180. MR **36** #6361.

19. ———, *On zeros of periodic Dirichlet functions*, Dokl. Akad. Nauk SSSR **192** (1970), 1214–1216 = Soviet Math. Dokl. **11** (1970), 802–805. MR **44** #413.

20. ———, *On the zeros of periodic Dirichlet functions defined by functional equations*, Internat. Congress Math. (Nice, 1970), 265 Communications Individuelles, p. 175. (Russian)

Translated by R. LENET

Trudy Mat. Inst. Steklov.
132 (1973)

Proc. Steklov Inst. Math.
132 (1973)

LARGEST PRIME FACTOR OF THE PRODUCT OF k CONSECUTIVE INTEGERS *

K. RAMACHANDRA

§ 1. Introduction

It is the object of this talk to present some of my results concerning lower bounds for the greatest prime factor $P=P(u, k)$ of the product $(u+1)\times$ $\times(u+2)\ldots(u+k)$, where u and k are natural numbers. *We assume that $k\geqslant k_0$ an (effective) absolute constant and also that the constants implied by the Vinogradov's symbol $\gg$ are absolute and effective (C, C_1, ... will denote absolute constants exceeding* 1).

Two results follow easily from well-known results. First, if $u\leqslant k^{3/2}$ one at least of the integers $u+1, \ldots, u+k$ is a prime number from the well-known deep results due to G. Hoheisel and A. E. Ingham. In other words $P\geqslant u+1$. On the other hand the work of A. Baker [2] on the equation $y^2=ax^3+bx^2+cx+d$ gives the lower bound $\gg\log\log u$ for the greatest prime factor of $(u+1)(u+2)$ and hence for P (see for instance [7] or the method of proving theorem 3 of [8]). In the present state of knowledge it seems difficult even to prove $P>k(\log k)^2$ for all $u\geqslant k^{3/2}$, though in various ranges for u it will be seen that we can improve on the result $P>{}^1/_3 k\log k$ $(u\geqslant k^{3/2})$, proved by a simple device by Erdös [4] (see also sections § 6 and § 7 for further comments and results).

§ 2. The range $k^{3/2}\leqslant u\leqslant k^{\log\log k}$

By combining a lemma of mine (lemma 1 below) with the sieve method of A. Selberg and a theorem of van der Corput I was able to prove (for details see [8])

Theorem 1. *If $k^{3/2}\leqslant u\leqslant k^{\log\log k}$ then*

$$P>k^{1+2^{-\lambda}}\quad where\quad \lambda=8+\frac{\log u}{\log k}. \tag{1}$$

Lemma 1. *Let $j=2$ for $k^{3/2}\leqslant u\leqslant k^3$ and more generally for $u\geqslant k^3$, $j=\frac{\log u}{\log k}-1$ or $\left[\frac{\log u}{\log k}\right]$ according as u is a power of k or not. Then*

AMS (MOS) subject classifications (1970). Primary 10A25.

**Editor's note.* This reprinting of the original paper incorporates suggestions made by the author.

$$\sum_{l \leqslant j-1} \sum_{n \leqslant uk^{-1}} \left\{ \vartheta\left(\left(\frac{u+k}{n}\right)^{1/l}\right) - \vartheta\left(\frac{u}{k}\right)^{1/l} \right\} = k \log \frac{u}{k} + O\left(\frac{k \log u}{\log \log k}\right), \quad \ldots \tag{2}$$

where as usual

$$\vartheta(x) = \sum_{p \leqslant x} \log p.$$

We now make two remarks about the proof of theorem 1 van der Corput's results [3] enable us to get the estimate

$$\sum_{1 \leqslant l \leqslant j-1} \sum_{u < np^l \leqslant u+k} \sum_{k < p \leqslant k^{1+\alpha_1}} \log p = O\left(\frac{k \log u}{\log \log k}\right) \text{ (where } \alpha_1 = 2^{-j-1}).$$

By using in addition A. Selberg's sieve method with careful handling of the error term we can get a similar estimate for the sum

$$\sum_{u < np \leqslant u+k} \sum_{k < p \leqslant k^{1+\alpha_2}} \log p \quad \text{(where } \alpha_2 = 2^{-j-7}).$$

These remarks will lead to the proof of theorem 1.

§ 3. The range $k^{\log \log k} \leqslant u \leqslant \exp\{(\log k)^2 (\log \log k)^{-1}\}$

Lemma 1 together with an adaptation of the method of H. Halberstam and K. F. Roth [5] enabled me to prove (see [9]) $P > (2-10^{-4})\, k \log k$ (the constant 10^{-4} is unimportant).

R. Tijdeman has a novel adaptation of the same method which gives the same lower bound in a bigger range and this will be explained in § 6. The results of [9] may be still of interest because of some lemmas contained therein.

§ 4. Some comments about the range $\exp\{(\log k)^2 (\log \log k)^{-1}\} \leqslant u \leqslant \exp\{(\log k)^{100}\}$

If we assume that $P \leqslant (2-10^{-4})\, k \log k$ it is not difficult to show that there exist two (distinct) numbers among $u+1, \ldots, u+k$ of the form mp^l, where $m \leqslant k^C$ and $k < p \leqslant 2k \log k$, the number l being the same for both. This at once raises the question of lower bounds for the quantity $L_1 = |\beta_1 \log \alpha_1 + \beta_2 \log \alpha_2|$ where α_1 and α_2 are multiplicatively independent positive rational numbers with height $\leqslant A$ and β_1, β_2 rational numbers not both zero with height $\leqslant (\log A)^{200}$. I obtained in [9] the lower bound $L_1 \gg \exp\{-(\log A)^{4.001}\}$. I later discussed this problem with A. Baker and we thought that some new ideas may be necessary to improve this. Combining the results of Tijdeman [12] with Baker's theory T. N. Shorey [11] recently succeeded in proving the estimate $L_1 \gg \gg \exp\{-(\log A)^3 (\log \log A)^3\}$. The best possible result in this direction seems to be very difficult to prove. Shorey's results show incidentally that $P > (2-10^{-4})\, k \log k$ for $u \geqslant \exp\{(\log k)^3 (\log \log k)^4\}$. My result gives this inequality in $u \geqslant \exp\{(\log k)^{4.001}\}$.

§ 5. The range $u \geqslant \exp\{(\log k)^{100}\}$

We begin by stating a lemma due to Erdös.

Lemma 2 (Erdös). *Let u and k be natural numbers and for $i=1, 2, \ldots, k$ let m_i denote the number obtained by omitting in the prime factor decomposition of $u+i$ all primes exceeding k. For each $p \leqslant k$ we select an i such that p divides m_i to the maximal power and call it an exceptional i. Then*

$$\Pi' m_i \leqslant k^k,$$

where Π' denotes the product over non exceptional i.

For the proof of this lemma see my paper [10].

Using this lemma we see that among $u+1, \ldots, u+k$ there exist two distinct numbers of the form $m'm''$, where all the prime factors of m' exceed k but the number of prime factors of m' does not exceed $\frac{5\pi(P)}{k}$ $\left(\pi(x)=\sum_{p\leqslant x} 1\right)$. Also m'' does not exceed k^4. This at once raises the question of lower bounds for the quantity $L=\left|\sum_{i=1}^{n} \beta_i \log \alpha_i\right|$, where $\alpha_1, \ldots, \alpha_n$ are multiplicatively independent positive rational numbers with heights not exceeding A and $\beta_1, \ldots, \beta_n$ are rational numbers not all zero with heights not exceeding B. If we follow Baker's work [1] closely we are led to the lower bound

$$L > \exp\{-(10^6 n^3 \log(70A))^{5n^4} (\log(70B))^{5n}\}$$

or slightly better ones. N. I. Feldman [6] has given much better results in a certain sense. But for our purposes it will be useful to have smaller orders of magnitude than $5n^4$ even at the cost of increasing the magnitude of $5n$ (see the remarks after theorem 3). In this connection I was led to an improvement [10] (with some slight changes in Baker's argument), namely

Theorem 2.

$$L > \exp\{-(10^6 n^3 \log(70A))^{5n^2} (\log(70B))^{5n}\}. \tag{3}$$

As a corollary one can deduce

Theorem 3. *If $u \geqslant k^{3/2}$, then*

$$P \gg k \log k \left(\frac{\log\log u}{\log\log k + \log\log\log u}\right)^{1/2}. \tag{4}$$

Remark 1. Theorem 3 is significant only with some condition of the kind $u \geqslant \exp\{(\log k)^{100}\}$ since the constant implied by $\gg$ is small. Also the estimate $P \gg \log\log u$ is better for $u \geqslant \exp\exp k^2$.

Remark 2. A result of the type

$$L > \exp\{-(\log(70A))(70B)^{\varepsilon(B) f(n)}\} \tag{5}$$

(where $\varepsilon(B)$ is a function depending only on B and tends to zero as B tends to infinity and where $f(n)$ is a function only of n), is very much desirable since it would imply better results than (4) viz $P > Ak \log k$ for every large constant A and $k \geqslant k_0(A)$ provided $u > k^{\log\log k}$ and by theorem 1 for all $u \geqslant k^{3/2}$. But (5) seems to be beyond the reach of the present methods though it appears to go only a little beyond the trivial result

$$L > \exp\{-70nB \log(70A)\}.$$

§ 6. Tijdeman's results [13] in the range $k^{\log\log k} \leqslant u \leqslant \exp\{(\log k)^{\log\log k}\}$

As already mentioned, a simple device due to Erdös gives $P > \frac{1}{3} k \log k$ if $u \geqslant k^{3/2}$. His method shows that if $u \geqslant k^{C_1}$, then $P > (1 - C_1^{-1}) k \log k$ and so theorem 1 imlies that if $u \geqslant k^{3/2}$ then $P > (1 - 10^{-4}) k \log k$. I succeeded in proving $P > (2 - 10^{-4}) k \log k$ for $u \geqslant k^{3/2}$ except when u satisfies $\exp\{(\log k)^2 (\log\log k)^{-1}\} \leqslant u \leqslant \exp\{(\log k)^{4.001}\}$. An improvement of Shorey's result with $(\log A)^{3/2}$ in place of $(\log A)^3 (\log\log A)^3$ would suffice to cover these exceptional values. But this seems to be difficult. R. Tijdeman closed the gap by proving

Theorem 4. *If u lies in the range specified in the title of* § 6, *then*

$$P > (2 - 10^{-4}) k \log k.$$

It is the object of this section to give a rough sketch of the proof of Tijdeman [13]. We may for convenience confine to the case $\exp\{(\log k)^{3/2}\} \leqslant u \leqslant \exp\{(\log k)^5\}$ by my results. Under the assumption that $P \leqslant (2-10^{-4}) k \log k$ it is not difficult to show that there exist $\geqslant k (\log k)^{-10}$ distinct numbers among $u+1, \ldots, u+k$ of the form mp^l, where $m \leqslant k^{C_2}$, $k < p \leqslant 2k \log k$ and l is the same for all the numbers. On the other hand Tijdeman shows that the number of such numbers is $\leqslant k^{1-\delta}$ for an absolute positive constant δ. His proof is as follows. Consider two such numbers $m_0 p_0^l$ and mp^l. We can write $mp^l - m_0 p_0^l = p^l (m (p/p_0)^l - m_0)$. The idea now is to approximate to x^l by rational functions $P(x)/Q(x)$ where $P(x)$ and $Q(x)$ are polynomials of degree $\leqslant t-1$, with integer coefficients whose absolute values are not too large. The requirement of the approximation is that $\frac{P(x)}{Q(x)} - x^l = (x-1)^{\vartheta} \frac{R(x-1)}{Q(x)}$, where ϑ is as large as possible and it turns out that $R(x-1)$ is polynomial with integer coefficients whose absolute values are not too large. This can be done by using the well-known lemma (due to Siegel, see for instance Lemma 1 of [1]) on the estimate for the solution of a system of linear homogeneous equations with integer coefficients.

Hence we have

$$mP\left(\frac{p}{p_0}\right) - m_0 Q\left(\frac{p}{p_0}\right) = m\left(\frac{p}{p_0} - 1\right)^{\vartheta} R\left(\frac{p}{p_0} - 1\right) + (mp^l - m_0 p_0^l)\, p_0^{-l} Q\left(\frac{p}{p_0}\right).$$

It follows that if $|p - p_0| \leqslant 2p_0$, $m|p - p_0|^{\vartheta} \leqslant 1/3 (12lt)^{-4t(l+t)(2t-\vartheta)^{-1}} p_0^{(\vartheta - t + 1)}$ and further $|mp^l - m_0 p_0^l| \leqslant 1/3 (12lt)^{-4t(l+t)(2t-\vartheta)^{-1}} p_0^{l-t+1}$. Then $mP(p/p_0) =$

$=m_0 Q(p/p_0)$. It will be convenient to set $\vartheta=[\alpha l]$, $t=[\beta l]$ where α and β are constants satisfying $\frac{1}{2}<\beta<1$ and $2\beta>\alpha>1$. We divide the range $k<p\leqslant 2k\log k$ into disjoint intervals I of equal length Y (Y suitably chosen). If $m_0p_0^l$ is a number such that p_0 is in some interval and mp^l is another number with p in the same interval it will follow that $mP(p/p_0)=$ $=m_0Q(p/p_0)$. It is not difficult to obtain the bound $\ll(\log k)^{10}$ for all mp^l with p belonging to I. Summing up over all the I's we are led to the result stated.

§ 7. Concluding remarks

Let $n_1<n_2<n_3\ldots$ be all the natural numbers which have at least one prime factor $>k$. Our results are nearly equivalent to bounds for $n_{i+1}-n_i$ in various ranges for n_i. We have $n_{i+1}-n_i\leqslant 2$ for $n\geqslant n_0(k)$. Erdös raised in [4] the question of estimating $F(k)=\max\limits_i(n_{i+1}-n_i)$. Erdös proved that $F(k)\leqslant 3\pi(k)(1+o(1))$ and proposed the problem of proving $F(k)\leqslant\pi(k)$. I proved that $F(k)\leqslant\pi(k)(1+o(1))$. Combining Theorems 1 and 3 of mine with his results in § 6, Tijdeman was in the first to prove that $F(k)\leqslant{}^1/_2\pi(k)(1+o(1))$. This settles the question of Erdös. Much remains to be done in this derection and a result of the type (5) together with my theorem 1 would imply that $F(k)=$ $=o(\pi(k))$. Actually it was conjectured by Erdös that $F(k)=O((\log k)^2)$.

Acknowledgements. I am thankful to the Soviet Academy for inviting me to lecture, to Professor P. Erdös for his constant interest in my work, and to Professor R. Tijdeman for communicating his proof sketched in § 6. My thanks are also due to Tata Institute of Fundamentall Research for enabling me to attend the Conference.

Remark.** After my return from the conference, T. N. Shorey and myself (see Acta Arith. XXIV (1973), 99–111) have used the results of [2] together with a suitable adaptation of Baker's well-known method to prove

$$F(k)=O\left(\frac{k}{\log k}\left(\frac{\log\log\log\log k}{\log\log\log k}\right)^{1/2}\right).$$

Subsequently Shorey refined the adaptation further, and this enables us to replace the exponent ½ by 1. Finally, M. Jutila has improved the results of [2] in a remarkable way, and this enables us to deduce by the method of Shorey and myself that

$$F(k)=O\left(\frac{k\log\log\log k}{\log k\log\log k}\right).$$

***Added in reprinting.*

BIBLIOGRAPHY

1. Alan Baker, *Linear forms in the logarithms of algebraic numbers.* I, Mathematika **13** (1966), 204–216. MR **36** #3732.

2. ———, *The Diophantine equation* $y^2 = ax^3 + bx^2 + cx + d$, J. London Math. Soc. **43** (1968), 1–9. MR **38** #111.

3. J. G. van der Corput, *Zahlentheoretische Abschätzungen mit Anwendung auf Gitterpunktprobleme.* II, Math. Z. **28** (1928), 301–310.

4. P. Erdös, *On consecutive integers,* Nieuw Arch. Wisk. (3) **3** (1955), 124–128. MR **17**, 461.

5. H. Halberstam and K. F. Roth, *On gaps between consecutive k-free integers,* J. London Math. Soc. **26** (1951), 268–273. MR **13**, 208.

6. N. I. Fel′dman, *Improved estimate of a linear form of the logarithms of algebraic numbers,* Mat. Sb. **77 (119)** (1968), 423–436 = Math. USSR Sb. **6** (1968), 393–406. MR **38** #1059.

7. M. Keates, *On the greatest prime factor of a polynomial,* Proc. Edinburgh Math. Soc. (2) **16** (1968/69), 301–303. MR **41** #1689.

8. K. Ramachandra, *A note on numbers with a large prime factor.* II, J. Indian Math. Soc. **34** (1970), 39–48 (1971). MR **45** #8616.

9. ———, *A note on numbers with a large prime factor.* III, Acta Arith. **19** (1971), 49–62. MR **45** #8617.

10. ———, *Application of Baker's theory to two problems considered by Erdös and Selfridge,* J. Indian Math. Soc. (to appear).

11. T. N. Shorey, *On a theorem of Ramachandra,* Acta Arith. **20** (1972), 215–221.

12. R. Tijdeman, *An auxiliary result in the theory of transcendental numbers,* J. Number Theory **5** (1973), 80–94.

13. ———, *On the maximal distance of numbers with a large prime factor,* J. London Math. Soc. (2) **5** (1972), 313–320.

Trudy Mat. Inst. Steklov.
132 (1973)

Proc. Steklov Inst. Math.
132 (1973)

ON THE UNIFORM BOUNDEDNESS OF THE TORSION OF ELLIPTIC CURVES OVER ALGEBRAIC NUMBER FIELDS

V. A. DEM'JANENKO

Dedicated to Academician Ivan Matveevič Vinogradov on his Eightieth Birthday

ABSTRACT. One of the results: the torsion of elliptic curves whose invariants are defined over a given field k is uniformly bounded over the set of all algebraic number fields of fixed degree.

Bibliography: 12 items.

In this lecture I wish to sketch the results which I have obtained on the torsion of elliptic curves over algebraic number fields. These results are published in [7]–[12].

Let k be an algebraic number field of finite degree n, X an elliptic curve over k and Φ the group of k-points of finite order on X.

In this survey [1] Cassels expressed the following conjecture.

CONJECTURE. *For a given k (in particular, for $k = Q$) the order of the groups Φ is bounded.*

Investigations by a number of authors were devoted to proving this. Thus, for example, in his elegant paper [2] Manin proved the following theorem.

Let p be a fixed prime number. There exists a constant c (depending only on p and k) such that the order of the group of p-torsion k-points of an elliptic curve defined over k does not exceed c.

Later in [3] Paršin proved under some restrictions on the field k that if $m = \Pi_i p_i^{\alpha_i}$, then $\Sigma_i \alpha_i$ is bounded from above by a constant. At about the same time a similar result was obtained by Novodvorskiĭ and Pjateckiĭ-Šapiro [4].

At the basis of my investigation of the torsion of elliptic curves is the interconnection of fields possessing points of finite order with Kummer fields.

Consider the following situation. Suppose that the Jacobi curve $Y: v^2 = u^4 + au^2 + b$ is defined over the field Q of rational numbers and that it has a primitive

AMS (MOS) subject classifications (1970). Primary 14G25, 14H25, 10B10; Secondary 12A35, 12A05, 14G05, 10B45.

point O_m of order m over Q. Put the coordinates u_t, v_t of the points tO_m into correspondence with U_t, V_t, W_t, where $u_t = U_t/W_t$, $v_t = V_t^2/W_t^2$ and U_t, V_t, W_t are defined from the recurrence relations

$$u_1 = U_1, \quad v_1 = V_1, \quad W_1 = 1,$$
$$U_t = U^4_{\frac{t}{2}} - bW^4_{\frac{t}{2}}, \quad W_t = 2U_{\frac{t}{2}}V_{\frac{t}{2}}W_{\frac{t}{2}},$$
$$V_t = V^4_{\frac{t}{2}} - (a^2 - 4b)\, U^4_{\frac{t}{2}} W^4_{\frac{t}{2}}$$

for $t \equiv 0 \pmod 2$ and

$$U_t U_1 = U^2_{\frac{t-1}{2}} U^2_{\frac{t+1}{2}} - bW^2_{\frac{t-1}{2}} W^2_{\frac{t+1}{2}},$$
$$V_t V_1 = V^2_{\frac{t-1}{2}} V^2_{\frac{t+1}{2}} - (a^2 - 4b)\, U^2_{\frac{t-1}{2}} U^2_{\frac{t+1}{2}} W^2_{\frac{t-1}{2}} W^2_{\frac{t+1}{2}},$$
$$W_t W_1 = U^2_{\frac{t-1}{2}} W^2_{\frac{t+1}{2}} - U^2_{\frac{t+1}{2}} W^2_{\frac{t-1}{2}}$$

for $t \not\equiv 0 \pmod 2$.

Then for any rational integers r and s, and any t which does not satisfy the congruence $2t \equiv 0 \pmod m$, the following equalities hold:

$$\begin{aligned} U_{r+s}U_{r-s} &= U_r^2U_s^2 - bW_r^2W_s^2, \\ V_{r+s}V_{r-s} &= V_r^2V_s^2 - (a^2 - 4b)\, U_r^2U_s^2W_r^2W_s^2, \\ W_{r+s}W_{r-s} &= U_s^2W_r^2 - U_r^2W_s^2, \\ U_{r\pm s}W_{r\mp s} &= U_rW_rV_s \mp U_sW_sV_r \end{aligned} \tag{1}$$

and

$$\frac{U_t}{W_t} = 2\varepsilon_t \prod_{m/d_i} \prod_{j=0}^{\left[\frac{d_i}{2}\right]} A_{d_i, j}^{\left\{\frac{tj}{d_i}\right\} d_i}, \qquad \frac{V_t}{U_tW_t} = \varepsilon'_t \prod_{m/d_i} \prod_{j=0}^{\left[\frac{d_i}{2}\right]} B_{d_i, j}^{\left\{\frac{tj}{d_i}\right\} d_i}$$

or $\qquad\qquad\qquad\qquad\qquad\qquad\qquad\qquad\qquad\qquad\qquad\qquad\qquad\qquad$ (2)

$$\frac{U_t}{W_t} = \varepsilon_t \prod_{m/d_i} \prod_{j=0}^{\left[\frac{d_i}{2}\right]} A_{d_i, j}^{\left\{\frac{tj}{d_i}\right\} d_i}, \qquad \frac{V_t}{U_tW_t} = 4\varepsilon'_t \prod_{m/d_i} \prod_{j=0}^{\left[\frac{d_i}{2}\right]} B_{d_i, j}^{\left\{\frac{tj}{d_i}\right\} d_i},$$

where $A_{d_i, j}$ and $B_{d_i, j}$ are relatively prime rational integers, $\{l\}$ is the distance from the number l to the nearest integer, and ϵ_t, ϵ'_t take the values ± 1.

We observe that in case of prime m the formulas (2) were also obtained by Hellegouarch [5].

Now if we assume m to be congruent to zero modulo 2^4, 3^3 or p^2, where p is a prime > 3, then by formulas (1) and (2) there must exist nontrivial rational points on the curves $x^4 - 1 = y^2$, $x^3 + y^3 = 1$ or $x^p + y^p = 1$.

The situation considered above is a special case of the following theorem.

THEOREM 1. *Let* k *be an algebraic number field of degree* n *over the field of rational numbers,* Y *a Jacobi curve over* k *and* O_m *a* k*-point of order* m *on* Y. *If* $m \equiv 0 \pmod{2^4}$, $m \equiv 0 \pmod{3^3}$ *or* $m \equiv 0 \pmod{p^2}$, *then the curves* $x^4 - 1 = y^2$, $x^3 + y^3 = 1$ *or* $x^p + y^p = 1$ *respectively have nontrivial points which are rational over* k.

This theorem indicates the interconnection of fields possessing points of finite order with Kummer fields.

Furthermore, let ρ_l $(l = 1, 2, 3)$ be roots of the equation $\rho^3 + r\rho + s = 0$, and let $tO_{M,\,l} = \{u_{t,\,l}, v_{t,\,l}\}$ and $tO'_{M,\,l} = \{u'_{t,\,l}, v'_{t,\,l}\}$ be points on the curves Y_l: $v^2 = u^4 - 6\rho_l u^2 - 4r - 3\rho_l^2$ and Y'_l: $v'^2 = u'^4 + 3\rho_l u'^2 + r + 3\rho_l^2$, corresponding to the point $tO_m = \{x_t, y_t\}$ of the curve X: $y^2 = x^3 + rx + s$ according to the following formulas:

$$u_{t,\,l} = \frac{y_t}{x_t - \rho_l}, \qquad \frac{v_{t,\,l}}{u_{t,\,l}} = \sqrt{x_{2t} - \rho_l},$$
$$u'_{t,\,l} = \frac{v_{t,\,l}}{u_{t,\,l}}, \qquad \frac{v'_{t,\,l}}{u'_{t,\,l}} = u_{2t,\,l}. \tag{3}$$

Obviously $M = m$ or $= m/2$, and Y_l, Y'_l, $tO_{M,\,l}$, $tO'_{M,\,l}$ are defined over the field $K_l = k(\rho_l)$.

We observe the following, the simplest properties of points of finite order on elliptic curves.

1. The set S of elliptic curves that are birationally equivalent to a given curve X over a finite extension of the field k contains only a finite number which possess points of finite order $m > 2$.

2. If $m/(m, t)$ contains at least two distinct prime divisors, or if $m/(m, t) \equiv 0 \pmod 8$, or if $m/(m, t) \geqslant 12n$, then the coordinates of the points tO_m, $tO_{M,\,l}$ and $tO'_{M,\,l}$ are integral.

3. Let K'_l be a field in which all divisors of K_l are principal. For any t for which $u_{t,\,l}, v_{t,\,l} \neq 0, \infty$ we have

$$\frac{U_{t,\,l}}{W_{t,\,l}} = 2\varepsilon_{t,\,l}\frac{P_l}{P_{(t,\,m),\,l}} \prod_{m/d_i} \prod_{j=0}^{\left[\frac{d_i}{2}\right]} A_{d_i,\,j,\,l}^{\left\{\frac{tj}{d_i}\right\} d_i},$$
$$\frac{V_{t,\,l}}{U_{t,\,l}W_{t,\,l}} = \varepsilon'_{t,\,l}\frac{P_l}{P_{(t,\,m),\,l}} \prod_{m/d_i} \prod_{j=0}^{\left[\frac{d_i}{2}\right]} B_{d_i,\,j,\,l}^{\left\{\frac{tj}{d_i}\right\} d_i} \tag{4a}$$

or

$$\frac{U_{t,l}}{W_{t,l}} = \varepsilon_{t,l}\frac{P_l}{P_{(t,m),l}} \prod_{m/d_i} \prod_{j=0}^{\left[\frac{d_i}{2}\right]} A_{d_i,j,l}^{\left\{\frac{tj}{d_i}\right\}d_i},$$

$$\frac{V_{t,l}}{U_{t,l}W_{t,l}} = 4\varepsilon'_{t,l}\frac{P_l}{P_{(t,m),l}} \prod_{m/d_i} \prod_{j=0}^{\left[\frac{d_i}{2}\right]} B_{d_i,j,l}^{\left\{\frac{tj}{d_i}\right\}d_i}, \tag{4b}$$

where $U_{t,l}$, $V_{t,l}$ and $W_{t,l}$ are defined by formulas (1), $A_{d_i,j,l}$, $B_{d_i,j,l}$, $P_{(t,m),l}$ and P_l are relatively prime integers of the field K'_l, $m \equiv 0 \pmod{P_{(t,m),l}}$ and $\epsilon_{t,l}$ and $\epsilon'_{t,l}$ are units of the field K_l.

4. For sufficiently large values of m there exist a sufficiently large number of points tO_m such that

$$c_0^{-1} < h_{t_1}/h_{t_2} < c_0, \qquad t_1, t_2 \in \{t\}, \tag{5}$$

where c_0 is an absolute constant, and

$$h_t = \ln\left[\prod_{i=1}^{3n} \max\{|(U_t/W_t)^{(i)}|, \ |(V_t/U_tW_t)^{(i)}|\}\right].$$

Already from these properties of the points of finite order of elliptic curves we can obtain definite results relating to the conjecture of Cassels expressed above. We give just two of these.

THEOREM A. *If* $m \equiv 0 \pmod{2p^2}$, *where* p *is a prime* > 3, *then there exists a constant* $c(p, k)$ *depending only on* p *and* k *such that* $m < c(p, k)$.

THEOREM B. *If* $m \equiv 0 \pmod{16}$ *or* $m \equiv 0 \pmod{18}$, *there exists an effectively calculable constant* $c(k)$, *depending only on* k, *such that* $m < c(k)$.

Note that in the proof of Theorem A we must apply an old result of Mumford [6] to the curve $x^p + y^p = 1$, and that in the proof of Theorem B we must use the fact that the sets of k-points on the curves $x^4 - 1 = y^2$ and $x^3 + y^3 = 1$ form groups with a finite number of generators.

We generalize formulas (2) and (4) in the following way. Let $o_{1,m}$ and $o_{2,m}$ be points on X, let p and q be rational integers whose least positive residues mod 2 are l_1 and l_2, and let α, β, γ and δ be respectively equal to $(p + l_1)/2$, $(q + l_2)/2$, $(p - l_1)/2$ and $(q - l_2)/2$.

Then the coordinates of the points $po_{1,m} + qo_{2,m} = \{x_{p,q}, y_{p,q}\}$ can be represented in the form $x_{p,q} = X_{p,q}/Z^2_{p,q}$, $y_{p,q} = Y_{p,q}/Z^3_{p,q}$, where $X_{p,q}$, $Y_{p,q}$ and $Z_{p,q}$ are defined from the recurrence relations

$$x_{1,0}=X_{1,0},\quad y_{1,0}=Y_{1,0},\quad Z_{1,0}=1,\qquad x_{0,1}=X_{0,1},\quad y_{0,1}=Y_{0,1}\quad Z_{0,1}=1,$$
$$X_{1,1}=-2y_{1,0}y_{0,1}+(x_{1,0}+x_{0,1})(x_{1,0}x_{0,1}+r)+2s,$$
$$Y_{1,1}=(x_{1,0}+3x_{0,1})(ry_{1,0}-x_{1,0}^2y_{0,1})+(3x_{1,0}+x_{0,1})(x_{0,1}^2y_{1,0}-ry_{0,1})$$
$$+4s(y_{1,0}-y_{0,1}),$$
$$Z_{1,1}=x_{0,1}-x_{1,0},$$
$$X_{p,q}X_{l_1,l_2}=(X_{\alpha,\beta}X_{\gamma,\delta}-rZ_{\alpha,\beta}^2Z_{\gamma,\delta}^2)^2-4s(X_{\alpha,\beta}Z_{\gamma,\delta}^2+X_{\gamma,\delta}Z_{\alpha,\beta}^2)Z_{\alpha,\beta}^2Z_{\gamma,\delta}^2,$$
$$Y_{p,q}Y_{l_1,l_2}=X_{\alpha,\beta}^3X_{\gamma,\delta}^3+r(X_{\alpha,\beta}X_{\gamma,\delta}-rZ_{\alpha,\beta}^2Z_{\gamma,\delta}^2)(X_{\alpha,\beta}^2Z_{\gamma,\delta}^4$$
$$+3X_{\alpha,\beta}X_{\gamma,\delta}Z_{\alpha,\beta}^2Z_{\gamma,\delta}^2+X_{\gamma,\delta}^2Z_{\alpha,\beta}^4)+s(X_{\alpha,\beta}Z_{\gamma,\delta}^2$$
$$+X_{\gamma,\delta}Z_{\alpha,\beta}^2)(X_{\alpha,\beta}^2Z_{\gamma,\delta}^4+8X_{\alpha,\beta}X_{\gamma,\delta}Z_{\alpha,\beta}^2Z_{\gamma,\delta}^2$$
$$+X_{\gamma,\delta}^2Z_{\alpha,\beta}^4-2rZ_{\alpha,\beta}^4Z_{\gamma,\delta}^4)-(r^3+8s^2)Z_{\alpha,\beta}^6Z_{\gamma,\delta}^6,$$
$$Z_{p,q}Z_{l_1,l_2}=X_{\gamma,\delta}Z_{\alpha,\beta}^2-X_{\alpha,\beta}Z_{\gamma,\delta}^2.$$

In correspondence with $X_{p,q}$, $Y_{p,q}$ and $Z_{p,q}$ we put

$$U_{p,q,l}=Y_{p,q}/\sqrt{X_{p,q}-\rho_lZ_{p,q}^2},$$
$$V_{p,q,l}=(X_{p,q}-\rho_lZ_{p,q}^2)^2-(r+3\rho_l^2)Z_{p,q}^4,$$
$$W_{p,q,l}=\sqrt{X_{p,q}-\rho_lZ_{p,q}^2}\,Z_{p,q}.$$

In this case for arbitrary rational integers α,β and γ,δ we have

$$U_{\alpha+\gamma,\beta+\delta,l}U_{\alpha-\gamma,\beta-\delta,l}=U_{\alpha,\beta,l}^2U_{\gamma,\delta,l}^2+(4r+3\rho_l^2)W_{\alpha,\beta,l}^2W_{\gamma,\delta,l}^2,$$
$$V_{\alpha+\gamma,\beta+\delta,l}V_{\alpha-\gamma,\beta-\delta,l}=V_{\alpha,\beta,l}^2V_{\gamma,\delta,l}^2-16(r+3\rho_l^2)U_{\alpha,\beta,l}^2U_{\gamma,\delta,l}^2W_{\alpha,\beta,l}^2W_{\gamma,\delta,l}^2,$$
$$W_{\alpha+\gamma,\beta+\delta,l}W_{\alpha-\gamma,\beta-\delta,l}=U_{\gamma,\delta,l}^2W_{\alpha,\beta,l}^2-U_{\alpha,\beta,l}^2W_{\gamma,\delta,l}^2,$$
$$U_{\alpha\pm\gamma,\beta\pm\delta,l}W_{\alpha\mp\gamma,\beta\mp\delta,l}=U_{\alpha,\beta,l}W_{\alpha,\beta,l}V_{\gamma,\delta,l}\mp U_{\gamma,\delta,l}W_{\gamma,\delta,l}V_{\alpha,\beta,l},\tag{6}$$

where

$$\left.\begin{aligned}
\frac{U_{p,q,l}}{W_{p,q,l}}&=2\varepsilon_{p,q,l}\frac{P_l}{P_{p,q,l}}\prod_{m/d_i}\prod_{j,s=0}^{\left[\frac{d_i}{2}\right]}A_{d_i,j,s,l}^{\left\{\frac{pj+qs}{d_i}\right\}d_i}B_{d_i,j,s,l}^{\left\{\frac{pj-qs}{d_i}\right\}d_i},\\
\frac{V_{p,q,l}}{U_{p,q,l}W_{p,q,l}}&=\varepsilon'_{p,q,l}\frac{P_l}{P_{p,q,l}}\prod_{m/d_i}\prod_{j,s=0}^{\left[\frac{d_i}{2}\right]}C_{d_i,j,s,l}^{\left\{\frac{pj+qs}{d_i}\right\}d_i}D_{d_i,j,s,l}^{\left\{\frac{pj-qs}{d_i}\right\}d_i}
\end{aligned}\right\}$$

or $\qquad$ (7)

$$\left.\begin{aligned}
\frac{U_{p,q,l}}{W_{p,q,l}}&=\varepsilon_{p,q,l}\frac{P_l}{P_{p,q,l}}\prod_{m/d_i}\prod_{j,s=0}^{\left[\frac{d_i}{2}\right]}A_{d_i,j,s,l}^{\left\{\frac{pj+qs}{d_i}\right\}d_i}B_{d_i,j,s,l}^{\left\{\frac{pj-qs}{d_i}\right\}d_i},\\
\frac{V_{p,q,l}}{U_{p,q,l}W_{p,q,l}}&=4\varepsilon'_{p,q,l}\frac{P_l}{P_{p,q,l}}\prod_{m/d_i}\prod_{j,s=0}^{\left[\frac{d_i}{2}\right]}C_{d_i,j,s,l}^{\left\{\frac{pj+qs}{d_i}\right\}}D_{d_i,j,s,l}^{\left\{\frac{pj-qs}{d_i}\right\}d_i},
\end{aligned}\right\}$$

where $A_{d_i, j, s, l}, \cdots, P_l$ are relatively prime intergers of the field K'_l, $m \equiv 0 \pmod{P_{p,q,l}}$, and $\epsilon_{p,q,l}$ and $\epsilon'_{p,q,l}$ are units of the field K_l.

We extend the field k to a field k_1 such that in k_1 the curve X will have primitive points of order $2^t m$. Then without losing generality we may assume that X has a group of k_1-points whose generators are $o_{2^t m}$ and $o'_{2^t m}$ which satisfy the equality $2^t o_{2^t m} = 2^t o'_{2^t m}$. Taking the points $o_{2^t m}$ and $o'_{2^t m}$ as $o_{1,m}$ and $o_{2,m}$, and applying the arguments of Theorem B to formulas (6) and (7), we will obtain the following theorem.

THEOREM 2. *There exists an effectively calculable constant* $c(k)$ *such that* $m < c(k)$.

I. R. Šafarevič has raised the following question.

Is the torsion of elliptic curves uniformly bounded over the set of all algebraic number fields of fixed degree n?

In connection with this question we have proved:

THEOREM 3. *The torsion of elliptic curves whose invariants are defined over a given field* k *is uniformly bounded over the set of all algebraic number fields of fixed degree.*

We note that for a more complete study of the torsion of elliptic curves it would be of interest to resolve the following three problems:

1) *Determine all the groups of torsion Q-points of elliptic curves.*

2) *For a given algebraic number field* k *find concrete estimates bounding the torsion from above.*

3) *Give an answer to the question of I. R. Šafarevič stated above.*

BIBLIOGRAPHY

1. J. W. S. Cassels, *Diophantine equations with special reference to elliptic curves,* J. London Math. Soc. **41** (1966), 193–291. MR **33** #7299.

2. Ju. I. Manin, *The p-torsion of elliptic curves is uniformly bounded,* Izv. Akad. Nauk SSSR Ser. Mat. **33** (1969), 459–465 = Math. USSR Izv. **3** (1969), 433–438. MR **42** #7667.

3. A. N. Paršin, *Isogenies and torsion of elliptic curves,* Izv. Akad. Nauk SSSR Ser. Mat. **34** (1970), 409–424 = Math. USSR Izv. **4** (1970), 415–430. MR **43** #218.

4. M. E. Novodvorskiĭ and I. I. Pjateckiĭ-Šapiro, *Some remarks on the torsion of elliptic curves,* Mat. Sb. **82** (124) (1970), 309–316 = Math. USSR Sb. **11** (1970), 283–290. MR **41** #8420.

5. Y. Hellegouarch, *Une propriété arithmétique des points exceptionnels rationnels d'ordre pair d'une cubique de genre* 1, C. R. Acad. Sci. Paris **260** (1965), 5989–5992. MR **31** #5862a.

6. D. Mumford, *A remark on Mordell's conjecture,* Amer. J. Math. **87** (1965), 1007–1016. MR **32** #4083.

7. V. A. Dem'janenko, *Points of finite order on elliptic curves,* Izv. Akad. Nauk SSSR Ser. Mat. **31** (1967), 1327–1340 = Math. USSR Izv. **1** (1967), 1271–1284. MR **36** #1440.

8. V. A. Dem'janenko, *Estimate of the remainder term in Tate's formula,* Mat. Zametki **3** (1968), 271–278 = Math. Notes **3** (1968), 173–177. MR **37** #2751.

9. ———, *Some properties of the group of torsion points of elliptic curves,* Abstracts of Reports, Tenth All-Union Algebra Colloq., Novosibirsk, 1969, vol. 2, p. 146. (Russian)

10. ———, *On torsion points of elliptic curves,* Izv. Akad. Nauk SSSR Ser. Mat. **34** (1970), 757–774 = Math. USSR Izv. **4** (1970), 765–783. MR **43** #217.

11. ———, *Torsion of elliptic curves,* Izv. Akad. Nauk SSSR Ser. Mat. **35** (1971), 280–307 = Math. USSR Izv. **5** (1971), 289–318. MR **44** #2755.

12. ———, *The points of finite order of elliptic curves,* Acta Arith. **19** (1971), 185–194. (Russian) MR **45** #5133.

Translated by J. S. JOEL

Trudy Mat. Inst. Steklov.
132 (1973)

Proc. Steklov Inst. Math.
132 (1973)

ON THE PARTITION FUNCTION OF POSITIVE DEFINITE MATRICES*

T. MITSUI

1. In this note I shall consider a problem in the additive theory of positive definite matrices, namely, the partition problem of positive definite matrices, which is a generalization of usual partition of rational integers.

Let $\mathfrak{M}$ be the set of the positive definite matrices n by n with the entries in the set of rational integers.

We take an element M in $\mathfrak{M}$ and assume that the M is represented as the sum of $M_1, M_2, \ldots, M_s$ in $\mathfrak{M}$; $M=M_1+\cdots+M_s$.

The number of such representations of M is finite. We denote by $P(M)$ this number, namely,

$$P(M)=\sum_{M=M_1+\ldots+M_s,\ M_i\in\mathfrak{M}\ (i=1,\ldots,s)} 1$$

We call $P(M)$ the *partition function* for M. The object of this note is to prove the following theorem.

Assume that the characteristic values of M in $\mathfrak{M}$ are in the interval $[c_1t, c_2t]$, where t is sufficiently large number, c_1 and c_2 are positive constants. Then we have an asymptotic formula for $\log P(M)$ *as t tends to infinity*:

$$\log P(M) \sim \left(n+\frac{2}{1+n}\right)\left(\frac{1+n}{2}\zeta(1+\nu)A_n\right)^{\frac{1}{1+\nu}} |M|^{\frac{\nu}{n(1+\nu)}},$$

where $\nu=n(n+1)/2$, $\zeta(s)$ is the Riemann zeta function, $|M|$ is the determinant of M, and

$$A_n=\pi^{\frac{n(n-1)}{4}}\prod_{i=2}^{n+1}\Gamma(i/2).$$

If $n=1$, then $P(M)$ is the usual partition function, and we see

$$\log P(M)\sim\pi(2M/3)^{1/2}.$$

If $n=2$, then

$$\log P(M)\sim\frac{8\pi^{5/4}}{3(120)^{1/4}}|M|^{3/8}$$

and for $n=3$, we have

$$\log P(M)\sim\frac{7}{2}(\pi^2\zeta(7))^{1/7}|M|^{2/7}.$$

AMS (MOS) subject classifications (1970). Primary 10J20, 10M20.

**Editor's note.* The body of this article is an unaltered reprinting of the original paper. Details of the bibliography have been checked and put in standard form, and MR citations have been added.

2. Let X be a positive definite matrix with entries in $\mathbf{R}$, the set of real numbers. We put

$$f(X)=\prod_{T\in\mathfrak{M}}(1-\exp(-\sigma(TX)))^{-1},$$

where $\sigma(*)$ is the trace of $*$. This infinite product is convergent, and we can expand this product into an infinite series

$$f(X)=1+\sum_{M\in\mathfrak{M}}P(M)\exp(-\sigma(MX)).$$

This means that $f(X)$ is the *generating function* of the $P(M)$.

Now we denote by E_{jl} a matrix whose $j-l$ entry is equal to 1 and the others are 0. We take ν matrices as follows

$$E\,(=\text{unit matrix}),\qquad E+E_{jj}\;(j=2,\ldots,n),$$
$$E+E_{ii}+E_{ij}+E_{ji}\qquad(1\leqslant i<j\leqslant n).$$

We write these matrices $T_1, T_2, \ldots, T_\nu$. They belong to $\mathfrak{M}$, and, as the vectors in $\mathbf{R}^\nu$, they are linearly independent over $\mathbf{R}$ and span a parallelotope with volume 1. Hence, we can write any symmetric matrix Y as a linear combination of $T_1, \ldots, T_\nu$

$$Y=\sum_{i=1}^{\nu}u_iT_i \tag{1}$$

in unique way with real coefficients $u_1, \ldots, u_\nu$. In particular, if T is an integral matrix, then

$$T=\sum_{i=1}^{\nu}m_iT_i,$$

where $m_1, \ldots, m_\nu$ are rational integers.

3. Now, we define by $D(T)$ for any integral matrix T a set in $\mathbf{R}^\nu$

$$D(T)=\left\{Y\,\middle|\,Y=T+\sum_{i=1}^{\nu}u_iT_i,\quad 0\leqslant u_i<1\;(i=1,\ldots,\nu)\right\}$$

and consider the integral

$$\int_{D(T)}\exp(-\sigma(YX))\,dY\qquad\left(dY=\prod_{i\leqslant k}dy_{ik}\right),$$

where X is positive definite. Since the equation (1) can be regarded as a linear transformation from $\{y_{ik}\}$ into $u_1, \ldots, u_\nu$ for which the value of Jacobian is 1, we have for $T=\sum_{i=1}^{\nu}m_iT_i$

$$\int_{D(T)}\exp(-\sigma(YX))\,dY=\prod_{i=1}^{\nu}\int_{m_i}^{m_i+1}\exp(-u_i\sigma(T_iX))\,du_i$$
$$=\prod_{i=1}^{\nu}\frac{1-\exp(-\sigma(T_iX))}{\sigma(T_iX)}\exp(-\sigma(TX))=i(X)\,e^{-\sigma(TX)}.$$

Hence, putting

$$F(X)=i(X)(f(X)-1),$$

we have

$$F(X)=\sum_{M\in\mathfrak{M}}\int_{D(M)} P(M)\exp(-\sigma(YX))\,dY.$$

Furthermore, if we define $P(Y)$ for $Y\in D$, where D is the set of real positive definite matrices, as follows

$$P(Y)=\begin{cases} P(M) & \text{if } Y\in D(M)\ (M\in\mathfrak{M}), \\ 0 & \text{if } Y\in D-\bigcup_{M\in\mathfrak{M}} D(M), \end{cases}$$

then we have

$$F(X)=\int_D P(Y)\exp(-\sigma(YX))\,dY. \tag{2}$$

In the similar way, if we introduce a function

$$G(X)=\sum_{T\in\mathfrak{M}}\exp(-\sigma(TX)),$$

then we have

$$G(X)=i(X)^{-1}\int_D a(Y)\exp(-\sigma(YX))\,dY,$$

where

$$a(Y)=\begin{cases} 1 & \text{if } Y\in D(T)\ (T\in\mathfrak{M}), \\ 0 & \text{if } Y\in D-\bigcup_{T\in\mathfrak{M}} D(T). \end{cases}$$

4. Now, it is known that

$$\int_D \exp(-\sigma(YX))\,dY=A_n|X|^{-(n+1)/2} \tag{3}$$

(*C. L. Siegel*. Über die analytische Theorie der quadratischen Formen. Ann. Math., 1935, **36**, Hilfssatz 37). Making use of this formula, we have

$$G(X)\leqslant i(X)^{-1}\int_D \exp(-\sigma(YX))\,dY\leqslant A_n|X|^{-(n+1)/2}\prod_{i=1}^{\nu}(1+\sigma(T_iX)). \tag{4}$$

On the other hand,

$$\int_D a(Y)\exp(-\sigma(YX))\,dY\geqslant\int_{Y-T_0\in D}\exp(-\sigma(YX))\,dY,$$

where $T_0=\sum_{i=1}^{\nu}T_i$. Hence

$$G(X)\geqslant A_n|X|^{-(n+1)/2}\,i(X)^{-1}e^{-\sigma(T_0X)}\geqslant A_n|X|^{-(n+1)/2}e^{-\sigma(T_0X)}. \tag{5}$$

From (4), (5) and a formula

$$\log f(X) = \sum_{k=1}^{\infty} \frac{1}{k} G(kX),$$

it follows that, if $\sigma(X)$ is sufficiently small, then

$$\log F(X) = \zeta(1+\nu) A_n |X|^{-(n+1)/2} \{1 + O(\sigma(X) \log(1/|X|))\}. \tag{6}$$

5. Now we put

$$\varphi(X) = \zeta(1+\nu) A_n |X|^{-(n+1)/2}$$

and

$$z_{ik} = -\frac{1}{d_{ik}} \frac{\partial \varphi(X)}{\partial x_{ik}} \qquad (X = (x_{ik})),$$

where

$$d_{ik} = \begin{cases} 1 & \text{if } i = k, \\ 2 & \text{if } i \neq k. \end{cases}$$

We define a symmetric matrix

$$Z = (z_{ik})_{1 \leqslant i,\ k \leqslant n}.$$

Then, we easily see that

$$Z = \frac{n+1}{2} \zeta(1+\nu) A_n |X|^{-(n+1)/2} X^{-1}.$$

Hence we can write X as the function of Z;

$$X = \Psi(Z) = B_n^{\frac{1}{1+\nu}} |Z|^{\frac{1+n}{2(1+\nu)}} Z^{-1},$$

where

$$B_n = \frac{n+1}{2} \zeta(1+\nu) A_n.$$

Now we take $Z = (z_{ik})$ whose characteristic values are in the interval $[c_1 t,\ c_2 t]$, where c_1 and c_2 are positive constants and t is sufficiently large.

If we define X by $X = (x_{ik}) = \Psi(Z)$, then the characteristic values of X are in the interval $\left[c_3 t^{-\frac{1}{1+\nu}},\ c_4 t^{-\frac{1}{1+\nu}}\right]$, where c_3 and c_4 are suitably chosen positive constants, and we have from (6)

$$F(X) \leqslant \exp\{\varphi(X)(1+\varepsilon_1)\} \tag{7}$$

(we denote by $\varepsilon_1, \ldots, \varepsilon_7$ sufficiently small positive numbers).

On the other hand, we have by (2)

$$F(X) \geqslant \exp\{-\sigma((Z+T_0)X)\} \int_D P(Y+Z+T_0) \exp(-\sigma(YX))\, dY$$

$$\geqslant \exp\{-\sigma((Z+T_0)X)\} \int_D P(Z) \exp(-\sigma(YX))\, dY. \tag{8}$$

Therefore we have, by (3), (7) and (8),

$$P(Z) \leqslant A_n^{-1} |X|^{(n+1)/2} \exp\{\sigma((Z+T_0)X) + \varphi(X)(1+\varepsilon_1)\}$$
$$\leqslant \exp[(1+\varepsilon_2)\{\varphi(\Psi(Z)) + \sigma(Z\Psi(Z))\}]. \qquad (9)$$

6. If we define, for $Z=(z_{ik})$ and small δ,

$$D_0 = \left\{ Y=(y_{ik}) \sum_{i \leqslant k} (y_{ik} - z_{ik})^2 \leqslant \delta^2 t^2 \right\}$$

and divide the integral (2) into two parts

$$F(X) = \int_D P(Y) \exp(-\sigma(YX))\, dY = \int_{D-D_0} + \int_{D_0},$$

then we can prove that

$$\int_{D-D_0} \leqslant \exp\{\varphi(X)(1-\varepsilon_3)\}.$$

Since we have by (6)

$$F(X) \geqslant \exp\{\varphi(X)(1-\varepsilon_4\},$$

it follows that

$$\int_{D_0} > \exp\{\varphi(X)(1-\varepsilon_5)\}.$$

On the other hand,

$$\int_{D_0} \leqslant P((1+c_5\delta)Z) \int_{D_0} \exp(-\sigma(YX))\, dY \leqslant P((1+c_5\delta)Z) \exp\{-\sigma(XZ)(1-\varepsilon_6)\}.$$

Hence

$$P((1+c_5\delta)Z) > \exp\{(\varphi(X) + \sigma(XZ))(1-\varepsilon_7)\}. \qquad (10)$$

From (9) and (10) it follows that

$$\log P(Z) \sim \varphi(\Psi(Z)) + \sigma(\Psi(Z)Z) \qquad (t \to \infty).$$

By some calculations we have

$$\varphi(\Psi(Z)) = \frac{2}{1+n} B_n^{\frac{1}{1+\nu}} |Z|^{\frac{1+n}{2(1+\nu)}}, \qquad \sigma(\Psi(Z)Z) = nB_n^{\frac{1}{1+\nu}} |Z|^{\frac{1+n}{2(1+\nu)}}.$$

Hence we obtain our theorem.

Trudy Mat. Inst. Steklov.
132 (1973)

Proc. Steklov. Inst. Math.
132 (1973)

ÜBER DAS NORMENRESTSYMBOL EINER LOKALEN UNVERZWEIGTEN ERWEITERUNG VON 2-POTENZGRAD*

H. KOCH

Sei k eine endliche Erweiterung des Körpers Q_p der rationalen p-adischen Zahlen vom Grad m und K/k eine endliche normale Erweiterung mit Galoisscher Gruppe G. Der Körper K enthalte die p^s-ten Einheitswurzeln. $K^\times$ bezeichne die multiplikative Gruppe von K. Die Faktor-gruppe $K^\times/K^{\times p^s}$ ist ein freier $\mathbf{Z}/p^s\mathbf{Z}$-Modul von Rang $m+2$, auf dem das Normenrestsymbol eine nichtausgeartete, schiefsymmetrische Bilinearform induziert. In der vorliegenden Arbeit wird der Zusammenhang dieser Bilinearform mit der Struktur von $K^\times/K^{\times p^s}$ als G-Modul in dem Fall, daß K/k eine unverzweigte Erweiterung von 2-Potenzordnung f ist, untersucht. Die benötigten Tatsachen über lokale Körper findet man in [4] und [9].

1. Wir entwickeln zunächst die genaue Problemstellung. Sei $\mathfrak{R}_s$ der Gruppenring von G über $\mathbf{Z}/p^s\mathbf{Z} \cdot K^\times/K^{\times p^s} = A = A^{(s)}$ läßt sich in offensichtlicher Weise als $\mathfrak{R}_s$-Modul betrachten. Wir zeichnen eine primitive p^s-te Einheitswurzel ζ_{p^s} in K aus und erhalten einen Isomorphismus φ der additiven Gruppe von $\mathbf{Z}/p^s\mathbf{Z}$ auf die Gruppe der p^s-ten Einheitswurzeln durch die Zuordnung

$$\varphi(\nu) = \zeta_{p^s}^{\nu}, \qquad \nu \in \mathbf{Z}/p^s\mathbf{Z}.$$

Wir definieren einen Homomorphismus χ von G in $\mathbf{Z}/p^s\mathbf{Z}$ durch

$$\zeta_{p^s}^{\chi(\eta)} = \zeta_{p^s}\eta \quad \text{für} \quad \eta \in G.$$

Das Normenrestsymbol $(\alpha, \beta)_K = (\alpha, \beta)_s = (\alpha, \beta)$, $\alpha, \beta \in K$, zur p^s-ten Potenz über K induziert auf A eine schiefsymmetrische, nichtausgeartete Bilinearform $\Phi_s = \Phi$ mit Werten in $\mathbf{Z}/p^s\mathbf{Z}$:

$$\Phi(\bar{\alpha}, \bar{\beta}) = \varphi^{-1}(\alpha, \beta), \qquad \bar{\alpha} = \alpha K^{\times p^s}, \qquad \bar{\beta} = \beta K^{\times p^s}.$$

Die Transformationsformel

$$(\alpha\eta, \beta\eta) = (\alpha, \beta)\eta, \qquad \eta \in G, \quad \alpha, \beta \in K^\times,$$

für das Normenrestsymbol übersetzt sich in die Verträglichkeitsforderung

$$\Phi(a\eta, b\eta) = \chi(\eta)\Phi(a, b) \quad \text{für} \quad a, b \in A, \quad \eta \in G. \tag{1}$$

Weiter gilt für $1 \leqslant t \leqslant s$

$$(\alpha, \beta)_t = (\alpha, \beta)_s^{p^{s-t}}$$

AMS (MOS) subject classifications (1970). Primary 12B10, 10C20.

**Editor's note.* The body of this article is an unaltered reprinting of the original paper. Details of the bibliography have been checked and put in standard form, and MR citations have been added.

und daher, wenn $\zeta_{p^t} = \zeta_{p^s}^{r^{s-t}}$ gesetzt wird

$$\Phi_t(\bar{a}, \bar{b}) = \varphi_t \Phi_s(a, b), \tag{2}$$

wobei $\bar{a}$, $\bar{b}$ die Klasse von a, $b \in A_s$ in A_t und φ_t den Homomorphismus von $\mathbf{Z}/p^s\mathbf{Z}$ auf $\mathbf{Z}/p^t\mathbf{Z}$ bezeichnet, der von der identischen Abbildung $\mathbf{Z} \to \mathbf{Z}$ induziert wird.

Sei schließlich K'/K eine beliebige endliche Erweiterung. Dann gilt

$$(\alpha, \beta)_{K'} = (n_{K'/K}\alpha, \beta)_K \quad \text{für} \quad \alpha \in K', \ \beta \in K. \tag{3}$$

Sei insbesondere K^H Fixkörper einer Untergruppe H von G und $\zeta_{p^s} \in K^H$, die Norm von K nach K^H werde mit n_H bezeichnet. Dann gilt

$$(\alpha, \alpha)_K = (\alpha, -1)_K = (n_H\alpha, -1)_{K^H},$$

$$(\alpha, n_H\alpha)_K = (n_H\alpha, n_H\alpha)_{K^H} = (n_H\alpha, -1)_{K^H},$$

d. h.

$$(\alpha, \alpha)_K = (\alpha, n_H\alpha)_K \quad \text{für} \quad \alpha \in K. \tag{4}$$

Das ergibt

$$\Phi(a, a) = \Phi(a, a\,\mathrm{sp}_H) \quad \text{für} \quad a \in A, \tag{5}$$

wobei die Untergruppe H von G im Kern von χ liegt und $\mathrm{sp}_H = \sum_{\eta \in H} \eta$ gesetzt ist.

Die Erweiterung K/k definiert also ein objekt $U(K/k)$, bestehend aus einem unitären $\mathfrak{R}_s$-Rechtsmodul A, einem Homomorphismus χ von G in die Gruppe $(\mathbf{Z}/p^s\mathbf{Z})^\times$ der primen Restklassen modulo p^s und einer nichtausgearteten schiefsymmetrischen Bilinearform Φ mit der Verträglichkeitsbedingung (1). Wir bezeichnen ein solches Objekt $(A, \chi, \Phi) = (A, \Phi)$ als symplektischen $\mathfrak{R}_s$-Modul. Für jedes t mit $1 \leqslant t \leqslant s$ ist (A, χ, Φ) ein symplektischer $\mathfrak{R}_t$-Modul $(A^{(t)}, \lambda^{(t)}, \Phi_{(t)})$ zugeordnet: $A_{(t)} = A/p^tA$

$$\chi^{(t)}(\eta) = \varphi_t\chi(\eta), \qquad \Phi^{(t)}(\bar{a}, \bar{b}) = \varphi_t\Phi(a, b) \quad \text{für} \quad a, b \in A.$$

Wir sagen, daß (A, χ, Φ) der Spurenbedingung genügt, wenn für alle $(A^{(t)}, \chi^{(t)}, \Phi^{(t)})$, $1 \leqslant t \leqslant s$ und alle Untergruppen H von C mit $\chi^{(t)}(H) = 1$ die Bedingung

$$\Phi^{(t)}(a, a) = \Phi^{(t)}(a, a\,\mathrm{sp}_H) \quad \text{für} \quad a \in A$$

gilt.

Nach dem oben gesagten genügt $U(K/k)$ der Spurenbedingung. Ein Morphismus von (A, Φ) in (A', Φ') ist ein $\mathfrak{R}_s$-Homomorphismus ψ von A in A' mit

$$\Phi'(\psi a, \psi b) = \Phi(a, b) \quad \text{für} \quad a, b \in A.$$

Es ergibt sich das Problem, die Struktur des symplektischen Moduls $U(K/k)$ aufzuklären. Dieses Problem wurde in den Arbeiten [1—3, 5, 6, 8] für Spezialfälle behandelt. Das Problem is besonders interessant für zahnverzweigte Erweiterungen K/k, da seine Lösung in diesem Fall eine wesentliche Voraussetzung für die Bestimmung der Struktur der Galloisschen

Gruppe der algebraischen Abschließung von k ist (siehe hierzu [5]). Für $p \neq 2$ wurde das Problem in den Arbeiten [8, 5] gelöst. Im folgender wird der Fall $p=2$ und K/k unverzweigt von 2-Potenzgrad f behandelt.

Im folgenden bezeichnet G immer eine zyklische Gruppe von 2-Potenz Ordnung f mit der Erzeugenden σ.

Wir setzen in der ganzen Arbeit voraus, daß f gerade und $\chi(\sigma^{f/2})=1$ ist. Diese Forderung ist immer erfüllt, wenn $f \geqslant 2^s$ ist. Im Hinblick auf eine Anwendung zur Bestimmung der Struktur der Caloisschen Gruppe der algebraischen Abschließung von k stellt die Forderung keine Einschränkung dar, weil es dabei nur auf hinreichend große f ankommt.

Die Bedingung, daß die Bilinearform Φ nicht ausgeartet ist, kann folgendermaßen formuliert werden: Jedem $a \in A$ ist die Linearform Φ_a mit

$$\Phi_a(b) = \Phi(a, b) \quad \text{für} \quad b \in A$$

zugeordnet. Φ ist genau dann nicht ausgeartet, wenn diese Zuordnung eine Bijektion von A auf Hom $(A, \mathbf{Z}/p^s\mathbf{Z})$ ist. Sei L_Φ die Linearform mit

$$L_\Phi(a) = \Phi(a, a) \quad \text{für} \quad a \in A.$$

Es gibt also genau ein Element b von A mit $L_\Phi = \Phi_b$. Dieses Element wird im folgenden mit a_Φ bezeichnet. In den oben genannten Arbeiten wurde eine Reihe von Begriffen im Zusammenhang mit symplektischen Moduln eingeführt. Wir benötigen hier nur den Begriff des vollständigen Teilmoduls eines symplektischen Moduls (A, Φ), das ist ein Teilmodul B von A für den die Beschränkung von Φ auf B nicht ausgeartet ist, sowie den Begriff direkte Summe von symplektischen Moduln (A_ν, Φ_ν) (bei festem χ), $\nu = 1, \ldots, h$. Das ist der Modul $A = \sum_{\nu=1}^{k} A_\nu$ mit der Bilinearform Φ:

$$\Phi\left(\sum_{\nu=1}^{h} a_\nu, \sum_{\nu=1}^{h} b_\nu\right) = \sum_{\nu=1}^{h} \Phi_\nu(a_\nu, b_\nu) \quad \text{für} \quad a_\nu, b_\nu \in A_\nu.$$

2. In diesem Abschnitt betrachten wir den Fall $s=1$, $a_\Phi \neq 0$.

Dann ist χ trivial und ein symplektischer Modul (A, Φ) genügt wegen (1) genau dann der Spurenbedingung, wenn

$$\Phi(a, a\sigma^{f/2}) = 0 \quad \text{für} \quad a \in A \tag{6}$$

gilt.

Bei den folgenden Beweisen benutzen wir ausgiebig den Operator $v = 1+\sigma$, der auch in [3] die Hauptrolle spielt. Es gilt, wie man leicht sieht,

$$v^{f-1} = \mathrm{sp}_G = \mathrm{sp}, \qquad v^f = 0 \tag{7}$$

und die Potenzen $1, v, \ldots, v^{f-1}$ bilden eine $\mathbf{Z}/2\mathbf{Z}$-Basis von $\mathfrak{R}_1$. Aus (1) folgt

$$\Phi(av, b) = \Phi(a, bv^*) \quad \text{für} \quad a, b \in A \quad \text{mit} \tag{8}$$

$$v^* = v + v^2 + \ldots + v^{f-1}.$$

Satz 1. *Sei A_m freier $\mathfrak{R}_1$-Modul vom Rang m und (A_m, Φ) symplektischer Modul mit $a_\Phi \neq 0$, A_m genüge der Spurenbedingung. Dann gibt es Elemente $a_1, \ldots a_m$ in A_m mit*

$$\Phi(a_\mu, a_\mu) = 1, \qquad \mu = 1, \ldots, m, \tag{9}$$

so daß A_m in die direkte Summe der vollständigen Teil-moduln $(a_\mu \mathfrak{R}_1, \Phi)$ zerfällt.

Satz 2. *Sei A ein $\mathfrak{R}_1$-Modul mit einer Erzeugenden und (A, Φ) ein symplektischer Modul mit Spurenbedingung und $a_\Phi \neq 0$.*

Dann gibt es ein $a \in A$ mit folgenden Eigenschaften:

1. *$\{a\sigma^\nu \mid \nu = 0, \ldots, f-1\}$ ist eine $\mathbf{Z}/2\mathbf{Z}$-Basis von A, d. h. $a\mathfrak{R}_1$ ist freier $\mathfrak{R}_1$-Modul.*

2. *In Falle $f = 2$ ist*

$$\Phi(a\sigma^\nu, a\sigma^\mu) = \begin{cases} 1 & \text{für } \nu \equiv \mu \pmod{f}, \\ 0 & \text{für } \nu \not\equiv \mu \pmod{f}. \end{cases}$$

Im Falle $f = 4$ ist

$$\Phi(a\sigma^\nu, a\sigma^\mu) = \begin{cases} 1 & \text{für } \nu - \mu \equiv 0 \pmod{f}, \\ i & \text{für } \nu - \mu \equiv 1, 3 \pmod{f}, \\ 0 & \text{für } \nu - \mu \equiv 2 \pmod{f}. \end{cases}$$

Im Falle $f \geqslant 8$ ist

$$\Phi(a\sigma^\nu, a\sigma^\mu) = \begin{cases} 1 & \text{für } \nu - \mu \equiv 0 \pmod{f}, \\ i & \text{für } \nu - \mu \equiv 2,\ 3,\ f-2,\ f-3 \pmod{f}, \\ 0 & \text{in allen anderen Fällen.} \end{cases}$$

Dabei ist

$$i = \Phi(b, bv^{f-3}) + 1$$

für alle $b \in A$ mit $\Phi(b, b) = 1$.

Wir bezeichnen i als Invariante des symlektischen Moduls (A, Φ).

Satz 3. *Sei A freier $\mathfrak{R}_1$-Modul vom Rang m und (A, Φ) ein symplektischer Modul mit Spurenbedingung und $a_\Phi \neq 0$*

$$(A, \Phi) = \sum_{\mu=1}^{fn} (A^\mu, \Phi)$$

sei entsprechend Satz 1 eine Zerlegung von (A, Φ) in die direkte Summe von vollständigen, freien $\mathfrak{R}_1$-Moduln (A^μ, Φ) mit den Invarianten i_μ.

Das Element a_Φ liegt in A sp. Für jedes b mit $b\,\mathrm{sp} = a_\Phi$ gilt

$$\Phi(b, bv^{f-3}) = \sum_{\mu=1}^{m} i_\mu + m.$$

Wir bezeichnen $\sum_{\mu=1}^{m} i_\mu$ als Invariante von A.

Zwei symplektische Moduln (A, Φ), (A, Φ') mit Spurenbedingung und $a_\Phi \neq 0$, $a'_\Phi \neq 0$ sind genau dann isomorph, wenn ihre Invarianten gleich sind.

Bemerkung. Die Sätze 1—3 geben eine volle Übersicht über alle symplektischen Moduln (A, Φ) zu freien $\mathfrak{R}_1$-Modul A mit Spurenbedingung und $a_\Phi \neq 0$. Zu festem Modul A gibt es genau zwei Isomorphieklassen von symplektischen Moduln.

3. In diesem Abschnitt betrachten wir den Fall $s = 1$, $a_\Phi = 0$. Im Gegen-

satz zu dem Ergebnis des vorigen Abschnittes sind in diesem Fall alle symplektischen Moduln mit Spurenbedingung über gegebenem freien $\mathfrak{R}_1$-Modul zueinander isomorph.

Satz 4. *Sei A_m freier $\mathfrak{R}_1$-Modul vom Rang m und (A_m, Φ) symplektischer Modul mit $a_\Phi = 0$.*

Dann ist m gerade, und (A_m, Φ) zerfällt in die direkte Summe von $m/2$ vollständigen, freien Teilmoduln vom Rang 2.

Satz 5. *Sei A ein freier $\mathfrak{R}_1$-Modul vom Rang 2 und (A, Φ) symplektischer Modul mit Spurenbedingung und $a_\Phi = 0$. Dann gibt es in A Erzeugende a, b mit*

$$\Phi(a\sigma^\nu, a\sigma^\mu) - \Phi(b\sigma^\nu, b\sigma^\mu) = 0 \text{ für alle } \nu, \mu$$

$$\Phi(a\sigma^\nu, b\sigma^\mu) = \begin{cases} 1 & \text{für } \mu = \nu, \\ 0 & \text{für } \mu \neq \nu, \quad \nu\mu = 0, \ldots, f-1. \end{cases}$$

4. Wir betrachten jetzt den Fall $s > 1$ und nehmen zusätzlich an, daß $\chi(\sigma^{f/2}) = 1$ ist, um für einen symplektischen Modul (A, Φ) die Spurenbedingung anwenden zu können, woraus dann

$$\Phi(a, a\sigma^{f/2}) = 0 \quad \text{für } a \in A$$

folgt. Aus der Schiefsymmetrie von Φ folgt $2\Phi(a, a) = 0$ für $a \in A$. Daher gilt bei $t < s$ für den induzierten Modul $(A^{(t)}, \Phi^{(t)})$

$$a_\Phi(t) = 0.$$

Wir setzen δ_Φ gleich 1 oder 0 je nach dem ob $a_\varphi \neq 0$ oder $a_\Phi = 0$ ist.

Satz 6. *Sei A, freier $\mathfrak{R}_s$-Rechtsmodul vom Rang m und $s > 1$. Weiter sei (A_m, Φ) symplektisher Modul mit Spurenbedingung. Dann ist m gerade, und (A_m, Φ) zerfällt in die direkte Summe von $m/2$ vollständigen, freien Teilmoduln vom Rang 2 mit Erzeugenden a_ν, b_ν, wobei*

$$\Phi(a_\nu, a_\nu) = \Phi(b_\nu, b_\nu) = 2^{s-1}\delta_\Phi,$$

$$\Phi^{(1)}(\bar{a}_\nu \mathrm{sp}, \bar{b}_\nu) = 1, \qquad \nu = 1, \ldots, m/2,$$

gilt (der Querstrich bezeichnet die Restklasse in $A_{(1)}$).

Satz 7. *Sei A freier $\mathfrak{R}_s$-Rechtsmodul vom Rang 2, $s > 1$ und (A, Φ) symplektischer Modul mit Spurenbedingung. Dann gibt es Erzeugende $a, b \in A$ mit*

$$\Phi(a\sigma^\nu, a\sigma^\mu) = \Phi(b\sigma^\nu, b\sigma^\mu) = \begin{cases} 2^{s-1}\delta_\Phi & \text{für } \nu = \mu, \\ 0 & \text{für } \nu \neq \mu; \end{cases}$$

$$\Phi(a\sigma^\nu, bs^\mu) = \begin{cases} \chi(\sigma)_\nu & \text{für } \nu = \mu, \\ 0 & \text{für } \nu \neq \mu. \end{cases} \quad \nu, \mu = 0, \ldots, f-1$$

5. Wir kommen nun zum eigentlichen Gegenchtand dieser Arbeit, der Beschreibung von $U(K/k)$ für eine unverzweigte Erweiterung K/k von 2-Potenzgrad f. Wir erinnern daran, daß wir stets $\chi(\sigma^{f/2}) = 1$ voraussetzen.

Satz 8. *Sei Π ein Primelement von k. Es gibt ein 2^s-primäres Element $\beta_0 \in K$ mit*

$$(\Pi, \beta_0)_s = \zeta_2 s.$$

Es gilt

$$(\Pi, \Pi)_s = (\beta_0, \beta_0)_s = 1. \tag{10}$$

Die Elemente $a = \Pi K^{\times 2^s} b = \beta_0 K^{\times 2^s}$ *erzeugen einen vollständigen* $\mathfrak{R}_s$*-Teilmodul B von* $U(K/k)$ *der Ordnung* 2^{2s}, *dessen Transformationsverhalten bei G durch*

$$a\eta = a, \quad b\eta = \chi(\eta) b, \quad \eta \in G$$

gegeben ist, wobei χ *der zu* $U(K/k)$ *gehörige Homomorphismus ist.*

Der zu B orthogonale Teilmodul B' *ist freier Modul vom Rang* $[k : Q_2]$.

Für die zu $U(K/k)$ *gehörige Bilinearform* Φ *ist* $a_\Phi = 0$ *genau dann, wenn die* 2^{s+1}*-ten Einhetswurzeln in K liegen.*

Bemerkung. Für $s > 1$, oder $s = 1$, $a_\Phi = 0$ beschreibt Satz 8 zusammen mit Satz 4 bis Satz 7 die Struktur von $U(K/k)$ vollständig. Dazu ist nur zu beachten, daß wegen (10) $a_\Phi(U(K/k)) = a_\Phi(B', \Phi)$ ist.

Für den Fall $s = 1$, $a_\Phi \neq 0$, hat man neben den Sätzen 1—3 noch den folgenden Satz zu berücksichtigen, um zu einer vollständigen Strukturbestimmung von $U(K/k)$ zu gelangen.

Satz 9. *Sei* $s = 1$, $a_\Phi \neq 0$ *und* $f \equiv 0 \pmod 4$. *Der Grundkörper k habe über* Q_2 *den Trägheitsgrad* f_0 *und den Verzweigungsexponenten* e_0. *Die Invariante i des in Satz 8 definierten symplektischen Moduls* (B', Φ) *ist*

$$i = e_0(f_0 + 1) + 1.$$

Zu den Beweisen der vorstehenden Sätze siehe demnächst in den Mathematischen Nachrichten erscheinende Arbeit [10] mit dem gleichen Titel.

BIBLIOGRAPHY

1. Z. I. Borevič, *A symplectic space with p-groups of operators over a field of characteristic p*, Mat. Zametki 6 (1969), 181–186 = Math. Notes 6 (1969), 567–569. MR 40 #5641.

2. ———, *The symplectic spaces with groups of operators*, Ural. Gos. Univ. Mat. Zap. 7 (1969/70), tetrad' 3, 36–50. (Russian) MR 44 #1743.

3. D. K. Faddeev, *On the structure of the reduced multiplicative group of a cyclic extension of a local field*, Izv. Akad. Nauk SSSR Ser. Mat. 24 (1960), 145–152. (Russian) MR 27 #2496.

4. H. Hasse, *Zahlentheorie*, Zweite Auflage, Akademie Verlag, Berlin, 1963. MR 27 #3621.

5. A. V. Jakovlev, *The Galois group of the algebraic closure of a local field*, Izv. Akad. Nauk SSSR Ser. Mat. 32 (1968), 1283–1322 = Math. USSR Izv. 2 (1968), 1231–1270. MR 38 #4453.

6. ———, *Symplectic spaces with operators over commutative rings*, Vestnik Leningrad. Univ. 25 (1970), no. 19, 58–64 = Vestnik Leningrad Univ. Math. 1970 (to appear). MR 43 #6255.

7. M. Kneser, *Zum expliziten Reziprozitätsgesetz von I. R. Šafarevič*, Math. Nachr. 6 (1951), 89–96. MR 13, 726.

8. H. Koch, *Über Darstellungsräume und die Struktur der multiplikativen Gruppen eines p-adischen Zahlkörpers*, Math. Nachr. 26 (1963), 67–100. MR 28 #3985.

9. J.-P. Serre, *Corps locaux*, Publ. Inst. Math. Univ. Nancago, 8, Actualités Sci. Indust., no. 1296, Hermann, Paris, 1962. MR 27 #133.

10. H. Koch, *Über das Normenrestsymbol einer lokalen unverweigten Erweiterung von 2-Potenzgrad*, Math. Nachr. 52 (1972), 355–369.

Trudy Mat. Inst. Steklov.
132 (1973)

Proc. Steklov Inst. Math.
132 (1973)

DISTRIBUTION PROBLEMS OF ARITHMETIC FUNCTIONS

Ĭ. P. KUBILIUS

ABSTRACT. This paper is a survey of the most important results in probabilistic number theory.

Bibliography: 26 items.

We shall consider real arithmetic functions $\psi(m)$, i.e. sequences of real numbers $\psi(1), \psi(2), \cdots$.

Most of the arithmetic functions considered in the theory of numbers are additive or multiplicative.

An arithmetic function $g(m)$ is called multiplicative if, for any relatively prime pair m, n,

$$g(mn) = g(m)\, g(n).$$

Examples of such functions are the Möbius function $\mu(m)$, the Liouville function $\lambda(m)$, the Euler function $\varphi(m)$, the number $\tau(m)$ of all divisors of m, the sum $\sigma(m)$ of all divisors of m, the so-called singular series in additive problems, and so on.

An arithmetic function $f(m)$ is called additive if

$$f(mn) = f(m) + f(n)$$

for all m and n satisfying the condition $(m, n) = 1$. Examples of additive functions are the number $\omega(m)$ of distinct prime divisors of m, the number $\Omega(m)$ of all prime divisors of m taking multiplicity into account, the logarithms of positive multiplicative functions, and so on.

The values of additive as well as multiplicative functions are distributed in a very complicated way. If one traces the variation in the values of such functions, when the argument runs through successive natural numbers, one gets a very chaotic picture, which is ordinarily observed by considering additive characteristics of integers jointly with the multiplicative ones. It is therefore natural to examine their distribution statistically.

It has been observed that, of late, greater interest is being taken in the problems of the so-called probabilistic theory of numbers.

In classical investigations, for examining the distribution of arithmetic functions, a study was usually made of the asymptotic behavior of the sum

---

AMS (MOS) subject classifications (1970). Primary 10K20; Secondary 60F05, 60G50

$$M_n = \frac{1}{n} \sum_{m=1}^{n} \psi(m),$$

which represents the mean value of the function $\psi(m)$ on the segment $\{1, \cdots, n\}$ of natural numbers. However, in general, M_n gives little information on the fluctuations of the function $\psi(m)$. Even in the case of comparatively simple arithmetic functions, the values of $\psi(m)$ may deviate considerably from M_n. Thus, in the case of $\omega(m)$, as one can easily calculate,

$$M_n \sim \ln \ln n,$$

whereas

$$\liminf_{n\to\infty} \omega(m) = 1, \qquad \limsup_{n\to\infty} \omega(m) \frac{\ln \ln m}{\ln m} = 1.$$

Moreover, it is found that very large or very small values are encountered, on the whole, rather infrequently. It is therefore natural to raise the question of to what extent the values of these functions deviate from the mean value for an overwhelming majority of values of the argument.

Some information on the fluctuations of the function $\psi(m)$ is given by a sum of the form

$$S_n = \frac{1}{n} \sum_{m=1} (\psi(m) - K_n)^2,$$

where K_n are appropriately chosen constants. For each n, the sum S_n assumes the least value

$$\frac{1}{n} \sum_{m=1} \psi^2(m) - M_n^2,$$

if K_n is chosen equal to M_n.

The values of additive and multiplicative functions are completely determined by specifying their values at powers of primes. It is therefore natural to formulate the various conditions in problems on the distribution of values of such functions in terms of $\psi(p^\alpha)$ (from now on the letter p will denote primes, and $\alpha = 1, 2, \cdots$). For an extensive class of additive functions $f(m)$,

$$M_n \approx \sum_{p^\alpha \leqslant n} \frac{f(p^\alpha)}{p^\alpha}.$$

By supposing

$$A_n = \sum_{p^\alpha \leqslant n} \frac{f(p^\alpha)}{p^\alpha}, \quad B_n^2 = \sum_{p^\alpha \leqslant n} \frac{f^2(p^\alpha)}{p^\alpha},$$

$$\frac{1}{n} \sum_{m=1} (f(m) - A_n)^2 = \varkappa_n B_n^2,$$

we succeeded in proving in 1955 [1], [2] that κ_n is bounded by an absolute constant.

Recently, using the method of the large sieve, Elliott [3] has shown that $\kappa_n \leqslant 51$ for sufficiently large n. It was found that our original method of proof gives the better result

$$\varkappa_n \leqslant \beta + c_1 \left(\frac{\ln \ln n}{\ln n}\right)^{1/2},$$

where

$$\beta = 1 + \left[\ln^2 2 - 2 \int_0^{1/2} \frac{\ln (1-y)}{y}\, dy + \left(\sum_p \frac{1}{p^2 (p-1)}\right)^2 + \sum_p \frac{2p^3 - 6p^2 + 2p + 1}{p^4 (p^2 - 1)}\right]^{1/2} = 2.29\ldots,$$

while in the case of functions with a constant sign

$$\varkappa_n \leqslant 1 + c_2 \left(\frac{\ln \ln n}{\ln n}\right)^{1/2},$$

where, in general, 1 cannot be replaced by a smaller constant; c_1 and c_2 are absolute constants.

Hence, in particular, follows the existence of an absolute constant n_0 such that, for $n \geqslant n_0$, for any real additive function $f(m)$, arbitrary $t > 0$ and for all $m \leqslant n$, with the exception of $< 3nt^{-2}$ numbers (possibly different for each function), one has the inequality

$$|f(m) - A_n| < tB_n. \tag{1}$$

By $N(\cdot\cdot\cdot)$ we shall denote the number of positive integers m satisfying the conditions indicated inside the parentheses.

Wishing to characterize more precisely the distribution of values of arithmetic functions, we are led to consider the asymptotic behavior of the frequency

$$\frac{1}{n} N(m \leqslant n, \ \psi(m) \in E) \tag{2}$$

as $n \to \infty$ for a sufficiently wide class of numerical sets E, say Borel sets. Our problem can be given a probability-theoretic interpretation.

As the set of elementary events we choose the segment $\Omega_n = \{1, \cdots, n\}$ of natural numbers; and as the set of random events $\mathfrak{A}_n$, the system of all subsets of Ω_n. We define the probability measure for all $A \in \mathfrak{A}_n$ as the frequency

$$\nu_n(A) = \frac{1}{n} N(m \in A).$$

The triple $(\Omega_n, \mathfrak{A}_n, \nu_n)$ is a probability space. The function $\psi(m)$, $m \leqslant n$, is a random variable defined on this space.

Let us consider the case of additive functions. For simplicity, we restrict ourselves to the so-called strongly additive functions. An additive function $f(m)$ is called strongly additive if $f(p^\alpha) = f(p)$ for all prime p and for all $\alpha = 2, 3, \cdots$. We introduce the functions

$$f^{(p)}(m) = \begin{cases} f(p), & \text{if} \quad p \mid m, \\ 0 & \text{otherwise.} \end{cases}$$

Then $f(m)$ can be represented in the form

$$f(m) = \sum_p f^{(p)}(m).$$

Thus the function $f(m)$ is a sum of random variables $f^{(p)}(m)$, where $f^{(p)}(m)$ takes two values: $f(p)$ with probability $[n/p]/n$ and 0 with probability $1 - [n/p]/n$. For studying the behavior of (2), it is desirable for us to employ limit theorems of sums of random variables. The random variables $f^{(p)}(m)$ are not, in general, independent. In fact, for prime p and q, $p \neq q$,

$$\nu_n\{f^{(p)}(m) = f(p),\ f^{(q)}(m) = f(q)\} = \frac{1}{n}\left[\frac{n}{pq}\right],$$

$$\nu_n\{f^{(p)}(m) = f(p)\} = \frac{1}{n}\left[\frac{n}{p}\right], \quad \nu_n\{f^{(q)}(m) = f(q)\} = \frac{1}{n}\left[\frac{n}{q}\right],$$

where $[n/pq]/n$ is not, in general, equal to

$$\frac{1}{n}\left[\frac{n}{p}\right]\frac{1}{n}\left[\frac{n}{q}\right].$$

However,

$$\frac{1}{n}\left[\frac{n}{pq}\right] = \frac{1}{n}\left[\frac{n}{p}\right]\frac{1}{n}\left[\frac{n}{q}\right] + \frac{\theta}{n},$$

where $|\theta| < 2$. This suggests that the dependence of the random variables must be, in a certain sense, weak, when p and q are not too large in comparison with n.

Let $r = r(n)$ be a function of n satisfying the conditions $r(n) \geqslant 2$ and $\ln r(n) = o(\ln n)$ for $n \to \infty$, and let $p_1, \cdots, p_k$ be an arbitrary set of primes not exceeding r. By the sieve method, one can prove that

$$\nu_n\{f^{(p_1)}(m) = \varepsilon(p_1), \ldots, f^{(p_k)}(m) = \varepsilon(p_k)\} = \prod_{j=1}^{k} \nu_n\{f^{(p_j)}(m) = \varepsilon(p_j)\} + o(1),$$

where $\epsilon(p)$ is equal to 0 or $f(p)$. Thus the "truncated" function

$$f(m)_r = \sum_{p \leqslant r} f^{(p)}(m)$$

is a sum of weakly dependent random variables.

Limit theorems for sums of weakly dependent random variables are usually proved by reducing them to sums of independent random variables. Hence it is preferable to reduce the problem of the behavior of (2) to problems in the better developed theory of sums of independent random variables. The interpretation mentioned above suggests that in order that the functions $f^{(p)}(m)$ could be considered as independent random variables, it is necessary to modify the probability space constructed above by taking, as the set of elementary events Ω_n, the entire sequence of natural numbers instead of only a finite segment, and, as probability measure,

$$\lim_{n\to\infty} \frac{1}{n} N(m \leqslant n, \ m \in A),$$

the asymptotic density of the set A, and, moreover, by choosing the σ-algebra of sets $\mathfrak{A}$ such that its sets possess asymptotic density.

At this point we run into difficulties. The union (as also the intersection and difference) of two sets having asymptotic density does not always possess this property. Besides, asymptotic density is not fully (countably) additive.

In order to avoid these difficulties, we proceed in the following manner. Taking $r = r(n)$, $r(n) \to \infty$ and $\ln r(n) = o(\ln n)$, we form two sequences of probability spaces: $\{\Omega_n, \mathfrak{L}_n, \nu_n\}$ and $\{\Omega_n, \mathfrak{L}_n, P_n\}$. Here, as before, $\Omega_n = \{1, \cdots, n\}$. Let Q_p be the set of positive integers $m \leqslant n$, divisible by p. The system of random events $\mathfrak{L}_n$ is the smallest algebra of sets with identity Ω_n containing all sets Q_p with $p \leqslant r(n)$. For each set $A \in \mathfrak{L}_n$, along with the frequency $\nu_n(A)$, we introduce another measure $P_n(A)$ such that $P_n(Q_p) = 1/p$ for all $p \leqslant r(n)$. This condition determines $P_n(A)$ uniquely. The sieve method allows one to prove that

$$\nu_n(A) - P_n(A) = o(1)$$

for $n \to \infty$, uniformly for all $A \in \mathfrak{L}_n$. Besides, the estimate $o(1)$ can be replaced by the more precise

$$O\left(\exp\left(-c \frac{\ln n}{\ln r(n)} \ln \frac{\ln n}{\ln r(n)}\right)\right)$$

with some constant $c > 0$. In addition, $f^{(p)}(m)$, $m \leqslant n$, are independent random variables relative to the measure P_n; and $f^{(p)}(m) = f(p)$ with probability $1/p$ and $f^{(p)}(m) = 0$ with probability $1 - 1/p$. Thus for any Borel set E

$$\frac{1}{n} N\left(m \leqslant n, \ \sum_{p\leqslant r} f^{(p)}(m) \in E\right) = P_n\left(\sum_{p\leqslant r} f^{(p)}(m) \in E\right) + o(1).$$

The use of the theory of sums of independent random variables allows one to study the asymptotic behavior of the frequency

$$\frac{1}{n} N\left(m \leqslant n, \ \sum_{p\leqslant r} f^{(p)}(m) \in E\right) \tag{3}$$

for the truncated function $f(m)_r$. For a very extensive class of functions $f(m)$, frequency (3) gives a good approximation for the frequency

$$\frac{1}{n} N(m \leqslant n, \ f(m) \in E). \tag{4}$$

Analogous arguments show that multiplicative functions can be interpreted as the products of certain random variables.

Among the asymptotic laws for (2), laws of two special types are of the greatest interest. In the first case one studies the asymptotic behavior of the frequency

$$F_n(x) = \frac{1}{n} N(m \leqslant n, \quad \psi(m) < C_n + D_n x) \tag{5}$$

as $n \to \infty$ with suitably chosen C_n and D_n (the integral laws). Since $F_n(x)$ is a distribution function in the probability-theoretic sense, it is natural to consider the convergence of $F_n(x)$ for $n \to \infty$ to some distribution function $F(x)$ at all points of continuity of the latter. In the second case one studies the asymptotic behavior as $n \to \infty$ of the frequency

$$\frac{1}{n} N(m \leqslant n, \quad \psi(m) = k) \tag{6}$$

(the local laws), where it is usually assumed that the values of $\psi(m)$ belong to some arithmetic progression.

The problem of the asymptotic behavior of (2), in particular of (5) and (6), is conveniently reduced to a study of the behavior of the so-called characteristic functions of Fourier transforms:

$$\varphi_n(t) = \int_{-\infty}^{\infty} \exp itx \, dF_n(x) = \exp\left(-\frac{itC_n}{D_n}\right)\frac{1}{n}\sum_{m=1}^{n} \exp\frac{it\psi(m)}{D_n}. \tag{7}$$

As is well known, there is a one-to-one correspondence between distribution functions and characteristic functions. This correspondence is continuous in the following sense. If the sequence of distribution functions $F_n(x)$ $(n = 1, 2, \cdots)$ converges to some distribution function $F(x)$ at all points of continuity of the latter, then the sequence of the corresponding characteristic functions $\varphi_n(t)$ converges to the characteristic function $\varphi(t)$ of $F(x)$; moreover, this convergence is uniform in every finite interval of values of the argument t. Conversely, if the sequence of characteristic functions $\varphi_n(t)$ $(n = 1, 2, \cdots)$ converges to the function $\varphi(t)$, continuous at the origin, then the sequence of the corresponding distribution functions $F_n(x)$ converges to some distribution function $F(x)$ at all its points of continuity; moreover, $\varphi(t)$ is the characteristic function of $F(x)$.

In this way the problem of the asymptotic behavior of the frequency $F_n(x)$ defined by (5) is reduced to a study of the trigonometric sum $\varphi_n(t)$ given by (7). If $\psi(m)$ is a polynomial or a function similar to a polynomial, we can apply the methods evolved by I. M. Vinogradov and his school.

If $\psi(m)$ is an additive function, then $\exp(it\psi(m)/D_n)$ is a complex-valued multiplicative function (although at the outset of this report we defined only real multiplicative functions, that definition is also suitable, without any change, for complex-valued multiplicative functions). Consequently our problem is reduced to a study of the behavior of the mean value of a complex-valued multiplicative function.

In the case of multiplicative functions it is found that, in place of Fourier transforms, it is more appropriate to utilize a certain modification of Mellin transforms, suggested by V. M. Zolotarev [4] for studying the multiplication of independent random variables.

For any distribution function $F(x)$ let

$$w_k(t) = \int_{-\infty}^{\infty}{}' |x|^{it} \operatorname{sgn}^k x dF(x) \qquad (k = 0, 1), \tag{8}$$

where the prime indicates that the point $x = 0$ is excluded from the range of integration. The functions $w_0(t)$ and $w_1(t)$ are defined for all real t. They are usually called characteristic transformations. Properties of characteristic transformations are analogous to the properties of characteristic functions.

There exists a one-to-one correspondence between distribution functions and pairs of characteristic transformations. This correspondence, as in the case of characteristic functions, is continuous in the same sense, only this time, instead of the correspondence of the distribution functions $F_n(x)$ to the limit function $F(x)$ at all its points of continuity, it is necessary to require, in addition, convergence at the origin; i.e. $F_n(0) \to F(0)$ and $F_n(+0) \to F(+0)$ (if, or course, 0 is a point of discontinuity of $F(x)$).

For the distribution law $N(m \leqslant n;\, \psi(m) < x)/n$ the characteristic transformations are

$$\frac{1}{n} \sum_{\substack{m=1 \\ \psi(m) \neq 0}}^{n} |\psi(m)|^{it} \operatorname{sgn}^k \psi(m) \qquad (k = 0, 1). \tag{9}$$

If $\psi(m)$ is a multiplicative function, we again have to deal with the asymptotic behavior of the mean value of a complex-valued multiplicative function.

Thus, in the case of additive as well as multiplicative functions, our problem reduces to an investigation of the behavior as $n \to \infty$ of sums of the form

$$\frac{1}{n} \sum_{m=1}^{n} h(m), \tag{10}$$

where $h(m)$ is a complex-valued multiplicative function with $|h(m)| \leqslant 1$ (in fact, $|h(m)| = 1$ in the case of additive functions and $|h(m)| = 1$ or 0 in the case of multiplicative functions).

Such sums can be investigated by various methods; in particular, in the case of additive and multiplicative functions one can apply the probability-theoretic interpretation mentioned above. However, the method of generating Dirichlet series has been found to be the most powerful. One considers the Dirichlet series

$$Z(s) = \sum_{m=1}^{\infty} \frac{h(m)}{m^s},$$

where $s = \sigma + i\tau$ is a complex variable. For $\sigma > 1$ the sum of this series is an analytic function. The sum (10) can be computed by the method of contour integration, well known in analytic number theory. But for this one must know the behavior of the function $Z(s)$ near the line $\sigma = 1$. From the Euler product it follows that for $\sigma > 1$

$$Z(s) = \prod_p \left(1 + \frac{h(p)}{p^s} + \frac{h(p^2)}{p^{2s}} + \dots\right) = H(s) \exp\left(\sum_p \frac{h(p)}{p^s}\right),$$

where $H(s)$ is an analytic function for $\sigma > \frac{1}{2}$ and is bounded for $\sigma > \frac{1}{2} + \epsilon$. The problem reduces to the behavior of the function

$$\sum_p \frac{h(p)}{p^s}.$$

Let us consider the series

$$\sum_p \frac{1 - \operatorname{Re} h(p)\, p^{-i\tau}}{p^\sigma}$$

for $\sigma > 1$. The sum of this series is a monotonic function of σ, and therefore it has a finite or infinite limit as $\sigma \to 1 + 0$. There are two possibilities: either for all τ this limit is infinite, or else there exists a $\tau = \tau_0$ for which it is finite. Let us take λ equal to τ_0 in the second case and equal to 0 (or any fixed number) in the first case. For $\sigma \to 1 + 0$, the function $Z(s)$ can be represented in the form

$$Z(s) = \frac{CL\left(\frac{1}{\sigma - 1}\right)}{s - (1 + i\lambda)} + o\left(\frac{1}{\sigma - 1}\right),$$

where the estimate of the remainder term is uniform on every finite segment $|\tau| \leqslant K$. Here C is a complex constant, while the function $L(u)$ has modulus 1 and varies slowly:

$$|L(u)| = 1, \quad \frac{L(u_1)}{L(u)} \to 1$$

for $u \to \infty$, uniformly for $u_1 \in [u, 2u]$.

Utilizing this fact and certain highly ingenious devices, G. Halász [5] succeeded in proving that

$$\frac{1}{n} \sum_{m=1}^{n} h(m) = \frac{n^{i\lambda} Z(a + i\lambda)}{(1 + i\lambda)\, \zeta(a)} + o(1) = \frac{n^{i\lambda}}{1 + i\lambda} CL(\ln n) + o(1)$$

for $n \to \infty$, where $a = 1 + (\ln n)^{-1}$ and $\zeta(a)$ is the Riemann zeta-function. The validity of the last result was conjectured by E. Wirsing [6].

However, the result of Halász is still inadequate for our purpose. The fact is that, in calculating the sums (7) and (9), $h(m)$ depends on the parameters n and t; we require estimates which are uniform in these parameters. Besides, λ depends on n

and t, while the method of Halász demonstrates only the existence of the function $\lambda = \lambda(t, n)$. Nevertheless, upon introducing certain additional restrictions, this method is found to be highly effective.

We mention some results on integral and local laws for additive and multiplicative functions.

Let us first consider additive functions. If $f(m)$ is an additive function and for some C_n and D_n the frequency

$$G_n(C_n + D_n x) = \frac{1}{n} N(m \leqslant n,\ f(m) < C_n + D_n x) \tag{11}$$

has nondegenerate (i.e. not concentrated at one point) limit distribution for $n \to \infty$, then D_n necessarily must tend to a finite nonzero or an infinite limit [7], [8]. In the case of a finite limit it suffices to restrict our discussion to limit distributions for $G_n(C_n + x)$. The following assertion is true:

In order that for $n \to \infty$ the frequency $G_n(C_n + x)$ have a nondegenerate limit distribution with some C_n, it is necessary and sufficient that $f(x)$ have the form

$$f(m) = b \ln m + v(m),$$

where b is a constant, and

$$\sum_{|v(p)| \geqslant 1} \frac{1}{p} < \infty, \qquad \sum_{|v(p)| < 1} \frac{v^2(p)}{p} < \infty.$$

In addition, the centering constants C_n must be given by

$$C_n = b \ln n + \sum_{\substack{p \leqslant n \\ |v(p)| < 1}} \frac{v(p)}{p} + C + o(1),$$

where C is a constant. The choice of the numbers C_n is unique apart from the summands $C + o(1)$. The limit distribution is discrete if and only if

$$\sum_{f(p) \neq 0} \frac{1}{p} < \infty.$$

This result was obtained, independently of each other, by several authors: Elliott and Ryavec [7], Levin and Timofeev [8], H. Delange (written communication) and also by the present author (not published), and is an extension of a theorem of Erdös and Wintner [9], proved in 1939, giving necessary and sufficient conditions for the existence of a limit law for $G_n(x)$: in order that $G_n(x)$ have a limit distribution, it is necessary and sufficient that the series

$$\sum_{|f(p)| \geqslant 1} \frac{1}{p}, \qquad \sum_{|f(p)| < 1} \frac{f(p)}{p}, \qquad \sum_{|f(p)| < 1} \frac{f^2(p)}{p} \tag{12}$$

converge (the analogue of the probability-theoretic three series criterion). Some of these sums may simply be finite sums. It is assumed that p runs through the primes in their natural order of succession.

The case $D_n \to \infty$ was found to be more difficult and has not been completely investigated up till now. The first result of this kind was obtained in 1939 by Erdös and Kac [10] for strongly additive functions $f(m)$ with the condition $|f(p)| \leqslant 1$. It was proved that if

$$C_n = \sum_{p \leqslant n} \frac{f(p)}{p} \text{ and } D_n^2 = \sum_{p \leqslant n} \frac{f^2(p)}{p} \to \infty,$$

then

$$G_n(C_n + D_n x) \to \frac{1}{\sqrt{2\pi}} \int_{-\infty}^{x} e^{-\frac{u^2}{2}} du. \tag{13}$$

In 1954, using the probability-theoretic interpretation mentioned above, the limit relation (13) was proved by us [1], [11] for arbitrary additive functions $f(m)$ with

$$C_n = \sum_{p \leqslant n} \frac{f(p)}{p}, \qquad D_n^2 = \sum_{p^\alpha \leqslant n} \frac{f^2(p^\alpha)}{p^\alpha} \tag{14}$$

under the assumption that the analogue of the Lindeberg condition is valid, viz. for every fixed $\epsilon > 0$

$$\frac{1}{D_n^2} \sum_{\substack{p^\alpha \leqslant n \\ |f(p^\alpha)| > \varepsilon D_n}} \frac{f^2(p^\alpha)}{p^\alpha} \to 0. \tag{15}$$

If (15) is satisfied, then D_n is a slowly varying function of $\ln n$ in the sense of Karamata. Further, if D_n is such a function, then condition (15) is necessary for the validity of (13) with C_n and D_n defined by (14).

This theorem was extended [1], [11] to a wider class of additive functions. Suppose that the quantity D_n defined by (14) is a slowly varying function, in the sense of Karamata, of $\ln n$. In order that $G_n(C_n + D_n x)$, with C_n and D_n defined by (14), converge to a limit distribution with variance 1, it is necessary and sufficient that there exist a nondecreasing function $V(u)$, $-\infty < u < \infty$, $V(-\infty) = 0$, $V(\infty) = 1$, such that as $n \to \infty$, for all u with the possible exception of $u = 0$, the quantity

$$\frac{1}{D_n^2} \sum_{\substack{p^\alpha \leqslant n \\ f(p^\alpha) < u D_n}} \frac{f^2(p^\alpha)}{p^\alpha}$$

converges to $V(u)$. The characteristic function $\varphi(t)$ of the limit law, in case it exists, is determined by the formula

$$\varphi(t) = \exp\left(\int_{-\infty}^{\infty} (e^{itu} - 1 - itu) u^{-2} dV(u) \right).$$

The method of proof of this theorem is the following. Instead of the function $f(m)$ we consider, on the set $\{1, \cdots, n\}$, the "truncated" function $f(m_r)$, where

m_r contains only those prime factors of m which do not exceed $r = r(n)$, $r(n) \to \infty$, $\ln r(n) = o(\ln n)$. According to the interpretation mentioned above, the distribution of the function $f(m_r)$ is well approximated by the distribution of the sum of certain independent random variables, and to this sum one can apply the limit theorems of probability theory. Finally the difference of the distributions of $f(m_r)$ and $f(m)$ is estimated with the help of (1).

The idea of truncation was found to be highly fruitful. Levin and Faĭnleĭb [13] succeeding in generalizing this method by introducing the approximation of $f(m)$ by means of the function $f(m_r) + f_0(m/m_r)$, where $f_0(m)$ is an additive function for which

$$\frac{1}{\ln n} \sum_{\substack{p \leqslant n \\ f_0(p) < u}} \frac{\ln p}{p}$$

converges as $n \to \infty$ to some distribution function. The authors developed a method which made it possible to reduce the investigation of the distribution of the function $f(m_r) + f_0(m/m_r)$ to the summation of independent random variables, whence the distribution of the function $f(m)$ itself is obtained with the help of (1). In this way they succeeded in widening the class of functions which have a limit distribution.

The method of truncation works well when the truncated function and the function $f(m)$ itself, under appropriate normalizations, have the same limit distributions if, of course, the latter exist. However, this does not always happen. If we wish to obtain more general results, we have to manage without truncation and use the method of generating Dirichlet series, of which we have spoken above.

We cite one of the results obtained in this way [8]. *Suppose that* $f(m)$ *is an additive function,* $D_n \to \infty$ *and there exists a* $t(n)$ *such that for every fixed* $\tau > 0$ *we have*

$$\frac{1}{\ln n} \sum_{\substack{p \leqslant n \\ |f(p) - t(n)\ln p| > \tau D_n}} \frac{\ln p}{p} \to 0$$

and

$$\sum_{p \leqslant n} \frac{1}{p} \frac{(f(p) - t(n)\ln p)^2}{D_n^2 + (f(p) - t(n)\ln p)^2} \to A$$

for $n \to \infty$, *where* A *is a nonzero constant. In order that, under these assumptions, the frequency* $G_n(C_n + D_n x)$, *given in* (11), *with some* C_n *have a nondegenerate limit distribution, it is necessary and sufficient that*

$$\Phi_n(u) = \sum_{\substack{p \leqslant n \\ f(p) - t(n)\ln p < uD_n}} \frac{1}{p} \frac{(f(p) - t(n)\ln p)^2}{D_n^2 + (f(p) - t(n)\ln p)^2} + cD_n + o(D_n)$$

tend as $n \to \infty$ *to some nondecreasing function* $V(u)$, $-\infty < u < \infty$, $0 = V(-\infty) <$

$V(\infty) = A$, at all its points of continuity with the possible exception of $u = 0$, and for every $\epsilon > 0$ there exist $u_0(\epsilon)$ and $n_0(\epsilon)$ such that

$$\Phi_n(\infty) - \Phi_n(u) + \Phi_n(-u) < \varepsilon$$

for all $u > u_0$ and $n > n_0$. Here C_n is determined by the formula

$$C_n = t(n) \ln n + \sum_{\substack{p \leqslant n \\ |f(p) - t(n) \ln p| < D_n}} \frac{f(p) - t(n) \ln p}{p} + cD_n + o(D_n),$$

where c is a constant.

We pass on to the consideration of limit laws for multiplicative functions $g(m)$. Let

$$H_n(C_n + D_n x) = \frac{1}{n} N(m \leqslant n,\ g(m) < C_n + D_n x). \tag{16}$$

If $g(m)$ is positive for all m, by taking $\log g(m)$ we reduce the problem to that of additive functions. The case of negative functions is not directly reducible to that of additive functions, but Erdös [12], [14] succeeded in finding necessary and sufficient conditions for the existence of the limit law for (16) in the case $C_n \equiv 0$, $D_n \equiv 1$. The case of functions with arbitrary sign is even more complicated. The fundamental step for studying such functions was taken by A. Bakštys [15], who employed the characteristic transformations (8) for investigating such functions. He obtained necessary and sufficient conditions for the convergence of (16) to the asymmetric distribution law in the case $C_n \equiv 0$, $D_n \equiv 1$, and also certain sufficient conditions in the case $D_n \to \infty$. For considering the symmetric case, the results of Halász–in particular, necessary and sufficient conditions for the convergence of (10) to 0–are lacking. The use of the method of generating Dirichlet series [16], [17] made it possible to make considerable advances.

It was found that $H_n(C_n + D_n x)$ tends to some nondegenerate distribution function $F(x)$ at each of its points of continuity, and

$$H_{n_k}(0) \to F(c), \quad H_{n_k}(+0) \to F(c+0)$$

at least for one finite c such that $C_{n_k}/D_{n_k} \to c$, if such a c exists, if and only if the following conditions are satisfied:

$$\sum_{g(p)=0} \frac{1}{p} < \infty;$$

there exist numbers b and d such that

$$\sum_{g(p) \neq 0} \frac{1}{p} \frac{(\ln |g(p)| - b \ln p)^2}{1 + (\ln |g(p)| - b \ln p)^2} < \infty,$$

$$D_n = dn^b \exp\left\{ \sum_{\substack{p \leq n,\ g(p) \neq 0 \\ |\ln|g(p)| - b\ln p| < 1}} \frac{\ln|g(p)| - b\ln p}{p} \right\} (1 + o(1)),$$

$$C_n = cD_n + o(D_n).$$

In particular, $H_n(x)$ tends to a limit law $F(x)$ at the points of continuity of $F(x)$, and $H_n(0) \to F(0)$ and $H_n(+0) \to F(+0)$, if and only if the series

$$\sum_{g(p)=0} \frac{1}{p}, \quad \sum_{\substack{g(p) \neq 0 \\ |\ln|g(p)|| \geq 1}} \frac{1}{p}, \quad \sum_{\substack{g(p) \neq 0 \\ |\ln|g(p)|| < 1}} \frac{\ln|g(p)|}{p}, \quad \sum_{\substack{g(p) \neq 0 \\ |\ln|g(p)|| < 1}} \frac{\ln^2|g(p)|}{p}$$

converge. Hence it follows that $H_n(x)$ has a limit distribution if and only if

$$\sum_{g(p)=0} \frac{1}{p} < \infty \tag{17}$$

and the additive function $\ln|g^*(m)|$, where $g^*(m)$ is the multiplicative function obtained from $g(m)$ by replacing $g(p^\alpha) = 0$ by arbitrary numbers, has a limit distribution with null centralization and unit normalization.

For multiplicative functions one can have another normalization:

$$K_n(x, C_n, D_n) = \frac{1}{n} N(m \leq n \mid C_n g(m)|^{1/D_n} \operatorname{sgn} g(m) < x).$$

It turns out that in this case the problem reduces to that of additive functions. Suppose that the function $g(m)$ assumes at least one value different from $-1, 0, 1$, and $D_n \to \infty$. Let us consider the convergence of the frequency $K_n(x, C_n, D_n)$ to a distribution function $F(x)$ at all the points of continuity of $F(x)$ and of $K_n(\pm 0, C_n, D_n)$ to $F(\pm 0)$ with the assumption that $F(x)$ does not belong to the class of functions having points of growth possibly only at the origin and at two points symmetric with respect to the origin. Under these conditions $K_n(x, C_n, D_n)$ has a limit distribution function if and only if (17) holds and any additive function $\ln|g^*(m)|$, where $g^*(m)$ is the multiplicative function given above, has a nondegenerate limit distribution with centralization $\ln|C_n|$ and normalization D_n.

We now turn to local laws. We restrict ourselves to integer-valued functions $\psi(m)$. If for arbitrary integral k the quantity $N(m \leq n;\ \psi(m) = k)/n$ tends as $n \to \infty$ to some limit λ_k, where the convergence is uniform in k and $\Sigma_{-\infty}^{\infty} \lambda_k = 1$, then we shall say that $\psi(m)$ has a local limit law of distribution.

Apparently the first nontrivial example of an additive function with a local limit law of distribution was found by Rényi [18]. This was the function $\Omega(m) - \omega(m)$. Rényi's method was elementary-combinatorial and was based on specific properties of the function $\Omega(m) - \omega(m)$; in particular, use was made of the property that $\Omega(p) - \omega(p) = 0$ for all prime p. A combination of number-theoretic and probability-theoretic considerations enabled us [1] to prove a local limit law for any integer-valued function $f(m)$ with the condition that $f(p) = 0$ for all prime p. Shortly afterwards it was found that this law remains valid if the condition $f(p)=0$ is violated for not too large a part of the primes; viz. if one requires that the series

$$\sum_{f(p)\neq 0} \frac{1}{p} \tag{18}$$

converge. Later on [19] it was found that the last condition is necessary. Thus an integer-valued additive function has a local limit law of distribution if and only if the series (18) converges. Moreover, if this series converges, one has the equality

$$\sum_{k=-\infty}^{\infty} \lambda_k \exp itk = \prod_p \left(1-\frac{1}{p}\right) \sum_{\alpha=0}^{\infty} \frac{\exp itf(p^\alpha)}{p^\alpha}.$$

The convergence of (18) in the case of integer-valued additive functions $f(m)$ is equivalent to the convergence of the series (12), which in turn is equivalent to the existence of an integral limit law of distribution for (11) with $C_n \equiv 0$ and $D_n \equiv 1$. The following more general theorem is valid:

An integer-valued arithmetic function $\psi(m)$ *has a local limit law of distribution if and only if* $\psi(m)$ *has an asymptotic integral law of distribution with centering constants* 0 *and normalizing constants* 1, *i.e.*

$$\frac{1}{n} N(m \leqslant n,\ \psi(m) < x)$$

converges to some distribution function $F(x)$ *at its points of continuity. If a limit law exists, the numbers* λ_k *are the Fourier coefficients of the characteristic function of the limit law* $F(x)$:

$$\int_{-\infty}^{\infty} \exp itx dF(x) = \sum_{k=-\infty}^{\infty} \lambda_k \exp itk.$$

Analogous results can be obtained for multiplicative functions too. The necessary and sufficient condition for the existence of a local limit law for an integer-valued multiplicative function $g(m)$ is the convergence of the series

$$\sum_{g(p)\neq 1} \frac{1}{p}.$$

We have presented only a small part of the results now known in the theory of distributions of the values of additive and multiplicative functions. The framework of the report did not allow us to dwell on many interesting and important results, nor on other methods. In particular, we did not touch upon the question of rapidity of convergence to the limit laws and questions of the joint distribution of several functions. We have also not considered the distribution of these functions when the argument runs through some sequence of natural numbers which does not coincide with the entire natural sequence (positive values of integer-valued polynomials, "shifted" primes and the like). One can also consider additive and multiplicative functions whose domains of definition are constituted by the integers of an algebraic number field or semigroup with regular normalization. Additional information may be found, besides the literature cited above, in the books and articles [20] – [26].

BIBLIOGRAPHY

1. J. P. Kubilius, *Probabilistic methods in the theory of numbers*, Uspehi Mat. Nauk **11** (1956), no. 2 (68), 31–66; English transl., Amer. Math. Soc. Transl. (2) **19** (1962), 47–85. MR **18**, 17; **24** #A1266.

2. ———, *Probabilistic methods in the theory of numbers*, Gos. Izdat. Politič. i Naučn. Lit. Litovsk. SSR, Vilna, 1959; rev. ed., 1962; English transl., Transl. Math. Monographs, vol. 11, Amer. Math. Soc., Providence, R. I., 1964. MR **23** #A134; **26** #3691; **28** #3956.

3. P. D. T. A. Elliott, *The Turán-Kubilius inequality, and a limitation theorem for the large sieve*, Amer. J. Math. **92** (1970), 293–300. MR **41** #8360.

4. V. M. Zolotarev, *On a theorem of multiplication of independent random variables*, Dokl. Akad. Nauk SSSR **142** (1962), 788–791 = Soviet Math. Dokl. **3** (1962), 166–169. MR **24** #A2402.

5. G. Halász, *Über die Mittelwerte multiplikativer zahlentheoretischer Funktionen*, Acta Math. Acad. Sci. Hungar. **19** (1968), 365–403. MR **37** #6254.

6. E. Wirsing, *Das asymptotische Verhalten von Summen über multiplikative Funktionen.* II, Acta Math. Acad. Sci. Hungar. **18** (1967), 411–467. MR **36** #6366.

7. P. D. T. A. Elliott and C. Ryavec, *The distribution of the values of additive arithmetical functions*, Acta Math. **126** (1971), 143–164. MR **43** #3224.

8. B. V. Levin and N. M. Timofeev, *An analytic method in probabilistic number theory*, Vladimir. Gos. Ped. Inst. Učen. Zap. **38** (1971), 57–150. (Russian) MR **46** #1740.

9. P. Erdös and A. Wintner, *Additive arithmetical functions and statistical independence*, Amer. J. Math. **61** (1939), 713–721. MR **1**, 40.

10. P. Erdös and M. Kac, *The Gaussian law of errors in the theory of additive number-theoretic functions*, Amer. J. Math. **62** (1940), 738–742. MR **2**, 42.

11. J. P. Kubilius, *On an application of the theory of summation of random variables to the theory of additive arithmetic functions*, Trudy Fiz.-Tehn. Inst. Akad. Nauk Litovsk. SSR **1** (1955), 5–24. (Russian) RŽMat **1957** #116.

12. P. Erdös, *Some remarks about additive and multiplicative functions*, Bull. Amer. Math. Soc. **52** (1946), 527–537. MR **7**, 507.

13. B. V. Levin and A. S. Faĭnleĭb, *Multiplicative functions and probabilistic number theory*, Izv. Akad. Nauk SSSR Ser. Mat. **34** (1970), 1064–1109 = Math. USSR Izv. **4** (1970), 1071–1118. MR **43** #6176.

14. P. Erdös, *Some remarks and corrections to one of my papers*, Bull. Amer. Math. Soc. **53** (1947), 761–763. MR **9**, 12.

15. A. Bakštys, *Limit laws of a distribution of multiplicative arithmetic functions.* I, II, III, Litovsk. Mat. Sb. **8** (1968), 5–20, 201–219, 643–680. (Russian) MR **40** #4231, #4232.

16. B. V. Levin, N. M. Timofeev and S. T. Tuljaganov, *Distribution of values of multiplicative functions*, Vladimir. Gos. Ped. Inst. Učen. Zap. **38** (1971), 282–288. (Russian) RŽMat **1972** #3A103.

17.* S. T. Tuljaganov, *Distribution of values of multiplicative functions*, Vladimir. Gos. Ped. Inst. Učen. Zap. 1971, no. 2, 287.

Editor's note.* This reference is reproduced essentially as it appears in the original, although it conflicts with the preceding citation. Possibly the author intended his paper of the same title in Izv. Akad. Nauk UzSSR Ser. Fiz.-Mat. Nauk **1972, no. 5, 30–38. (Russian)

18. A. Rényi, *On the density of certain sequences of integers*, Acad. Serbe Sci. Publ. Inst. Math. **8** (1955), 157–162. MR **17**, 944.

19. J. P. Kubilius, *On asymptotic local laws for arithmetic functions*, Litovsk. Mat. Sb. **11** (1971), 809–816. (Russian)

20. M. Kac, *Probability methods in some problems of analysis and number theory*, Bull. Amer. Math. Soc. **55** (1949), 641–665. MR **11**, 161.

21. ———, *Statistical independence in probability, analysis and number theory*, Carus Math. Monographs, no. 12, Math. Assoc. Amer.; distributed by Wiley, New York, 1959. MR **22** #996.

22. A. Rényi, *Probabilistic methods in number theory*, Proc. Internat. Congr. Math. (Edinburgh, 1958), Cambridge Univ. Press, New York, 1960, pp. 529–539. MR **22** #9478.

23. Ju. V. Linnik, *Les nombres entiers se prêtent-ils aux jeux du hasard*? Atomes **22** (1967), 441–446.

24. J. P. Kubilius, *On the distribution of number-theoretic functions*, Séminaire Delange-Pisot-Poitou. 11e année: 1969/70. Théorie des nombres, fasc. 2, Exposé 23, Secrétariat mathématique, Paris, 1970.

25. J. Galambos, *Distribution of arithmetical functions. A survey*, Ann. Inst. H. Poincaré Sect. B. **6** (1970), 281–305. MR **45** #6780.

26. W. Phillipp, *Mixing sequences of random variables and probabilistic number theory*, Mem. Amer. Math. Soc. No. 114 (1971).

Translated by V. N. SINGH

Trudy Mat. Inst. Steklov.
132 (1973)

Proc. Steklov Inst. Math.
132 (1973)

ON SUMS OF SQUARES

J. W. S. CASSELS

ABSTRACT. Results are formulated, and several problems posed, on the representation of functions of fixed sign in the form of a sum of squares.

Bibliography: 11 items.

Let $f \in \mathbf{R}[x_1, \cdots, x_n]$, where $\mathbf{R}$ is the field of real numbers, and $x_1, \cdots, x_n$ are independent variables. It is evident that if f is a sum of squares of polynomials from $\mathbf{R}[x_1, \cdots, x_n]$, then f is a semidefinite function. Hilbert [1] asked the question: Is the converse assertion correct? It is very easy to prove for $n = 1$. Hilbert himself constructed a very complex counterexample for $n = 2$. Motzkin [2] recently gave the following simple counterexample. The polynomial

$$f = 1 + x^2y^4 + x^4y^2 - 3x^2y^2$$

is semidefinite, since x^2y^2 is the geometric mean of the values $1, x^2y^4$ and x^4y^2.

Let us assume that

$$f = \sum_\lambda g_\lambda^2, \tag{1}$$

where $g_\lambda \in \mathbf{R}[x, y]$. It is evident that the g_λ are polynomials of degree not higher than the third. Let us write

$$g_\lambda = a_\lambda + b_\lambda + c_\lambda + d_\lambda,$$

where $a_\lambda, b_\lambda, c_\lambda$ and d_λ are homogeneous forms of the zeroth, first, second and third degrees.

Comparing the terms of the sixth degree in (1), we get

$$\sum_\lambda d_\lambda^2 = x^2y^4 + x^4y^2, \qquad \sum_\lambda p_\lambda^2 = 0,$$

where p_λ is the coefficient of x^3 in the form d_λ. Therefore all the coefficients p_λ vanish. Consequently every form d_λ is divisible by x, and, similarly, d_λ is divisible by y, i.e. $d_\lambda = xyd'_\lambda$, $d'_\lambda \in \mathbf{R}[x, y]$. Comparing the terms of the fourth degree, we find that the form

$$\sum_\lambda c_\lambda^2 = -2\sum_\lambda b_\lambda d_\lambda - 3x^2y^2 \tag{2}$$

AMS (MOS) subject classifications (1970). Primary 10J05.

is divisible by xy, from which it follows that $c_\lambda = xyc'_\lambda$, $c'_\lambda \in \mathbf{R}[x, y]$. In a similar way we get

$$\sum_\lambda b_\lambda^2 = -2 \sum a_\lambda c_\lambda \equiv 0 \pmod{xy}$$

and $b_\lambda = xyb'_\lambda$, $b'_\lambda \in \mathbf{R}[x, y]$. Since b_λ is a form of the first degree, we have $b_\lambda = 0$. From equation (2) it follows that

$$\sum_\lambda c_\lambda^2 = -3x^2y^2.$$

A contradiction!

We note that the polynomial f is only semidefinite, not strictly definite. However, the proof does not depend on the free term. Therefore we cannot represent the *strictly* definite polynomial

$$1 + f = 2 + x^2y^4 + x^4y^2 - 3x^2y^2$$

as a sum of squares of polynomials.

In his well-known lecture in Paris, Hilbert stated the conjecture that every semidefinite rational function $f \in \mathbf{R}(x_1, \cdots, x_n)$ can be represented by a sum of squares. The first general proof of this conjecture, given by Artin [3], was completely nonconstructive. After this, Robinson [4] gave a nonconstructive proof for the existence of a constructive proof of Artin's theorem, soon after which Kreisel (see [5] and in more detail [6]) gave a constructive proof. Habicht's [7] simple proof for strictly definite functions appears in the second edition of [8], because Littlewood thought that it needed to be better known.

None of these proofs speak about the number of squares necessary for a representation. Recently Pfister [9] proved that every semidefinite function $f \in \mathbf{R}(x_1, \cdots, x_n)$ can be represented by a sum of not more than 2^n squares. More exactly, he proved that every sum of squares can be represented by such a sum, and then he applied Artin's theorem.

On the other hand, I proved that $1 + x_1^2 + \cdots + x_n^2$ cannot be represented by the sum of n squares.

A problem arises naturally: *to find the smallest number* $t(n)$ *such that every indefinite function* $f \in \mathbf{R}(x_1, \cdots, x_n)$ *can be represented by a sum of not more than* $t(n)$ *squares.*

It follows from the above that $n + 1 \leqslant t(n) \leqslant 2^n$.

For $n = 1$ we have $n + 1 = 2 = 2^n$. But for $n = 2$ we have only $3 \leqslant t(2) \leqslant 4$. Ellison, Pfister, and I recently proved that, indeed, $t(2) = 4$.

For this we established the following connection between the theory of sums of squares and the theory of elliptic curves.

LEMMA. *Let R be a formally real field, and let*

$$f = 1 + ay^2 + by^4, \tag{3}$$

where y is a variable, and $a, b \in R$, $b \neq 0$ and $a^2 \neq 4b$. The following condition is necessary and sufficient for f to be a sum of three squares in R: there exist $\xi, \eta \in R$ such that

$$\text{(i)}\quad -\eta^2 = \xi(\xi^2 - 2a\xi + a^2 - 4b), \tag{4}$$

(ii) ξ *and* $-(\xi^2 - 2a\xi + a^2 - 4b)$ *are sums of two squares in* R.

PROOF. We have $f = f_1^2 + f_2^2 + f_3^2$, where $f_j \in R(y)$. It follows from one of my theorems that

$$f = g_1^2 + g_2^2 + g_3^2, \qquad g_j \in R[x]. \tag{5}$$

Setting

$$g_j = \alpha_j + \beta_j y + \gamma_j y^2, \tag{6}$$

we have $\alpha_1^2 + \alpha_2^2 + \alpha_3^2 = 1$. Therefore, after an orthogonal transformation with coefficients in the field R, we may suppose that

$$\alpha_1 = 1, \qquad \alpha_2 = \alpha_3 = 0. \tag{7}$$

It follows from the equalities (3)–(7) that

$$\beta_1 = 0, \qquad \beta_2^2 + \beta_3^2 = a - 2\gamma_1,$$
$$\beta_2\gamma_2 + \beta_3\gamma_3 = 0, \qquad \gamma_2^2 + \gamma_3^2 = b - \gamma_1^2$$

and consequently

$$(a - 2\gamma_1)(b - \gamma_1^2) = (\beta_2^2 + \beta_3^2)(\gamma_2^2 + \gamma_3^2) = (\beta_2\gamma_2 + \beta_3\gamma_3)^2 + (\beta_2\gamma_3 - \beta_3\gamma_2)^2 = (\beta_2\gamma_3 - \beta_3\gamma_2)^2.$$

Therefore the numbers

$$\xi = a - 2\gamma_1, \qquad \eta = 2(\beta_2\gamma_3 - \beta_3\gamma_2)$$

satisfy conditions (i) and (ii).

The converse is easy.

COROLLARY. *If the function*

$$f = 1 + x^2(x^2 - 3)y^2 + x^2y^4 \tag{8}$$

is a sum of three squres in the field $\mathbf{R}(x, y)$, then there exist $\xi, \eta \in \mathbf{R}(x)$ such that

$$-\eta^2 = \xi(\xi - x^2(x^2 - 3) - 2x)(\xi - x^2(x^2 - 3) + 2x), \tag{9}$$

where ξ is a semidefinite function.

Ellison, Pfister, and I proved that there are no such functions ξ and η.

The curve (9) is elliptic over the field of constants $\mathbf{R}(x)$. The theory of elliptic curves over fields of algebraic numbers has been well worked out. The classical method for determining a group of points is inapplicable to fields of functions, since the Tate-Šafarevič group may be infinitely divisible. We needed to find another method.

We consider the group G of points over the curve (9), which is definite over the field $\mathbf{C}(x)$, where $\mathbf{C}$ is the field of complex numbers. From an analogue of the Mordell-Weil theorem, we know that the group G is finitely generated. From this it is easy to conclude that its points are defined over the field $K(x)$, where the field K is a certain finite extension of the field $\mathbf{Q}$ of rational numbers. If the field K is minimal, then it is obviously a Galois extension of $\mathbf{Q}$, and the action of its Galois group Γ on the group G of points must be proper (i.e. without a kernel).

Using classical methods, it is not very difficult to find all the points of finite order in Γ and to prove that it has no more than three generators of infinite order. It follows from what has been said that one can compute a bound for the order of the Galois group Γ. This is a decisive step.

It turns out that there are no points of the required kind on the curve (9), and therefore the function (8) does not have a representation as a sum of three squares. The details of the proof are in the article [10].

In passing, we note that the Tate-Šafarevič group of the curve (9) is infinite.

PROBLEMS. (i) *Find a strictly definite function in the field* $\mathbf{R}(x, y)$ *which is not a sum of three squres.*

(Motzkin's function (8) is only semidefinite.) It is very probable that this problem can be solved by the method in [10]. The problem is for graduate students.

(ii) *Find* $t(n)$ *for* $n > 2$.

We have only inequalities (2 bis). I do not see any method for solving this problem.

(iii) *Investigate representations by a sum of squares in the field* $\mathbf{Q}(x_1, \cdots, x_n)$, *where* $\mathbf{Q}$ *is the field of rational numbers.*

It follows from Artin's general theorem that every indefinite function is such a sum.

For $n \geqslant 2$, it is at present unknown whether there exists a bound for the number of squares that depends only on the number of variables n.

For $n = 1$, it has long been known that every semidefinite function is a sum of eight squares.* The young French mathematician Pourchet [11] recently proved that the exact bound is five.

Editor's note.* See E. Landau, *Über die Darstellung definiter Funktionen durch Quadrate,* Math. Ann. **62 (1906), 272–285.

BIBLIOIGRAPHY

1. D. Hilbert, *Über die Darstellung definiter Formen als Summe von Formenquadraten,* Math. Ann. **32** (1888), 342–350.

2. T. S. Motzkin, *The arithmetic-geometric inequality,* Proc. Sympos. Inequalities (Wright-Patterson Air Force Base, Ohio, 1965), Academic Press, New York, 1967, pp. 205–224. MR **36** #6569.

3. E. Artin, *Über die Zerlegung definiter Funktionen in Quadrate*, Abh. Math. Sem. Univ. Hamburg **5** (1927), 100–115.

4. A. Robinson, *On ordered fields and definite functions,* Math. Ann. **130** (1965), 257–271, 405–409. MR **17**, 822; 1180.

5. G. Kreisel, *Mathematical significance of consistency proofs,* J. Symbolic Logic **23** (1958), 155–182. MR **22** #6710.

6. ———, *Sums of squares,* Summer Inst. Symbolic Logic, Cornell University, Summaries of talks presented, 1957, pp. 313–320; 2nd ed., Comm. Res. Div. Inst. Defense Anal., Princeton, N. J., 1960. J. Symbolic Logic **31**, 128.

7. W. Habicht, *Über die Zerlegung strikte definiter Formen in Quadrate,* Comment. Math. Helv. **12** (1940), 317–322. MR **2**, 119.

8. G. H. Hardy, J. E. Littlewood and G. Pólya, *Inequalities,* 2nd ed., Cambridge Univ. Press, New York, 1952. MR **13**, 727.

9. A. Pfister, *Zur Darstellung definiter Funktionen als Summe von Quadraten,* Invent. Math. **4** (1967), 229–237. MR **36** #5095.

10. J. W. S. Cassels, W. J. Ellison and A. Pfister, *On sums of squares and on elliptic curves over function fields,* J. Number Theory **3** (1971), 125–149. MR **44** #1863.

11. Y. Pourchet, *Sur la représentation en somme de carrés des polynômes à une indéterminée sur un corps de nombres algébriques,* Acta Arith. **19** (1971), 89–104.

Translated by D. PIRANIAN

Trudy Mat. Inst. Steklov.
132 (1973)

Proc. Steklov Inst. Math.
132 (1973)

ON ($\infty \times p$)-ADIC COVERINGS OF CURVES (THE SIMPLEST EXAMPLE)

Y. IHARA

ABSTRACT. This paper serves as an introduction to the ($\infty \times p$)-adic theory of algebraic function fields over finite fields of characteristic p.

Bibliography: 3 items.

This note is for an introduction to the $(\infty \times p)$-adic study of algebraic function fields over finite fields of characteristic p. We shall try to explain the close connection between the geometry of some $(\infty \times p)$-adic transformation spaces and the arithmetic of some types of infinite Galois extensions of such function fields, by taking the simplest possible example. The talk was on some observations and partial results related to more general cases, and I would like to ask the permission of the readers for replacing the notes of my talk by a very introductory exposition. The reason for this «going down» and taking a very elementary example as the object for discussion lies on my side, i. e., it was useful for the study of more general cases to review the simplest case.

Let $K=F_p(t)$ be the rational function field of one variable over the prime field of characteristic p. Let $\mathfrak{K}$ be the maximum tamely ramified Galois extension of K which is unramified outside the two points $t=0, \infty$. Then $\mathfrak{K}/K$ is an infinite Galois extension of an elementary type. In fact, it is obtained by the adjunction of all constants and n-th roots of t, for $n \not\equiv 0 \pmod p$. One generally looks at this extension only in connection with the multiplicative group G_m and its n-th power endomorphisms, but then it no longer has any interest for us. We shall look at $\mathfrak{K}/K$ in connection with the $(\infty \times p)$-adic transformation space $(X, \Gamma)=(Q, p^Z \cdot Z^{(p)})$. Here, $X=Q$ is the set of all rational numbers, and $\Gamma=p^Z \cdot Z^{(p)}$ is the semi-direct product of the multiplicative group p^Z (generated by p) and the additive group of the ring $Z^{(p)}=Z[p^{-1}]$, acting on X as $x \to p^Z x+Z^{(p)}$. This is the simplest example of the connections of an analogous type, and it is only accidental in our general picture that our extension $\mathfrak{K}/K$ correponds to the isogenies of algebraic groups, or that $\mathfrak{K}/K$ is an abelian extension modulo the constant extension. The connection may be briefly described as follows; X is the «universal

AMS (MOS) subject classifications (1973). Primary 14H05; Secondary 12B15, 10D10.

**Editor's note.* This reprinting of the original paper incorporates suggestions made by the author.

covering» associated with $\mathfrak{K}/K$, and Γ is the «covering group». Moreover, the $(\infty \times p)$-adic geometric presentation of (X, Γ) gives a precise account of the arithmetic of $\mathfrak{K}/K$. It is our purpose in this note to explain the connection in detail for this example.

In § 1, we formulate the connection just as a matter of fact. In § 2, we define the «$(\infty \times p)$-adic equivalent» $\mathfrak{M}/M$ of $\mathfrak{K}/K$, as the union of the fields of «automorphic functions» on X with respect to subgroups of Γ with finite indices. The «complex parts» of such «automorphic functions» are nothing more than the rational functions of $\exp 2\pi iz/n$, but the natural action of Γ on the complex part *does not* give its proper action on $\mathfrak{M}/M$ or $\mathfrak{K}/K$, and this is why we must consider some twisted automorphic functions. (For instance, the transform $z \to pz$ must be twisted so as to act trivially on $F_p(\exp 2\pi iz)$). By defining the field $\mathfrak{M}$ (and its reduction $\mathfrak{K}$) in this way, we can regard X and Γ naturally as the «universal covering» and the «covering group» associated with the extension $\mathfrak{K}/K$. The final section § 3 is for the converse of this procedure.

There is another class of similar examples, in which K is an elliptic function field over finite field and $\mathfrak{K}$ is the maximum unramified Galois extension of K. In this case, the transformation space (X, Γ) and its connection with $\mathfrak{K}/K$ are almost similar to our example. But in more general cases, $\mathfrak{K}/K$ is some essentially non-abelian extension, and Γ is an irreducible discrete subgroup of $G = PSL_2(R) \times PSL_2(k_\mathfrak{p})$ with finite volume quotient $\Gamma \backslash G$, where $k_\mathfrak{p}$ is a $\mathfrak{p}$-adic field. If T is the direct product of the orthogonal subgroup of $PSL_2(R)$ and the diagonal subgroup of $PSL_2(k_\mathfrak{p})$, then Γ acts on G/T in a natural manner, and X is defined to be the set of all fixed points of Γ. In these cases, the connections are, in general, only conjectured. Partial results along this line are given in [1—3]. But these are relatively old now, and our present review of the simplest example gives us some new insight for the general cases. On the other hand, there are some essential points which pertain to the general cases and which cannot be indicated by treating the simplest example only. So, we hope to continue this note to the (new) expositions of more general cases.

For any field k, $\bar{k}$ will denote its algebraic closure.

§ 1. The connections between (X, Γ) and $\mathfrak{K}/K$

1.1. The transformation space (X, Γ). We denote by X the set of all rational numbers, and by Γ the group of all linear transforms of X of the form

$$x \to p^d x + b; \quad d \in Z, \ b \in Z^{(p)}. \tag{1}$$

The transform (1) will be denoted by $[p^d, b]$. Put $\chi([p^d, b]) = d$. Then χ gives a surjective homomorphism of Γ onto Z, and the kernel Γ^1 consists of all translations $x \to x+b\,(b \in Z^{(p)})$. And if $\pi_0=[p,\ 0]$, we have $\pi_0^{-1}[1,\ b]\pi_0=[1,\ p^{-1}b]$, and Γ is a semi-direct product of Γ^1 and the free cyclic group π_0^Z generated by π_0.

For each $x \in X$, d_x will denote the smallest positive integer d with which $(p^d-1)x \in Z^{(p)}$. So, if n' is the denominator of the irreducible fraction for x, and if we decompose n' as $n'=p^k n$ with $n \not\equiv 0 \pmod p$, then d_x is nothing but the order of the class of p in the multiplicative group $(Z/nZ)^\times$, and $(p^d-1)x \in Z^{(p)}$ holds if and only if d is a multiple of d_x. Now let Γ_x denote the stabilizer of $x \in X$, and take any $\gamma=[p^d,\ b] \in \Gamma$. Then the condition $\gamma \in \Gamma_x$ is equivalent to $(p^d-1)x=-b$; hence b is uniquely determined by d, and the condition for d is that $(p^d-1)x \in Z^{(p)}$, i. e., $d \equiv 0 \pmod{d_x}$. Therefore, χ maps Γ_x isomorphically to $d_x Z$, and Γ_x is free cyclic, being generated by

$$\pi_x=[p^{d_x},\ -(p^{d_x}-1)\,x].$$

(The above notation $\pi_0=[p,\ 0]$ agrees with this definition.) Each element of Γ not contained in the translation group Γ^1 has a unique fixed point $x \in X$, and can be expressed uniquely as an integral power of π_x. Thus, by classifying the elements of $\Gamma-\Gamma^1$ with respect to their fixed points on X, we obtain the decomposition of $\Gamma-\Gamma^1$ into the following disjoint union:

$$\Gamma-\Gamma^1=\bigcup_{x \in X} \pi_x^{Z-\{0\}}.$$

Let Γ' be any subgroup of Γ with finite index. Put $D=(Z:\chi(\Gamma'))$ and $N=(\Gamma^1:\Gamma' \cap \Gamma^1)$. Then D is the smallest positive exponent for which there is some $B \in Z^{(p)}$ with $[p^D,\ B] \in \Gamma'$, and N is the smallest positive integer for which $[1,\ N] \in \Gamma'$. We call D (resp. N) the *degree* (resp. the *level*) of Γ'. Take any $B \in Z^{(p)}$ for which $[p^D,\ B] \in \Gamma'$. Then B is determined modulo N, and it is easy to see that Γ' is generated by the two elements $[p^D,\ B]$ and $[1,\ N]$. Conversely, we can define Γ' starting from any data $(D;\ N,\ B)$, where D, N are positive integers with $N \not\equiv 0 \pmod p$, and B is a representative of a class of $Z^{(p)}$ modulo $NZ^{(p)}$. In this way, Γ' and $(D;\ N,\ B)$ are in a one-to-one correspondence with each other, and hence we may write $\Gamma'=\Gamma_{(D;\,N,\,B)}$. Finally, we note that the congruence subgroup property is valid for Γ. In fact, if $(\Gamma:\Gamma')=n$, then Γ' contains $[p^n,\ 0]$ and $[1,\ n]$, from which follows immediately that Γ' contains all $[p^m,\ b] \in \Gamma$ with $[p^m,\ b] \equiv [1,\ 0] \bmod (n(p^n-1))$.

1.2. The extension $\mathfrak{K}$/K. Consider the rational function field of one variable $K=F_p(t)$. Denote by P_0 and P_∞ the two primes of K having 0 and ∞ as the residue classes of t. By adjoining to K the n-th roots of unity and of t *for all* $n \not\equiv 0 \pmod p$, we obtain an infinite Galois extension

$$\mathfrak{K}=\bar{F}_p(t^{1/n} \mid n \not\equiv 0 \pmod p)$$

of K. This extension is a tower $\mathfrak{K}/\bar{F}_p(t)/K$ of two abelian extensions, and can be characterized as the maximum tamely ramified Galois extension of K *unramified outside* P_0, P_∞. Denote by σ^* the p-th power automorphism of $\bar{F}_p$. Then the group of its integral powers σ^{*Z} is dense in the Galois group of $\bar{F}_p/F_p$, playing a more important role than the full Galois group. Let $\mathfrak{G}$ denote the group of all automorphisms of $\mathfrak{K}/K$ whose restrictions to $\bar{F}_p$ belong to σ^{*Z}. It is of course dense in the Galois group of $\mathfrak{K}/K$. Denote by $\mathfrak{X}$ the set of all primes of $\mathfrak{K}$ not lying on P_0, P_∞. Then both $\mathfrak{G}$ and $\mathfrak{X}$ carry the Krull topologies, and $\mathfrak{G}$ acts on $\mathfrak{X}$ continuously by $\xi \to g\xi$ ($g \in \mathfrak{G}$, $\xi \in \mathfrak{X}$), where $g\xi$ is the prime of $\mathfrak{X}$ defined by $g\xi(\alpha) = \xi(g^{-1}\alpha)$ ($\alpha \in \mathfrak{K}$). Two primes of $\mathfrak{K}$ lie on the same prime of K if and only if they are conjugate by some automorphism of $\mathfrak{K}/K$. By the definition of $\mathfrak{G}$, it is clear that such an automorphism can be taken from $\mathfrak{G}$. Thus, the quotient $\mathfrak{G}\backslash\mathfrak{X}$ can be identified with the set of all primes $\neq P_0, P_\infty$ of K. For each $\xi \in \mathfrak{X}$, its stabilizer $\mathfrak{G}_\xi$ is the decomposition group of ξ in $\mathfrak{G}$. It is generated by the Frobenius automorphism $\left(\frac{\mathfrak{K}/K}{\xi}\right)$, as an *abstract group*.

1.3. The embedding ι: $(X, \Gamma) \to (\mathfrak{X}, \mathfrak{G})$. Now we are going to show that $(\mathfrak{X}, \mathfrak{G})$ can be obtained by the completion of (X, Γ) with respect to the product of all l-adic topologies for $l \neq p$, and that (X, Γ) is «just as large as necessary» in describing the arithmetic of $\mathfrak{K}/K$ in a precise way.

We shall use the following conventions. Consider the direct sum $C \oplus \bar{Q}_p$ (C: the complex, Q_p: the p-adic fields), and embed the rational number field Q diagonally into $C \oplus \bar{Q}_p$. Fix one algebraic closure $\bar{Q}$ of Q in this direct sum. This is equivalent to fixing an isomorphism of the algebraic closure of Q in C onto that of Q in $\bar{Q}_p$. Thus, as far as algebraic numbers are concerned, a complex number gives a p-adic number, and vice versa. Also, we identify the residue field of $\bar{Q}_p$ with $\bar{F}_p$, and for each algebraic number α, complex or p-adic, we denote its residue class mod p by α^*; $\alpha^* \in \bar{F}_p \cup (\infty)$. (The same type of notations will also be used for some more general objects defined over $\bar{Q}$.)

Now, to define the embedding $\iota = (\iota_X, \iota_\Gamma)$ of (X, Γ) into $(\mathfrak{X}, \mathfrak{G})$, choose one n-th root t_n of t for each $n \not\equiv 0 \pmod p$ in such a way that $t_{nm}^m = t_n$ holds for all m, n; and write $e[z] = \exp 2\pi i z$ ($z \in C$). For each $a \in X$, let $\iota_X(a)$ be the unique prime of $\mathfrak{K}$ whose restriction to $\bar{F}_p(t_n)$ corresponds to the prime element $t_n - e[a/n]^*$ for every $n \not\equiv 0 \pmod p$. Then $\iota_X(a)$ does not lie on P_0 or P_∞; hence $\iota_X(a) \in \mathfrak{X}$. For a, $b \in Q$, $\iota_X(a)$ and $\iota_X(b)$ coincide on $F_{p^n}(t_{p^n-1})$ if and only if $e[a/(p^n-1)]^* = e[b/(p^n-1)]^{*p^{nk}}$ with some $k \in Z$; hence if and only if $a \equiv b \pmod{p^n - 1}$. Therefore, ι_X: $X \to \mathfrak{X}$ is injective, and the topology of X induced from the Krull topology of $\mathfrak{X}$ is equivalent to the product of l-adic topologies for all $l \neq p$. On the other hand, it is clear that $\iota_X(X)$ is dense in $\mathfrak{X}$. Hence we can identify $\mathfrak{X}$ with the completion of X with respect to this topology; i. e.,

$$\mathfrak{X} = \prod_{l \neq p}{}' Q_l,$$

where Π' denotes the restricted direct product. The embedding $\iota_\Gamma : \Gamma \to \mathfrak{G}$ will be defined as follows. For each $\gamma = [p^d, b] \in \Gamma$, define the automorphism $\iota_\Gamma(\gamma)$ of $\mathfrak{R}/K$ by

$$\iota_\Gamma(\gamma) t_n = \zeta_n^{-b} t_n; \quad \iota_\Gamma(\gamma) \zeta_n = \zeta_n^{p^d} \ (n \not\equiv 0 \ (\mathrm{mod}\ p)),$$

where $\zeta_n = e[1/n]^*$. This defines an automorphism of $\mathfrak{R}/K$ uniquely, since the two extensions $F_p(t_n \mid n \not\equiv 0 \ (\mathrm{mod}\ p))$ and $\bar{F}_p$ are linearly disjoint over F_p, and have $\mathfrak{R}$ as their composite. In particular, the translations $[1, b]$ define the automorphisms $t_n \to \zeta_n^{-b} t_n$ over $\bar{F}_p$, and π_0 defines the automorphism $\zeta_n \to \zeta_n^p$ over $F_p(t_n \mid n \not\equiv 0 \ (\mathrm{mod}\ p))$. Since the relation $\pi_0^{-1}[1, b]\pi_0 = [1, p^{-1}b]$ remains valid by applying ι_Γ, ι_Γ is a homomorphism of Γ into $\mathfrak{G}$. Since $\iota_\Gamma(\gamma)$ acts identically on $F_p(t_n, \zeta_n)$ if and only if $\gamma \equiv [1, 0] \ (\mathrm{mod}\ n)$, ι_Γ is injective, and the topology of Γ induced from the Krull topology of $\mathfrak{G}$ is equivalent to that topology of Γ defined with respect to the congruence subgroups, or equivalently, all subgroups with finite indices. But on the other hand, $\iota_\Gamma(\Gamma^1)$ is dense in the Galois group of $\mathfrak{R}/K \cdot \bar{F}_p$, and $\iota_\Gamma(\pi_0)$ induces the p-th power automorphism σ^* of $\bar{F}_p$; hence $\iota_\Gamma(\Gamma)$ is dense in $\mathfrak{G}$. Therefore, $\mathfrak{G}$ is contained in the completion of Γ with respect to the above topology, and the Galois theory is valid between finite extensions of K in $\mathfrak{R}$ and subgroups of Γ with finite indices. We have seen that such subgroups are determined by the data $(D; N, B)$. The corresponding field is given by $K_{(D;\, N,\, B)} = F_{p^D}(\zeta^B_{(p^D-1)N} t_N)$.

It can be checked easily that the embeddings ι_X and ι_Γ are compatible with the group actions, i. e., $\iota_\Gamma(\gamma)\, \iota_X(x) = \iota_X(\gamma x)$ $(\gamma \in \Gamma,\ x \in X)$. Therefore, the action of $\iota_\Gamma(\gamma)$ on $\mathfrak{X}$ is also given by $\xi \to p^d \xi + b$ $(\xi \in \mathfrak{X},\ \gamma = [p^d, b])$. Thus, $\mathfrak{G}$ can be identified with the group of all linear transforms of $\mathfrak{X}$ of the form

$$\xi \to p^d \xi + \beta; \quad d \in Z, \quad \beta \in \prod_{l \neq p} Z_l. \tag{2}$$

This element (2) of $\mathfrak{G}$ will be denoted by $[p^d, \beta]$.

1.4. Axioms satisfied by (X, Γ). Now we shall identify (X, Γ) with the embedded transformation space $(\iota_X(X), \iota_\Gamma(\Gamma))$, and shall list the basic properties of this transformation subspace (X, Γ) of $(\mathfrak{X}, \mathfrak{G})$ as the following axioms $(X\Gamma - 1) \sim (X\Gamma - 5)$.

Algebraic Axioms[1]

($X\Gamma$ — 1). *X is dense in $\mathfrak{X}$.*

($X\Gamma$ — 2). *For any two elements g, g' of $\mathfrak{G}$ the subspaces gX and $g'X$* of $\mathfrak{X}$ *are either disjoint or coincide totally, and Γ consists of all elements* of $\mathfrak{G}$ *leaving X fixed.*

In fact, for each $x \in X$ and $g \in \mathfrak{G}$, gx belongs to X if and only if $g \in \Gamma$. This follows immediately from (2). Since X is dense in $\mathfrak{X}$ and two elements of $\mathfrak{X}$ lying on the same prime of K are conjugate by $\mathfrak{G}$ (§ 1.2), every element of $\mathfrak{X}$ lies on gX for some $g \in \mathfrak{G}$. Hence $\mathfrak{X}$ can be decomposed into the disjoint union $\sum_{\bar{g} \in \mathfrak{G}/\Gamma} \bar{g}X$. (The index set $\mathfrak{G}/\Gamma$ is something very unpleasant; Γ is dense in $\mathfrak{G}$, and the cardinality of $\mathfrak{G}/\Gamma$ is $\aleph$-infinity. In fact, it can be identified with $(\prod_{l \neq p} Z_l)/Z^{(p)}$.)

($X\Gamma$—3). *Γ is dense in $\mathfrak{G}$, and a subgroup of Γ is open with respect to the Krull topology if and only if it is of finite index in Γ.*

Or in other words:

($X\Gamma$—3)*. *The Galois theory holds between finite extensions of K in $\mathfrak{K}$ and subgroups of Γ with finite indices.*

Let K' and Γ' be the corresponding finite extension of K and the subgroup of Γ, and let $\mathfrak{G}'$ be the Krull closure of Γ' in $\mathfrak{G}$. Then, from the above axioms we obtain immediately the following natural one-to-one correspondences:

$$\Gamma'\backslash X \underset{1:1}{\leftrightarrow} \mathfrak{G}'\backslash\mathfrak{X} \underset{1:1}{\leftrightarrow} \left\{\begin{matrix}\text{the primes of } K' \text{ not} \\ \text{lying on } P_0,\ P_\infty.\end{matrix}\right\}$$

Thus, we may say that «X covers the primes of K' (not on P_0, P_∞), with the covering group Γ'».

Arithmetic Axioms

($X\Gamma$ — 4). *For each point x of X, its stabilizer Γ_x is free cyclic, being generated by the Frobenius automorphism $\pi_x = \left(\frac{\mathfrak{K}/K}{x}\right)$ of x. In short,*

$$\Gamma_x = \pi_x^Z; \quad \pi_x = \left(\frac{\mathfrak{K}/K}{x}\right).$$

We know that $\Gamma_x = \pi_x^Z$, and so it remains to check that $\pi_x = \left(\frac{\mathfrak{K}/K}{x}\right)$. Put $\pi_x = [p^d, b]$ $(d = d_x,\ b = -(p^d - 1)x)$. Then the residue field of K at x is $F_p(e[-b/(p^d-1)]^*) = F_{p^d}$. On the other hand, π_x leaves x fixed, and induces on $\bar{F}_p$ the p^d-th power automorphism. Therefore, $\pi_x = \left(\frac{\mathfrak{K}/K}{x}\right)$.

---

[1]The following three algebraic axioms should be compared to the basic facts in the complex case, where the universal covering transformation space (C, Z) is associated to the extension $\mathfrak{K}^C/K^C$; $K = C(t)$, $\mathfrak{K}^C$: the maximum Galois extension of K^C unramified outside P_0, P_∞. On the other hand, arithmetic axioms are proper to objects over finite fields.

($X\Gamma$—5). $\Gamma_x \cap \Gamma_y = \{1\}$ *for* $x \neq y$.

This is obvious geometrically (see § 1). But as an arithmetic axiom, it requires that the Frobenius π_x determines x reciprocally; or going down to a finite extension K' of K with the corresponding subgroup Γ' of Γ, that the primes P' of K' not lying on P_0 or P_∞ are in one-to-one correspondence with the Γ'-conjugacy classes $\{\pi\}_{\Gamma'}$ defined by their Frobeniuses. The image $\{\pi\}_{\Gamma'}$ does not cover all Γ'-conjugacy classes, but it can be characterized group theoretically as the set of all positive and primitive Γ'-conjugacy classes. Here, an element $\pi = [p^d, b]$ of Γ' (or its Γ'-conjugacy class $\{\pi\}_{\Gamma'}$) is called positive if $d > 0$, and primitive if it is not a non-trivial power of other element (or conjugacy class). In this correspondence $P' \leftrightarrow \{\pi\}_{\Gamma'}$, the degree of P' is nothing but the *degree of* $\{\pi\}_{\Gamma'}$, i. e., the positive integer d/D, D being the degree of Γ'.

We note here that these arithmetic axioms are also satisfied by the stabilizers of any points ξ of $\mathfrak{X}$ in $\mathfrak{G}$. But $(\mathfrak{X}, \mathfrak{G})$ is unnecessarily large. In fact, (X, Γ) *is small but just as large as necessary in describing the arithmetic of* $\mathfrak{K}/K$, *and has a simple* ($\infty \times p$)*-adic geometric presentation which* $(\mathfrak{X}, \mathfrak{G})$ *does not have.*

§ 2. The ($\infty \times p$)-adic automorphic functions

2.1. (X, Γ) **as an** ($\infty \times p$)**-adic space.** As in § 1.1, let X be the set of all rational numbers, and Γ be the group of linear transforms of X of the form (1). We are going to show that the fields $\mathfrak{K}/K$ and the injections $\iota = (\iota_X, \iota_\Gamma)$ can be constructed *naturally* starting from (X, Γ). To begin with, consider the direct product $X_{\infty p} = R \times Q_p$ of the real number field R and the p-adic field Q_p. It has a ring structure, but we mainly look at its «($\infty \times p$)-adic» topological structure. Embed X diagonally into $X_{\infty p}$. Then X is dense in $X_{\infty p}$, and $X_{\infty p}$ can be considered as the completion of X with respect to the ($\infty \times p$)-adic topology. The action of Γ on X can be extended uniquely to that on $X_{\infty p}$ by the continuity, as

$$\gamma(x_{\infty p}) = (p^d x_\infty + b,\ p^d x_p + b),$$

where $x_{\infty p} = (x_\infty, x_p) \in X_{\infty p}$, $\gamma = [p^d, b] \in \Gamma$. It is clear that a fixed point of Γ is necessarily a point of X. Conversely, each point x of X is a fixed point; in fact, we know that $\Gamma_x \cong Z$. Therefore, X *is nothing but the set of all fixed points of* Γ *on* $X_{\infty p}$. (This is how X is defined in more general cases. Namely, in such cases, a transformation group acting on some ($\infty \times p$)-adic space is given, and X is defined to be the set of all fixed points on that space.) If we regard $X_{\infty p} = R \times Q_p$ as an additive topological group and $Z^{(p)}$ as a diagonally embedded subgroup, then $Z^{(p)}$ is a discrete subgroup with compact quotient. In fact, $Z^{(p)} \cap (R \times Z_p) = Z$, which implies that the restriction of $Z^{(p)}$ to a p-adically open subgroup of $X_{\infty p}$

is ∞-adically discrete; hence $Z^{(p)}$ is discrete. On the other hand, $Z^{(p)}+$ $+(R\times Z_p)=X_{\infty p}$; hence the quotient $Z^{(p)}\diagdown X_{\infty p}$ can be identified with $Z\diagdown(R\times Z_p)$; hence it is compact. This implies that the subgroup Γ^1 of Γ, consisting of all translations by $Z^{(p)}$, acts discontinuously on $X_{\infty p}$ and has a compact quotient. Since Γ is larger than Γ^1 by Z (i. e., $\Gamma/\Gamma^1\cong Z$), it follows that the action of Γ on $X_{\infty p}$ cannot be discontinuous. More precisely, if $x\notin X$ so that $\Gamma_x=\{1\}$, the Γ-orbit of x consists of infinitely many distinct Γ^1-orbits, and hence it cannot be discrete on $X_{\infty p}$. But, on the other hand, if $x\in X$, then the difference Z between Γ and Γ^1 is almost killed by the stabilizer $\Gamma_x\cong Z$, and the *Γ-orbit of x is discrete on $X_{\infty p}$*. In fact, it consists of exactly d_x distinct Γ^1-orbits

$$p^k x+Z^{(p)}\qquad(k=0,\ 1,\ \ldots,\ d_x-1).$$

(This of course corresponds to the fact observed in § 1 that the prime of K determined by the Γ-orbit of x is of degree d_x and hence splits into d_x distinct primes in the constant field extension $K\cdot\bar{F}_p$). In this sense, the action of Γ on X is discontinuous. These are the basic facts to be observed, but will be used only implicitly in the sequel.

2.2. The automorphic functions. The automorphic functions which we shall consider are functions defined over the set of rational numbers X having their values in $(\infty\times p)$-adic algebraic numbers, and having some invariance property with respect to subgroups of Γ with finite indices. The definition may sound strange, but it is natural. It is a definition which presupposes the reciprocity described by Γ.

For each positive integer d, let $Q_p^{(d)}$ denote the unique unramified extension of Q_p of degree d, and let $Q_p^{(\infty)}$ denote the composite of $Q_p^{(d)}$ for all d; i. e., the maximum unramified extension of Q_p. As in § 1. 3, fix an algebraic closure $\bar{Q}$ of Q in $C\oplus\bar{Q}_p$, and call ψ_∞ (resp. ψ_p) the projection $\bar{Q}\to C$ (resp. $\bar{Q}\to\bar{Q}_p$). Put

$$\mathfrak{Q}_d=\bar{Q}\cap(C\oplus Q_p^{(d)})\qquad(1\leqslant d\leqslant\infty).$$

Then the Galois groups of $\mathfrak{Q}_\infty/\mathfrak{Q}_1$, $Q_p^{(\infty)}/Q_p$ and of $\bar{F}_p/F_p$ can be identified in a natural manner. The automorphism of $\mathfrak{Q}_\infty/\mathfrak{Q}_1$ corresponding to the p-th power automorphism σ^* of $\bar{F}_p/F_p$ will be denoted by σ. Thus, if $\mathfrak{P}$ is the prime factor of p in $\bar{Q}$ defined by the projection ψ_p, then $\mathfrak{Q}_\infty$, $\mathfrak{Q}_1$ and σ are the inertia field, the decomposition field and the Frobenius automorphism of $\mathfrak{P}$ in $\bar{Q}/Q$ respectively.

Definition. *Let Γ' be a subgroup of Γ with finite index. A map*

$$f\colon X\to\mathfrak{Q}_\infty\cup(\infty)$$

is called an automorphic function with respect to Γ' if the following conditions (I), (II), (III) *are satisfied.*

(I) *Put* $X_0=X\cap(R\times Z_p)$; i. e., *the set of rational numbers without p-factors in the denominators. Then the restriction of f to X_0 is meromorphic, i. e., it can be extended to a meromorphic function $\tilde{f}$ on C.*

Here, we regard X_0 and $\mathfrak{Q}_\infty$ as subsets of C, the latter by the projection ψ_∞. Since X_0 is dense in R, such $\tilde{f}$ is uniquely determined by f.

(II) $f(\gamma x) = f(x)^{\sigma^{\chi(\gamma)}}$ *holds for all* $x \in X$ *and* $\gamma \in \Gamma'$.

Here, we use the convention $\infty^\sigma = \infty$. Since every Γ'-orbit in X meets X_0 (the intersection being a discrete subset of R), f is determined by its values on X_0, and hence by $\tilde{f}$. The condition (II) is satisfied if it is satisfied for the generators of Γ'. But if $\Gamma' = \Gamma_{(D;\, N,\, B)}$ (§ 1. 1), Γ' is generated by $[1, N]$ and $[p^D, B]$, so that (II) is equivalent to:

(II)* (a) $f(x+N) = f(x)$,

(b) $f(p^D x + B) = f(x)^{\sigma^D}$.

Now, (II)* (a) applied to the points of X_0 implies $\tilde{f}(z+N) = \tilde{f}(z)$ for all $z \in X_0$ and hence for all $z \in C$. Therefore, $\tilde{f}(z)$ is a meromorphic function of $\tilde{q}_N = e[z/N]$ on $0 < |\tilde{q}_N| < \infty$.

(III) *As a function of* $\tilde{q}_N$, $\tilde{f}$ *is meromorphic also at* $\tilde{q}_N = 0, \infty$.

This implies that $\tilde{f}$ is a rational function of q_N with complex coefficients.

2. 3. The field of automorphic functions. We shall write down all automorphic functions explicitly. Call $M(\Gamma')$ the set of all automorphic functions with respect to Γ'. First, a constant function $f(x) = c \in \mathfrak{Q}_\infty$ belongs to $M(\Gamma')$ if and only if it satisfies (II) * (b), since all other conditions are trivially satisfied. Therefore, it is automorphic if and only if $c^{\sigma^D} = c$; i. e., $c \in \mathfrak{Q}_D$. Secondly, to find a generating automorphic function, we note the following. For each $x_0 \in X_0$, $e[x_0]$ is a (complex) n-th root of unity for some $n \not\equiv 0 \pmod p$. By ψ_∞^{-1}, it can be considered as an element of $\bar{Q}$. But since $n \not\equiv 0 \pmod p$, it is contained in $\mathfrak{Q}_\infty$. For every such root of unity in $\mathfrak{Q}_\infty$, σ acts as the p-th power map, and hence we have $e[x_0]^\sigma = e[px_0]$. Therefore, if we define a function $q: X \to \mathfrak{Q}_\infty$ by

$$q(x) = e[x_0]^{\sigma^{-k}}$$

for $x = p^{-k}x_0$ ($x_0 \in X_0$, $k \geqslant 0$), it is independent of the way of this decomposition of x. Now, it is easy to check that $q(x)$ is an automorphic function with respect to Γ. In fact, $q(x_0) = e[x_0]$ on X_0; hence (I) and (III) are satisfied with $\tilde{q}(z) = e[z]$. As for (II) *, we have $q(x+1) = e[x_0 + p^k]^{\sigma^{-k}} = e[x_0]^{\sigma^{-k}} = q(x)$; hence (II) * (a); and $q(px) = e[px_0]^{\sigma^{-k}} = e[x_0]^{\sigma^{-k+1}} = q(x)^\sigma$; hence (II) * (b). Therefore, $q(x) \in M(\Gamma)$. We note that q is nothing but the function that defines an isomorphism $Q/Z^{(p)} \cong \nabla$, where ∇ is the group of all roots of unity in $\mathfrak{Q}_\infty$ whose orders are not divisible by p. But this trivial group-theoretic observation is not useful for our generalizations, and we do not care much about it. In any case, we have $q(x+y) = q(x)\,q(y)$, and $q(px) = q(x)^\sigma$ for $x, y \in X$. Now, to find a generating automorphic function for $\Gamma' = \Gamma_{(D;\, N,\, B)}$, consider the function $q'(x) = \varepsilon q(x/N)$, where ε

is some constant in $\mathfrak{Q}_\infty$. Then it satisfies the conditions (I), (II) * (a) and (III). The remaining condition (II) * (b) requires that $\varepsilon q(p^D x/N) q(B/N) = \varepsilon^{\sigma^D} q(x/N)^{\sigma^D}$, but since $q(p^d x/N) = q(x/N)^{\sigma^D}$, this is equivalent to $\varepsilon^{\sigma^D}/\varepsilon = q(B/N)$. Therefore, it is satisfied if $\varepsilon = q(B/N(p^D - 1))$. So, if we choose B to be an integer, $B \in Z$, the function

$$q'(x) = e[B/N(p^D - 1)]\, q(x/N)$$

gives a non-constant automorphic function with respect to $\Gamma' = \Gamma_{(D;\, N,\, B)}$.

Now, it is clear that the elements of $\mathfrak{Q}_D[q'(x)]$ and also the quotients of two relatively prime elements of $\mathfrak{Q}_D[q'(x)]$ are automorphic with respect to Γ'. Therefore, all rational functions of $q'(x)$ over $\mathfrak{Q}_D$ are automorphic functions. We shall check that these are in fact all. Take any $f(x) \in M(\Gamma')$. The corresponding meromorphic function $\tilde{f}(z)$ is then a rational function of $\tilde{q}'(z)$, but with complex coefficients. To show that the coefficients belong to $\mathfrak{Q}_\infty$ modulo a scalar multiple, it suffices to find an infinite set of points of C at which $\tilde{q}'$ takes infinitely many distinct values in $\mathfrak{Q}_\infty$ and $\tilde{f}$ also takes values in $\mathfrak{Q}_\infty$. But if z runs over all points of X_0 except the poles of $\tilde{f}$, then $\tilde{q}'(z)$ runs over almost all elements of ∇, and $\tilde{f}(z)$ for such z belongs to $\mathfrak{Q}_\infty$ (since $\tilde{f} = f$ on X_0). Therefore, $\tilde{f}$ is a rational function of $\tilde{q}'$ with coefficients in $\mathfrak{Q}_{nD}$ for some n. Let f_1 be the corresponding rational function of q'. Then f and f_1 are both automorphic functions with respect to $\Gamma'' = \Gamma_{(nD,\, N,\, B')}$, where B' is an element of $Z^{(p)}$ defined by $[p^D, B]^n = [p^{nD}, B']$. And we have $\tilde{f} = \tilde{f}_1$. Therefore, $f = f_1$, and hence f is a rational function of q' with coefficients in $\mathfrak{Q}_{nD}$. But since the functional operator $F(x) \to F(\gamma^{-1}x)^{\sigma^{\chi(\gamma)}}$ for $\gamma = [p^D, B]$ leaves f and q' invariant, and replaces elements of $\mathfrak{Q}_{nD}$ by their transforms by σ^D, we deduce easily (by the standard arguments in elementary algebra) that f must be a rational function of q' with coefficients in $\mathfrak{Q}_D$. We have thus obtained:

$$M(\Gamma') = \mathfrak{Q}_D(q'(x)).$$

For each $f \in M(\Gamma')$, define f^σ by $f^\sigma(x) = f(x)^\sigma$. For the constants $f = c \in \mathfrak{Q}_D$, f^σ equals c^σ in the previous sense. We have $q'^\sigma = q'^p$, since the values of q' belong to ∇. Therefore, σ defines an isomorphism of $M(\Gamma')$ into itself, and $[M(\Gamma') : M(\Gamma')^\sigma] = p$. Some important objects in $M(\Gamma')$ can be characterized by their σ-invariance. For instance, a prime of $M(\Gamma')$ is a *Teichmüller prime*, i. e., the residue class of q' with respect to that prime belongs to ∇, if and only if it is σ-invariant. Another important σ-invariant object is the following. Call $\mathfrak{P}_{\Gamma'}$ the unique class of places of $M(\Gamma')$ that extends $\mathfrak{P}$: $\mathfrak{Q}_D \to F_{p^D}$ and that maps q' to a transcendental element over F_{p^D}. Then it is the unique σ-invariant extension of $\mathfrak{P}$. In fact, if $\tilde{\mathfrak{P}}$ is any σ-invariant extension of $\mathfrak{P}$ to $M(\Gamma')$, then q' must be a $\tilde{\mathfrak{P}}$-adic unit, since $q'^{p-1} = q'^\sigma/q'$. Also, $q'^{p^d} - 1$ for any $d \geqslant 0$ must be a $\tilde{\mathfrak{P}}$-adic unit, since if we put $y = q'^{p^d}$, then $(y-1)^{p-1} \equiv (y^p - 1)/(y-1) = (y-1)^\sigma/(y-1) \pmod{\tilde{\mathfrak{P}}}$. Therefore, the residue class of q' mod $\tilde{\mathfrak{P}}$ must be transcendental, and hence $\tilde{\mathfrak{P}}$ must be equivalent to $\mathfrak{P}_{\Gamma'}$. It is obvious that σ induces the p-th power map of the residue field of $M(\Gamma')$. Finally, there is a differential ω of $M(\Gamma')$, unique up to $\mathfrak{Q}_1^\times$-multiples, such that $\omega^\sigma/\omega = p$. In fact, $\omega = cq'^{-1}dq'$ $(c \in \mathfrak{Q}_1^\times)$ are the solutions of this equation. We shall not make any explicit use of this differential here. But in the general cases, it is essential to consider such differentials [2, 3].

2.4. The Galois extension $\mathfrak{M}/M$. Now let $\mathfrak{M}$ be the union of $M(\Gamma')$ for all Γ', and put $M = M(\Gamma)$. Then $\mathfrak{M}$ is an infinite Galois extension of M, and Γ acts on $\mathfrak{M}$ by:

$$f(x) \to (\gamma f)(x) = f(\gamma^{-1}x)^{\sigma^{\chi(\gamma)}}.$$

It is easy to see that the Galois theory holds between subgroups of Γ with finite indices and finite extensions of M in $\mathfrak{M}$. The field corresponding to Γ' is of course, $M(\Gamma')$. The constant field of $\mathfrak{M}$ is $\mathfrak{Q}_\infty$, on which $\gamma \in \Gamma$ acts as $\sigma^{\chi(\gamma)}$. Each point x of X defines a prime of $\mathfrak{M}$ in a natural way. In fact, in terms of places, by $\mathfrak{M} \ni f \to f(x) \in \mathfrak{Q}_\infty \cup (\infty)$. The place of $\mathfrak{M}$ defined by γx $(\gamma \in \Gamma)$ maps f to $f(\gamma x) = (\gamma^{-1}f)(x)^{\sigma^{\chi(\gamma)}}$. This is equivalent to the place of $\mathfrak{M}$ defined by $f \to (\gamma^{-1}f)(x)$; hence the geometric action of Γ on X corresponds to the arithmetic action of Galois automorphisms of $\mathfrak{M}/M$ on primes of $\mathfrak{M}$.

The isomorphism $\sigma : f(x) \to f(x)^\sigma$ of each $M(\Gamma')$ defines an isomorphism of $\mathfrak{M}$ into itself, and we have $[\mathfrak{M} : \mathfrak{M}^\sigma] = p$. This isomorphism σ commutes with every element of Γ, and hence also with every element of the group $\mathfrak{G}^{\infty p}$ of all such automorphisms of $\mathfrak{M}/M$ that act as integral powers of σ on $\mathfrak{Q}_\infty$. On the other hand, it is easy to see that X is a dense subset of the space $\mathfrak{X}^{\infty p}$ of all Teichmüller primes of $\mathfrak{M}$, i. e., those primes of $\mathfrak{M}$ satisfying one of the following three equivalent conditions; (i) σ-invariant, (ii) the restriction to $M(\Gamma')$ is Teichmüller for every Γ', (iii) the restriction to M is Teichmüller. Finally, let $[\mathfrak{P}]$ be the unique class of places of $\mathfrak{M}$ whose restriction to each $M(\Gamma')$ coincides with $\mathfrak{P}_{\Gamma'}$. It is then the unique σ-invariant extension of $\mathfrak{P} : \mathfrak{Q}_\infty \to \bar{F}_p$ to $\mathfrak{M}$, and σ induces the p-th power map of the residue field. It is also invariant by $\mathfrak{G}^{\infty p}$.

Now we have obtained everything necessary. First, we can identify the residue field of $\mathfrak{M}$ modulo $[\mathfrak{P}]$ with $\mathfrak{K}$. The residue class of $q(x/N)$ is to correspond with t_N. Then, the reduction modulo $[\mathfrak{P}]$ induces the bijections $\mathfrak{X}^{\infty p} \to \mathfrak{X}$ and $\mathfrak{G}^{\infty p} \to \mathfrak{G}$. Finally, the embedding $(X, \Gamma) \to (\mathfrak{X}, \mathfrak{G})$ thus obtained is nothing but the embedding ι defined in § 1.3. Thus, we have constructed $\mathfrak{K}/K$ and ι by a natural method starting from (X, Γ).

2.5. Remarks. We have written down the field $M(\Gamma')$ explicitly. But the method we used does not indicate anything theoretical. So, we supplement here the following. Put $\mathfrak{C}' = C/NZ$, and let π' be the rational map of $\mathfrak{C}'$ onto itself induced by the transform $z \to p^D z + B$ of C. If we choose B

as an integer, such π' is uniquely determined. And a point of $\mathfrak{C}'$ comes from a point of X_0 if and only if it is a fixed point of some power of π'. Let Ψ' denote the set of all such points on $\mathfrak{C}'$. On the other hand, let $\tilde{f}$ be any rational function of $e[z/N]$ with complex coefficients, considered also as a function on $\mathfrak{C}'$. Then $\tilde{f}$ corresponds to some $f \in M(\Gamma')$ if and only if its restriction to Ψ' is $\mathfrak{Q}_\infty \cup (\infty)$ — valued and $\tilde{f} \circ \pi' = \tilde{f}^{\sigma^D}$ holds on Ψ'. Therefore, the existence of «sufficiently many» automorphic functions w. r. t. Γ' is equivalent to the existence of an algebraico-geometric model of $\{\mathfrak{C}', \pi\}$ defined over $\mathfrak{Q}_D$, with which each point of Ψ' is $\mathfrak{Q}_\infty$-rational and π' induces on Ψ' the conjugation by σ^D. In our case, we know that the isomorphism $\mathfrak{C}' \simeq P^1$ given by $\hat{q}'$ defines such a model, and our interest lies in the general methods for questions of this type. We hope to discuss this subject somewhere in near future.

Of course we can lower the field of constants of $\mathfrak{M}$ to the cyclotomic field $Q(\nabla)$, but we did not care about it because of the following reasons. First, we should not assume it to be known in advance, and secondly, it does not have much to do with $(\infty \times p)$-adic questions. So, the fields $\mathfrak{Q}_d$ $(1 \leqslant d \leqslant \infty)$ are probably the best fields with which to define the $(\infty \times p)$-adic automorphic functions in our case. In more general cases, we must take $\mathfrak{p}$-adic fields instead of Q_p as a base field on the p-part, and accordingly we must change the definitions of $\mathfrak{Q}_d$. The interrelation between (X, Γ) and the base $\mathfrak{p}$-adic field is a difficult question in the most general cases.

Finally, in the general cases, χ is not a character of Γ, but depends on two variables: $\chi = \chi(\gamma, x)\,(\gamma \in \Gamma,\ x \in X)$.

§ 3. The $(\infty \times p)$-adic universal covering

3. 1. The abstract version of $(\mathfrak{M}/M, \sigma)$. In § 2, we defined $\mathfrak{M}$ as the union of the fields of automorphic functions. The isomorphism σ of $\mathfrak{M}$ then came out naturally. We observed that the pair $\{\mathfrak{M}/M, \sigma\}$ is, in a sense, an $(\infty \times p)$-adic equivalent of $\{\mathfrak{K}/K, \sigma^*\}$, where σ^* is the p-th power isomorphism of $\mathfrak{K}$. Our purpose here is to explain the converse procedure, i. e., to construct the $(\infty \times p)$-adic transformation space (X, Γ) and the embeddings $\iota = (\iota_X, \iota_\Gamma)$, starting from the *abstractly given* pair $\{\mathfrak{M}/M, \sigma\}$. We leave out the discussion of the construction of $\{\mathfrak{M}/M, \sigma\}$ starting from $\{\mathfrak{K}/K, \sigma^*\}$, since this connection is too trivial in our example. We note here in advance that one essential point we use is that our lifting σ of σ^* leaves invariant the set of all primes of $\mathfrak{M}$ that are unramified in $\mathfrak{M}/M$.

To fix the notations, let q be an indeterminate, and define

$$M = \mathfrak{Q}_1(q), \qquad \mathfrak{M} = \mathfrak{Q}_\infty(q^{1/N} \mid N \not\equiv 0 \pmod{p}).$$

Then $\mathfrak{M}$ is an infinite Galois extension of M, and there is a unique extension of the automorphism σ of $\mathfrak{Q}_\infty$ to an isomorphism of σ into itself that maps every N-th root $q^{1/N}$ of q to its p-th power (for all $N \not\equiv 0 \pmod{p}$). Denote this extended isomorphism again by σ. Our purpose is to construct (X, Γ) and ι in some natural way, starting from $\{\mathfrak{M}/M, \sigma\}$.

For each $\nu \in Z$, let Φ_ν be the set of all places

$$\varphi_\nu : \mathfrak{M} \to \bar{Q} \cup (\infty),$$

whose restriction to $\mathfrak{Q}_\infty$ coincides with the automorphism σ^ν, and with which $\varphi_\nu(q) \neq 0, \infty$. For each $\varphi_\nu \in \Phi_\nu$, define $\varphi_\nu^\sigma \in \Phi_{\nu+1}$ by $\varphi_\nu^\sigma(f) = \varphi_\nu(f^\sigma)$ $(f \in \mathfrak{M})$. Then $\varphi_\nu \to \varphi_\nu^\sigma$ gives a *p-to-one* mapping of Φ_ν onto $\Phi_{\nu+1}$. In fact, take any $\varphi_{\nu+1} \in \Phi_{\nu+1}$, and let us look for the solution $\varphi_\nu \in \Phi_\nu$ of the equation $\varphi_\nu^\sigma = \varphi_{\nu+1}$. Since $(q^{1/N})^\sigma = (q^{1/N})^p$, it is clear that φ_ν is a solution if and only if $\varphi_\nu(q^{1/N})$ is a p-th root of $\varphi_{\nu+1}(q^{1/N})$ for all $N \not\equiv 0 \pmod{p}$. We can choose any one of the p-th roots of $\varphi_{\nu+1}(q)$ for $\varphi_\nu(q)$, but then $\varphi_\nu(q^{1/N})$ for each N is uniquely determined, since it must be the unique p-th root of $\varphi_{\nu+1}(q^{1/N})$ whose N-th power equals $\varphi_\nu(q)$. Thus, the mapping $\varphi_\nu \to \varphi_\nu^\sigma$ is p-to-one. We call $\varphi_\nu \in \Phi_\nu$ *Teichmüller* if φ_ν^σ is equivalent to φ_ν as a place, i. e., if $\varphi_\nu(f^\sigma) = \varphi_\nu(f)^{\tilde{\sigma}}$ $(f \in \mathfrak{M})$ holds for some automorphism $\tilde{\sigma}$ of $\bar{Q}$ (which must then be an extension of the automorphism σ of $\mathfrak{Q}_\infty$). It is easy to see that φ_ν is Teichmüller if and only if $\varphi_\nu(q^{1/N}) \in \nabla$ for all $N \not\equiv 0 \pmod{p}$, or if and only if $\varphi_\nu(q) \in \nabla$. Denote by Φ_ν^0 the set of all Teichmüller places in Φ_ν. Then, the p-to-one mapping $\varphi_\nu \to \varphi_\nu^\sigma$, restricted to Φ_ν^0, gives a *one-to-one* mapping $\Phi_\nu^0 \to \Phi_{\nu+1}^0$, since each element of ∇ has a unique p-th root in ∇. We may use the previous notation $\mathfrak{X}^{\infty p}$ for Φ_0^0, since Φ_0^0 consists of all Teichmüller primes of $\mathfrak{M}$, i. e., the Teichmüller places of $\mathfrak{M}$ that are the identity map on the constant field $\mathfrak{Q}_\infty$. We know that the reduction modulo $[\mathfrak{P}]$ of $\mathfrak{M}$ induces a bijection $\mathfrak{X}^{\infty p} \to \mathfrak{X}$.

3.2. The $(\infty \times p)$-adic coordinates. Now, consider the set Φ of all sequences $\varphi = (\varphi_\nu)_{\nu \in Z}$, with $\varphi_\nu \in \Phi_\nu$ and $\varphi_{\nu+1} = \varphi_\nu^\sigma$ *for each* $\nu \in Z$ and with $\varphi_\nu \in \Phi_\nu^0$ for *sufficiently large* ν. Then φ_ν determines $\varphi_{\nu+k}$ for all $k \geqslant 0$, but there are p distinct possibilities of $\varphi_{\nu-1}$ for each φ_ν. But if φ_ν is Teichmüller, there is a unique «Teichmüller choice» of $\varphi_{\nu-1}$. We call $\varphi^0 = (\varphi_\nu^0) \in \Phi$ Teichmüller if $\varphi_\nu^0 \in \Phi_\nu^0$ *for all* $\nu \in Z$. Call Φ^0 the set of all Teichmüller elements of Φ. Then each $\varphi_0^0 \in \Phi_0^0 = \mathfrak{X}^{\infty p}$ determines $\varphi^0 = (\varphi_\nu^0) \in \Phi^0$ uniquely, and elements of Φ^0 are in one-to-one correspondence with their 0-th coordinates in $\mathfrak{X}^{\infty p}$. On the other hand, for each $\varphi = (\varphi_\nu) \in \Phi$, there is a unique $\varphi^0 = (\varphi_\nu^0) \in \Phi^0$ with which $\varphi_\nu = \varphi_\nu^0$ holds for sufficiently large ν. In this case, we say that φ *belongs to* φ^0, or to the corresponding element ξ of $\mathfrak{X}^{\infty p}$. Thus, two elements $\varphi = (\varphi_\nu)$ and $\varphi' = (\varphi_\nu')$ of Φ belong to the same element of $\mathfrak{X}^{\infty p}$ if and only if $\varphi_\nu = \varphi_\nu'$ holds for sufficiently large ν.

Now fix any $\xi \in \mathfrak{X}^{\infty p}$, and call $\Phi(\xi)$ the set of all elements of Φ belonging to ξ. Then $\Phi(\xi)$ can be identified with Q_p in a natural way. To see this, recall that a solution φ_ν of the equation $\varphi_\nu^\sigma = \varphi_{\nu+1}$ is determined by $\varphi_\nu(q)$. Since ξ is fixed, the ν-th coordinates for sufficiently large ν are given for each $\varphi = (\varphi_\nu) \in \Phi(\xi)$. Hence each $\varphi \in \Phi(\xi)$ is determined by the sequence $\varphi(q)=(\varphi_\nu(q))$ of roots of unity. Let $\varphi_0=(\varphi_\nu^0)$ be the element of Φ^0 corresponding to ξ, and put $\eta=\varphi_0^0(q)\in\nabla$. Then $\varphi^0(q)=(\eta^{p^\nu})$; $\eta^{p^\nu}\in\nabla$. Now it is clear that $\alpha\in Q_p$ and $\varphi\in\Phi(\xi)$ correspond in a one-to-one manner by $\varphi(q)=(\eta^{p^\nu}e[\mathrm{tr}(p^\nu\alpha)])$, where $\mathrm{tr}(\alpha)$ for $\alpha\in Q_p$ denotes the image of the class of α modulo Z_p under the isomorphism $Q_p/Z_p\cong Z^{(p)}/Z$. In this way, each $\Phi(\xi)$ carries a p-adic structure. We call (ξ,α) the p-adic coordinate of φ, of which α is p-adic, but ξ is merely a «parameter» that runs over $\mathfrak{X}^{\infty p}$. The coordinate of the Teichmüller element φ^0 is $(\xi, 0)$. (In more general cases of $\{\mathfrak{M}/M, \sigma\}$, we use certain generalizations of the differential ω (§ 2.3) to define the p-adic coordinates.)

We preferred to define Φ as a set of sequences of places of $\mathfrak{M}$, but we may equivalently define it as follows. First, look at the subfield $\mathfrak{M}^\sigma$ of $\mathfrak{M}$ and the isomorphism $\sigma^{-1}:\mathfrak{M}^\sigma\to\mathfrak{M}$. Extend σ^{-1} to some isomorphism $\mathfrak{M}\to\mathfrak{M}^{\sigma^{-1}}$ in one way, $\mathfrak{M}^{\sigma^{-1}}$ being an overfield of $\mathfrak{M}$. Extend this again to an isomorphism $\sigma^{-1}:\mathfrak{M}^{\sigma^{-1}}\to\mathfrak{M}^{\sigma^{-2}}$ and proceeding step by step in this way, we obtain for each $\nu\in Z$ the field $\mathfrak{M}_\nu=\mathfrak{M}^{\sigma^\nu}$, the isomorphism $\sigma^{-1}:\mathfrak{M}_\nu\to\mathfrak{M}_{\nu-1}$, and the automorphisms σ^r $(r\in Z)$ of the injective union $\mathfrak{M}_{-\infty}=\bigcup_{\nu\in Z}\mathfrak{M}_\nu$. The constant field of $\mathfrak{M}_{-\infty}$ is still $\mathfrak{Q}_\infty$, since we did not extend σ^{-1} in all possible ways but in only one way at each step. (We may identify $\mathfrak{M}_{-\infty}$ with the field $\mathfrak{Q}_\infty(q^{1/M}\mid M\in Z)$, and then σ acts on the generators as $(q^{1/M})^\sigma=(q^{1/M})^p$ $(M\in Z)$). Now, Φ can be identified with the set of all places $\varphi:\mathfrak{M}_{-\infty}\to\bar{Q}\cup(\infty)$ which is the identity on the constant field $\mathfrak{Q}_\infty$ and which maps q to some root of unity (which may not belong to ∇). In fact, we can identify such a place φ of $\mathfrak{M}_{-\infty}$ with the sequence $(\varphi_\nu)\in\Phi$ defined by $\varphi_\nu(f)=\varphi(f^{\sigma^\nu})$ $(f\in\mathfrak{M}, \nu\in Z)$.

Now, to define the ∞-adic coordinates in Φ, extend $\mathfrak{M}_\nu$ to the function fields $\mathfrak{M}_\nu^C$ over C. Then $\mathfrak{M}_{-\infty}^C$ can be identified with the field $C(q^{1/M}\mid M\in Z)$, and hence it is an infinite Galois extension of $C(q)$ having no ramifications outside $q=0,\infty$. Let Φ^C be the set of all primes of $\mathfrak{M}_{-\infty}^C$ not lying on $q=0,\infty$. Then Φ^C carries a structure of a complex manifold isomorphic to $\sum_\lambda C_\lambda$, where each C_λ is the complex plane C, and $\sum_\lambda$ is a disjoint sum over ($\aleph$-infinity of) distinct connected components. In fact, each connected component of Φ^C can be identified with the universal covering of $P^1-\{0,\infty\}$, P^1 being the complex projective line. Now, embed Φ into Φ^C in a natural way. Then if we fix one isomorphism $\Phi^C\cong\sum_\lambda C_\lambda$, Φ can be considered as a subset of $\sum_\lambda C_\lambda$. Therefore, *each element $\varphi\in\Phi$ has two pairs of coordinates; the p-adic coordinate (ξ,α) and the complex coordinate (λ,β), where β is the point of $C_\lambda=C$ corresponding to φ.* And each of the two pairs (ξ,α) and (λ,β) determines φ uniquely.

Now fix one λ. Let φ and φ' be any two points of C_λ belonging to Φ, and let (ξ,α) and (ξ',α') be their p-adic coordinates. Then $\xi=\xi'$ umplies

$\alpha=\alpha'$, and vice versa. In fact, let $c(t)$ $(0\leqslant t\leqslant 1)$ be a path on C_λ connecting φ and φ', and suppose that $\xi=\xi'$. Then φ and φ' coincide on $\mathfrak{M}_\nu$ for some $\nu\in Z$. But since $\mathfrak{M}_\nu^C$ is a sufficiently large extension of $M_\nu^C= =C(q^{p^\nu})$, this implies that the projection of $c(t)$ to M_ν^C must be homotopic to 0. Hence $\varphi=\varphi'$. In the same way, by using $M^{\sigma^{-\infty}}$ instead of $\mathfrak{M}_\nu$, we see that $\alpha=\alpha'$ implies $\xi=\xi'$. Now, define X_λ to be the set of all elements of Φ lying on the fixed component C_λ of Φ^C. Take any $\varphi\in X_\lambda$, and let (λ, β) and (ξ, α) be the complex and the p-adic coordinates. Then, since one of ξ and α determines the other, the mappings $\varphi\to\beta$, $\varphi\to\alpha$, and $\varphi\to\xi$ define three injections $X_\lambda\to C$, $X_\lambda\to Q_p$, and $X_\lambda\to\mathfrak{X}^{\infty p}$. Hence X_λ can be considered as subsets of $C\times Q_p$, and also of $\mathfrak{X}^{\infty p}$. (Here, we note the following. We embed X_λ into $\mathfrak{X}^{\infty p}$, and hence each $\varphi\in X_\lambda$ corresponds to some $\xi\in\mathfrak{X}^{\infty p}$. But this does not imply that the place φ of $\mathfrak{M}_{-\infty}$ coincides with ξ on $\mathfrak{M}$. It only implies that φ and ξ coincide on $\mathfrak{M}_\nu$ for *some* ν; and more precisely, for such ν with which α belongs to $p^{-\nu}Z_p$. This is obvious by the definitions, but we must keep it in mind when we *embed* X_λ into $\mathfrak{X}^{\infty p}$.) Since each open subgroup of the Galois group of $\mathfrak{M}_{-\infty}^C/C(q)$ acts transitively on the set of all connected components of Φ^C, it follows immediately that X_λ is dense in $\mathfrak{X}^{\infty p}$. Take any λ, λ'. Then *the two subspaces X_λ and $X_{\lambda'}$ of $\mathfrak{X}^{\infty p}$ are either disjoint or coincide totally.* In fact, suppose that they meet at ξ, i. e., there are some φ and $\varphi'\in\Phi(\xi)$ lying on C_λ and $C_{\lambda'}$ respectively. Then since φ and φ' coincide on $\mathfrak{M}_\nu$ for some ν, and since $\mathfrak{M}_{-\infty}^C/\mathfrak{M}_\nu^C$ is a Galois extension, they are conjugate by some automorphism of $\mathfrak{M}_{-\infty}^C/\mathfrak{M}_\nu^C$. Such an automorphism leaves the complex structure of Φ^C invariant, and hence it maps the connected component C_λ of φ to the component $C_{\lambda'}$ of φ'. On the other hand, since such an automorphism is the identity on $\mathfrak{M}_\nu$, it leaves each $\Phi(\xi')$ $(\xi'\in\mathfrak{X}^{\infty p})$ invariant. Therefore, if ξ' belongs to X_λ, i. e., some $\psi\in\Phi(\xi')$ lies on C_λ, then there exists some $\psi'\in\Phi(\xi')$ lying on $C_{\lambda'}$, i. e., ξ' also belongs to $X_{\lambda'}$. This proves that the two subspaces X_λ and $X_{\lambda'}$ of $\mathfrak{X}^{\infty p}$ must coincide if they meet at all.

Now consider each X_λ as a (dense) subspace of $\mathfrak{X}^{\infty p}$. Then *elements of $\mathfrak{G}^{\infty p}$ map each X_λ to some $X_{\lambda'}$.* In fact, for those elements of $\mathfrak{G}^{\infty p}$ acting trivially on $\mathfrak{Q}_\infty$, the statement is obvious since they come from complex automorphisms. On the other hand, fix one N-th root q_N of q for each $N\not\equiv 0 \pmod p$ in such a way that $q_{MN}^M=q_N$ holds for all M, N. Let π_0 be the element of $\mathfrak{G}^{\infty p}$ that leaves each q_N fixed and acts as σ on $\mathfrak{Q}_\infty$. It suffices to check our statement for π_0, and this follows immediately from the fact that the automorphism of $\mathfrak{M}_{-\infty}^C$ over C defined by $q_N^{\sigma^\nu}\to q_N^{\sigma^{\nu-1}}$ leaves the complex structure of Φ^C invariant and that the effect of this automorphism on the coordinates ξ is the same as the effect of the automorphism π_0 on the primes ξ of $\mathfrak{X}^{\infty p}$.

Now, to define Γ, fix one λ, and embed $X=X_\lambda$ in $\mathfrak{X}^{\infty p}$. As a subgroup of $\mathfrak{G}^{\infty p}$, Γ is defined to be the group of all elements of $\mathfrak{G}^{\infty p}$ leaving X fixed. Then, it is easy to find for each $\gamma\in\Gamma$ a unique automorphism γ_C of $\mathfrak{M}_C^{-\infty}$ over C, that leaves C_λ fixed and whose action on the coordinates ξ of points on $\Phi\cap C_\lambda$ is the same as the action of γ on the primes ξ of $\mathfrak{X}^{\infty p}$. The action of γ_C on the coordinates β resp. α of points of $\Phi\cap C_\lambda$ defines the complex resp. the p-adic representations of Γ. We can check each expected property of (X, Γ) from these definitions and can normalize the choice of λ and the coordinates β in such a way that (X, Γ) thus obtained coincides precisely with the previously given one. Finally, it is easy to see that the two procedures of defining $\{\mathfrak{M}/M, \sigma\}$ from (X, Γ), and (X, Γ) from $\{\mathfrak{M}/M, \sigma\}$, are the inverse of each other. These are left to the readers.

BIBLIOGRAPHY

1. Y. Ihara, *On congruence monodromy problems.* Vols. 1, 2, Lecture Notes, nos. 1, 2, Dept. of Mathematics, Univ. of Tokyo, Tokyo, 1968, 1969. MR 44 #6706; #6707.

2. ———, *Non-abelian classfields over function fields in special cases,* Proc. Internat. Congress Math. (Nice, 1970), vol. 1, Gauthier-Villars, Paris, 1971, pp. 381–390.

3. ———, *Non-abelian invariant differentials and Schwarzian equations in p-adic theory of automorphic functions,* U. S.-Japan Seminar on Modern Methods in Number Theory, 1971 (available at Dept. Math. Univ. Tokyo). (Some other related papers and mimeographed notes are listed in the references of [3]).

Trudy Mat. Inst. Steklov.
132 (1973)

Proc. Steklov Inst. Math.
132 (1973)

SIEGEL FORMS AND ZETA-FUNCTIONS

A. N. ANDRIANOV

ABSTRACT. Results in the theory of Euler products corresponding to Siegel modular forms of rank Z are studied.

Bibliography: 9 items.

In this paper we present our results in the theory of Euler products corresponding to Siegel modular forms of rank 2. This theory turns out to be largely analogous to the classical theory of Hecke for ordinary modular forms. In the paper [1] analogous results have been obtained under some additional restrictions. Complete proofs will be published in Uspehi Mat. Nauk.

§1. Modular forms and Hecke operators

First we recall some known facts and definitions (cf. [2]–[4]). Let $\mathfrak{M}_k^n$ $(n, k \geqslant 1)$ be the space of all Siegel modular forms of rank n and weight k, i.e. the space of all functions $f(Z)$ of rank n which are regular on the Siegel upper half-plane

$$\mathfrak{Z}_n = \{Z = X + iY \in M_n(\mathbf{C}) \mid {}^t Z = Z, \quad Y > 0\}$$

satisfying for any transformation

$$M = \begin{pmatrix} A & B \\ C & D \end{pmatrix} \in \mathrm{Sp}_n(\mathbf{Z}) = \Gamma_n$$

(Siegel's modular group), the corresponding condition

$$f((AZ + B)(CZ + D)^{-1}) = \det (CZ + D)^k f(Z) \qquad (Z \in \mathfrak{Z}_n)$$

and which are furthermore bounded on all domains of the form $\{Z = X + iY \in \mathfrak{Z}_n;\ Y \geqslant cE_n \ (c > 0)\}$. It is well-known that $\dim \mathfrak{M}_k^n < \infty$. Each form $f \in \mathfrak{M}_k^n$ is expressible as a Fourier series of the form

$$f(Z) = \sum_N a(N) \exp (2\pi i \operatorname{Tr}(NZ)), \tag{1}$$

where $N = (n_{ij})$ runs through all symmetric half-integer matrices (i.e. $n_{ii}, 2n_{ij} \in \mathbf{Z}$) of order n satisfying the condition $N \geqslant 0$. Actually $a(N)$ depends only the equivalence class of the matrix $\mathfrak{N}$ in the narrow sense, i.e.

$$a(uN{}^t u) = a(N) \qquad (u \in SL_n(\mathbf{Z})). \tag{2}$$

AMS (MOS) subject classifications (1970). Primary 10D15; Secondary 10H05.

The forms f for which the equation $\det N = 0$ implies $a(N) = 0$ in the expansion (1) are called parabolic. The parabolic forms of rank n and weight k form a subspace $\mathfrak{N}_k^n$ of the space $\mathfrak{M}_k^n$.

Now we define Hecke's operators. We let

$$S_m = \left\{ M \in M_{2n}(\mathbf{Z}) \,\middle|\, {}^t M \begin{pmatrix} 0 & E_n \\ -E_n & 0 \end{pmatrix} M = m \begin{pmatrix} 0 & E_n \\ -E_n & 0 \end{pmatrix} \right\} \quad (m \geqslant 1).$$

Every set S_m splits into a finite number of left or right cosets modulo Γ_n. For $m \geqslant 1$ we define

$$T(m) f = m^{nk - \frac{n(n+1)}{2}} \sum_{M \in \Gamma_n \backslash S_m} f|_k M, \tag{3}$$

letting

$$f|_k M = \det (CZ + D)^{-k} f((AZ + B)(CZ + D)^{-1})$$

if $M = \begin{pmatrix} A & B \\ C & D \end{pmatrix} \in S_m$. The definition (3) does not depend on the choice of the system of representatives from the left cosets $\Gamma_n \backslash S_m$, and for each $m \geqslant 1$ the function $T(m)f$ is again contained in $\mathfrak{M}_k^n$. The operators $T(m)$ are called Hecke operators. The subspace $\mathfrak{N}_k^n$ of parabolic forms is invariant under all operators $T(m)$. Furthermore, the operators $T(m)$ commute with each other and satisfy the equation

$$T(m) T(m_1) = T(mm_1), \quad \text{if} \quad (m, m_1) = 1. \tag{4}$$

For every n and k there exists a base for the space $\mathfrak{N}_k^n$ of parabolic forms consisting of the eigenfunctions of all Hecke operators $T(m)$ $(m \geqslant 1)$. For small n (e.g., $n = 1, 2$) this has been shown also for the entire space $\mathfrak{M}_k^n$ (cf. [3]).

§2. Zeta-functions

For each eigenfunction of a Hecke operator there is a canonical correspondence between the Dirichlet series and an Euler product. If $f \in \mathfrak{M}_k^n$ and

$$T(m) f = \lambda_f(m) f \quad (m = 1, 2, \ldots), \tag{5}$$

then the Dirichlet series

$$Z_f(s) = \sum_{m=1}^{\infty} \frac{\lambda_f(m)}{m^s} \tag{6}$$

is expressible as an Euler product. (It is easy to see that it converges absolutely and uniformly if $\operatorname{Re} s$ is sufficiently large.) Indeed, by means of (4) one has

$$Z_f(s) = \prod_p \left(\sum_{j=0}^{\infty} \frac{\lambda_f(p^j)}{p^{js}} \right).$$

It was shown in [5] that for every prime number p the equation

$$\sum_{j=0}^{\infty} \frac{\lambda_f(p^j)}{p^{js}} = \frac{P_{f,p}(p^{-s})}{Q_{f,p}(p^{-s})}$$

holds, where $P_{f,p}(t)$ and $Q_{f,p}(t)$ are polynomials of degree $2^n - 2$ and 2^n respectively. Thus

$$Z_f(s) = \prod_p P_{f,p}(p^{-s}) Q_{f,p}(p^{-s})^{-1}. \tag{7}$$

The polynomials $P_{f,p}(t)$ and $Q_{f,p}(t)$ may be computed for every given n. For example,

$$\begin{aligned} P_{f,p}(t) &= 1, \\ Q_{f,p}(t) &= 1 - \lambda_f(p)\,t + p^{k-1}t^2, \end{aligned} \qquad (n=1)$$

$$\begin{aligned} P_{f,p}(t) &= 1 - p^{2k-4}t^2, \\ Q_{f,p}(t) &= 1 - \lambda_f(p)\,t + (\lambda_f(p)^2 - \lambda_f(p^2) - p^{2k-4})\,t^2 - \qquad (n=2) \\ &\quad - \lambda_f(p)\,p^{2k-3}t^3 + p^{4k-6}t^4. \end{aligned} \tag{8}$$

The following question arises naturally: What can be said about the analytic properties of the series $Z_f(s)$, and, in particular, does it possess an analytic continuation to the entire plane and does it satisfy any functional equation like the well-known Riemann zeta-function? Hecke's theory gives a complete answer to these questions in the case $n = 1$. Our results give a solution in the case $n = 2$.

For the sake of comparison we state first the basic theorems of Hecke's theory (cf. [6]).

Suppose the function

$$f(z) = \sum_{m=0}^{\infty} a(m) \exp(2\pi imz) \in \mathfrak{M}_k^1$$

is an eigenfunction of all Hecke operators $T(m)$, *i.e.*

$$T(m)\,f = \lambda_f(m)\,f \qquad (m = 1, 2, \ldots).$$

Then the following assertions hold:

I ($n = 1$; *relation between eigenvalues and Fourier coefficients of eigenfunctions*)

$$\sum_{m=1}^{\infty} \frac{a(m)}{m^s} = a(1) \sum_{m=1}^{\infty} \frac{\lambda_f(m)}{m^s} \qquad (\operatorname{Re} s \ \textit{sufficiently large}). \tag{9}$$

II ($n = 1$, *continuation and functional equation*). *The Dirichlet series* $Z_f(s)$ *possesses an analytic continuation into the entire s-plane in the form*

$$\Psi_f(s) = (2\pi)^{-s}\,\Gamma(s)\,Z_f(s), \tag{10}$$

where $\Gamma(s)$ *is the gamma-function, with* $\Psi_f(s)$ *having only a finite number of poles, and satisfies the equation*

$$\Psi_f(k - s) = i^k \Psi_f(s).$$

We observe that the proof of the second equation is based on the first. Indeed, the first equation yields an integral representation of the function $\Psi_f(s)$ in terms of $f(z)$ which implies the second equation. The same situation exists for $n = 2$.

Now let $n = 2$ and let the modular form

$$F(Z) = \sum_{N=\begin{pmatrix} a & b/2 \\ b/2 & c \end{pmatrix} \geqslant 0} a(N) \exp(2\pi i \operatorname{Tr}(NZ)) \in \mathfrak{M}_k^2 \tag{11}$$

be an eigenfunction of all Hecke operators $T(m)$, i.e.

$$T(m) F = \Lambda_F(m) F \qquad (m = 1, 2, \ldots). \tag{12}$$

The Fourier coefficients of the form F are computed as matrices of the form

$$N = \begin{pmatrix} a & b/2 \\ b/2 & c \end{pmatrix}$$

of an integral binary quadratic form with nonpositive discriminant $D(N) = b^2 - 4ac \leqslant 0$. The Hecke operators $T(m)$ and thus also the eigenvalues $\Lambda_F(m)$ are computed as integers. Therefore it would be unnatural to expect here a single identity of the type (9). Indeed, it can be shown that the identity (9) is in this case replaced by an infinite number of identities corresponding to the different discriminants $D(N) < 0$ and the different group characters for the classes of matrices with the discriminants under consideration. The reason for this situation lies in the fact that the Hecke operatos do not "mix" the Fourier coefficients with indices N unless their discriminants differ by a square factor.

Hence, let $D < 0$ be an arbitrary negative integer. For each half-integer primitive symmetric matrix

$$M = \begin{pmatrix} A & B/2 \\ B/2 & C \end{pmatrix}$$

with the discriminant $D(M) = B^2 - 4AC = D$ we let

$$R_{F,M}(s) = \sum_{m=1}^{\infty} \frac{a(mM)}{m^s}. \tag{13}$$

According to (2) the number $R_{F,M}(s)$ depends only on the equivalence class of M in the narrow sense. Let $D = df^2$, where d is the discriminant of the field $K = \mathbf{Q}(\sqrt{D})$ and let O_f be the subring of the ring of integers of index f in the field K. Each matrix M with the properties specified above is associated with the integral ideal $\mathfrak{A}(M) = \{A, (B - \sqrt{D})/2\}$ of the ring O_f, and thus we obtain a one-to-one correspondence between the set of all equivalence classes in the narrow sense of the given M and the set of all regular ideal-classes of the ring O_f (i.e. of ideals with the ring of divisors O_f) (cf. [7]–[9]). This enables us to introduce in the set of primitive classes of matrices M with discriminant D the structure of a group which is isomorphic to the group of regular ideal-classes in O_f. We denote this group by $H(D)$. For any character χ of the group $H(D)$ we let

$$R_F(s; \chi) = \sum_M \chi(M) R_{F,M}(s) \tag{14}$$

and furthermore

$$L_D(s;\ \chi) = \sum_{(\mathfrak{A},f)=1} \frac{\chi(\mathfrak{A})}{N(\mathfrak{A})^s}, \tag{15}$$

where M runs through a complete system of representatives of all classes from $H(D)$, and $\mathfrak{A}$ runs through the set of integral ideals of the ring O_f which are relatively prime to f (and thus regular), and $N(\mathfrak{A})$ denotes the norm of $\mathfrak{A}$.

On the other hand, we let

$$L_F(s) = \zeta(2s - 2k + 4) \sum_{m=1}^{\infty} \frac{\Lambda_F(m)}{m^s}, \tag{16}$$

where $\zeta(s)$ is the Riemann zeta-function and $\Lambda_F(m)$ the eigenvalue of the Hecke operator corresponding to F (compare (12)). The series (14)–(16) are convergent if $\operatorname{Re} s > 2k$. According to (7) and (8) the series $L_F(s)$ is expressible as an Euler product in the form

$$L_F(s) = \prod_p [1 - \Lambda_F(p)\, p^{-s} + (\Lambda_F(p)^2 - \Lambda_F(p^2) - p^{2k-4})\, p^{-2s} - \Lambda_F(p)\, p^{2k-3-3s} + p^{4k-6-4s}]^{-1}.$$

In the notation introduced above our results may be formulated as follows:

THEOREM I. ($n = 2$; *relation between eigenvalues and Fourier coefficients of eigenfunctions*). *For any* $D < 0$, $\chi \in \hat{H}(D)$ *and* $\operatorname{Re} s > 2k$ *the identity*

$$L_D(s - k + 2;\ \chi)\, R_F(s;\ \chi) = L_F(s) \left\{ \prod_{p \mid f} \left(1 - \frac{\Pi(p)}{p^{s-k+2}}\right)\left(1 - \frac{\Delta(p^{-1})}{p^{s-2k+3}}\right) \sum_M a(M)\, \chi(M) \right\} \tag{17}$$

holds, where M *runs through a complete system of representatives from the classes in* $H(D)$, $D = df^2$, *with* d *denoting the discriminant of the field* $\mathbf{Q}(\sqrt{D})$; p *runs through the prime divisors of* f, *and for each* p *the symbols* $\Delta(p^{-1})$ *and* $\Pi(p)$ *denote the operators acting linearly on functions of classes of half-integer symmetric matrices of order two in the following manner*:

$$\Delta(p^{-1})\, a(N) = \begin{cases} a\left(\frac{1}{p} N\right), & \text{if } \frac{1}{p} N \text{ is a half-integer,} \\ 0 & \text{otherwise.} \end{cases}$$

In the case under consideration (*i.e. if* M *is primitive*, $D(M) = df^2$ *and* $p \mid f$) *the operator* $\Pi(p)$ *has the form*

$$\Pi(p)\, a(M) = a(pM_p),$$

where M_p *is a primitive matrix with the discriminant* $D(M_p) = d(f/p)^2$ *whose class is the image of the class of* M *under the canonical epimorphism of the class-group* $H(D) \to H(D/p^2)$. (*In terms of regular ideals of the rings* O_f *and* $O_{f/p}$ *this epimorphism maps the* O_f*-ideal* $\mathfrak{A}$ *into the* $O_{f/p}$*-ideal* $\mathfrak{A} \cdot O_f$.) (*For the general definition and formulas for* $\Pi(p)$ *in other cases see* [1].)

REMARK. For given p the operators $\Delta(p^{-1})$ and $\Pi(p)$ do not commute and have to be applied in the exact order stated in (17). For different prime divisors p of f the order is immaterial.

THEOREM II ($n = 2$; *continuation and functional equation*). *The Dirichlet series* $L_F(s)$ (*cf.* (16)) *possesses an analytic continuation into the entire* s*-plane. The function*

$$\Psi_F(s) = (2\pi)^{-2s}\,\Gamma(s)\,\Gamma(s-k+2)\,L_F(s),$$

where $\Gamma(s)$ *is the gamma-function, has only a finite number of poles and satisfies the functional equation*

$$\Psi_F(2k-2-s) = (-1)^k\,\Psi_F(s).$$

BIBLIOGRAPHY

1. A. N. Andrianov, *Dirichlet series with Euler product in the theory of Siegel modular forms of genus* 2, Trudy Mat. Inst. Steklov. **112** (1971), 73–93 = Proc. Steklov Inst. Math. **112** (1971), 70–93.

2. C. L. Siegel, *Einführung in die Theorie der Modulfunktionen n-ten Grades*, Math. Ann. **116** (1939), 617–657. MR **1**, 203.

3. H. Maass, *Die Primzahlen in der Theorie der Siegelschen Modulfunktionen*, Math. Ann. **124** (1951), 87–122. MR **13**, 823.

4. *Séminaire H. Cartan.* 10iéme *année*: 1957/58. *Fonctions automorphes*, Secrétariat mathématique, Paris, 1958. MR **21** #2750.

5. A. N. Andrianov, *Spherical functions for* GL_n *over local fields, and summation of Hecke series*, Mat. Sb. **83 (125)** (1970), 429–451 = Math. USSR Sb. **12** (1970), 429–452. MR **44** #216.

6. A. Ogg, *Modular forms and Dirichlet series*, Benjamin, New York, 1969. MR **41** #1648.

7. B. A. Venkov, *Elementary number theory*, ONTI, Moscow, 1937; English transl., Wolters-Noordhoff, Groningen, 1970. MR **42** #178.

8. E. Hecke, *Vorlesungen über die Theorie der algebraischen Zahlen*, Akademische Verlagsgesellschaft (Geest & Portig), Leipzig, 1923; reprint, Chelsea, New York, 1948; Russian transl., GITTL, Moscow, 1940. MR **16**, 571.

9. Z. I. Borevič and I. R. Šafarevič, *Number theory*, "Nauka", Moscow, 1964; English transl., Pure and Appl. Math., vol. 20, Academic Press, New York, 1966. MR **30** #1080; **33** #4001.

Translated by B. VOLKMANN

Trudy Mat. Inst. Steklov.
132 (1973)

Proc. Steklov Inst. Math.
132 (1973)

APPLICATIONS OF THE METHOD OF TRIGONOMETRIC SUMS TO THE METRIC THEORY OF DIOPHANTINE APPROXIMATION OF DEPENDENT QUANTITIES

V. G. SPRINDŽUK

ABSTRACT. The method of trigonometric sums is used to study the extremality of some manifolds (in the sense of Diophantine approximation).

Bibliography: 21 items.

Because of the well-known applications of the method of trigonometric sums in quite different areas of analytic number theory, this method is regarded as one of the most powerful and universal tools in the theory of numbers.

However, we shall indicate an area for applying the method of trigonometric sums which has arisen recently and which is very attractive in view of both the precise and simple formulation of its basic problems and its intrinsic content which reveals deep aspects of the discrete and the continuous structure of numbers.

Let $\alpha_1, \cdots, \alpha_n$ be real numbers. We denote by $w(\alpha_1, \cdots, \alpha_n)$ the supremum of the set of those $w > 0$ for which the inequality

$$\|\alpha_1 a_1 + \dots + \alpha_n a_n\| < a^{-w}, \quad a = \max_{1 \leqslant i \leqslant n} |a_i| \neq 0, \tag{1}$$

has infinitely many solutions in integers $a_1, \cdots, a_n$ (the symbol $\|x\|$ denotes the distance between x and the nearest integer). It is well known that always $w(\alpha_1, \cdots, \alpha_n) \geqslant n$ and that there always exist numbers $\alpha_1, \cdots, \alpha_n$ with

$$w(\alpha_1, \dots, \alpha_n) = n. \tag{2}$$

For example, different rational powers of e, the basis of the natural logarithms, are such numbers. A point in the space $\mathbf{R}^n$, chose "at random", represents an n-tuple of numbers which satisfies the equation (2), i.e. this condition is valid for almost all points.

An n-tuple $\alpha_1, \cdots, \alpha_n$ satisfying (2) shall be said to be *badly approximable*.

Instead of the inequality (1) we may consider the "dual system" of inequalities

$$\max(\|\alpha_1 q\|, \dots, \|\alpha_n q\|) < q^{-v}, \quad v > 0, \tag{3}$$

AMS (MOS) subject classifications (1970). Primary 10K15.

where $q > 0$ is an integer obtained by Hinčin's "principle of transfer" (cf. [1] and also [2]). Let $v(\alpha_1, \cdots, \alpha_n)$ be the supremum of the set of those $v > 0$ for which the inequality (3) has infinitely many solutions in terms of integers $q > 0$. Then the equations (2) and

$$v(\alpha_1, \ldots, \alpha_n) = \frac{1}{n} \tag{4}$$

are equivalent.

Many authors have investigated the metric theory of the solutions of inequalities of the forms (1) and (3) in the case where the point $(\alpha_1, \cdots, \alpha_n)$ is chosen at random from the entire space $\mathbf{R}^n$. The first result in this direction was a theorem of Hinčin (cf. [3] and also [2]):

Let $\varphi(q)$ be a monotonically decreasing function defined for all integers $q > 0$. Then the inequality $\|\alpha q\| < \varphi(q)$ has infinitely many integer solutions $q > 0$ for almost all real α or at most finitely many solutions for almost all α according as the series $\Sigma_1^\infty \varphi(q)$ diverges or converges.

An analogous assertion is also true for systems of inequalities of the form $\max(\|\alpha_1 q\|, \cdots, \|\alpha_n q\|) < \varphi(q)$ for almost all n-tuples $(\alpha_1, \cdots, \alpha_n) \in \mathbf{R}^n$ (cf. [2]).

In this direction the theory was mainly developed by investigations of Schmidt [4] and [5], Gallagher [6] and [7] and the author [8], [9] and [10], thus reaching a state which has been considered satisfactory until now. In order to give an exposition of the results obtained we formulate the following theorem which covers a special case of a more general basic assertion (cf. [10], Russian p. 156 (English p. 162), Theorem 1):

Let $n \geqslant 2$ and with every vector $\mathbf{a} = (a_1, \cdots, a_n)$ with the g.c.d. $\mathbf{a} = (a_1, \cdots, a_n) = 1$ let there be associated a measurable set $A(\mathbf{a})$ from the unit interval $[0, 1)$. Then the condition $\alpha_1 a_1 + \cdots + \alpha_n a_n \in A(\mathbf{a}) \pmod 1$ has infinitely many solutions for almost all $(\alpha_1, \cdots, \alpha_n) \in \mathbf{R}^n$ or at most finitely many solutions for almost all $(\alpha_1, \cdots, \alpha_n) \in \mathbf{R}^n$ according as the series $\Sigma_{\mathbf{a}} |A(\mathbf{a})|$ diverges or converges. (The symbol $|A(\mathbf{a})|$ denotes the measure of $A(\mathbf{a})$, and the summation is extended over all integer vectors $\mathbf{a} = (a_1, \cdots, a_n)$ with g.c.d. $(a_1, \cdots, a_n) = 1$.)

The methods which have been used are related to the theory of orthogonal functions (Schmidt), ergodic theory (Gallagher) and the theory of measure-preserving transformations (author). The fact seems to be paradoxical that the only "unclaimed territory" should be the problem of extending Hinčin's theorem from which the entire theory originated to the case where the point $(\alpha_1 q, \cdots, \alpha_m q) \bmod 1$ is requested to be contained in a (possibly) more general domain. Such domains cannot be completely arbitrary as in the preceding theorem; even in the assumptions of Hinčin's theorem the condition of monotonicity of the function $\varphi(q)$ cannot be abandoned.

The problem assumes a completely different nature if we suppose that the points

$(\alpha_1, \cdots, \alpha_n)$ are chosen "at random" from a given manifold Γ within $\mathbf{R}^n$ with $\dim \Gamma < n$ rather than the space $\mathbf{R}^n$ itself.

PROBLEM. *Let* Γ *be a manifold in* $\mathbf{R}^n$ *with* $\dim \Gamma < n$. *What conditions on the* Γ *have to be satisfied in order that almost all its points correspond to badly approximable numbers ("almost all" with respect to a measure on* Γ)?

We shall call the manifold extremal if almost all its points are badly approximable (the term was introduced by Schmidt [11] for curves in $\mathbf{R}^n$). The first and so far the only example of such a problem is a conjecture of Mahler [12] which arose in the theory of transcendental numbers. This conjecture asserted that for any $n = 1, 2, \cdots$ the curve $\Gamma = (t, t^2, \cdots, t^n)$ in $\mathbf{R}^n$ is extremal. In other words, for any fixed n there exist, for almost all real numbers t, only finitely many integral polynomials $P(x)$ of degree n satisfying the inequality $|P(t)| < h_P^{-n-\epsilon}$, where $h_P \neq 0$ is the height of the polynomial $P(x)$ (i.e. the maximum of the absolute values of its coefficients) and $\epsilon > 0$ is arbitrary.

Mahler's conjecture was proved in 1964 by a special method based on a particular general idea (the introduction of "essential" and "inessential" domains) which came from the investigation of similar problems. This method also led to a solution of the complex case of Mahler's conjecture and of its analogue for locally compact nonarchimedean normed rings (cf. [10] and [13], [14], [15]). However, the attempt to apply this method to manifolds subject to sufficiently general conditions leads to considerable difficulties in view of the insufficiency of the information on the "structure" of these manifolds. The question whether it is possible to give a general characterization of extremality of manifolds under nonspecial conditions is answered affirmatively (at least for $n = 2$) by the following theorem of Schmidt [11]:

Let Γ *be a curve in* $\mathbf{R}^2, \Gamma = (f_1(t), f_2(t))$, *where the functions* f_1 *and* f_2 *are three times differentiable on an interval I, and* $(f_1'f_2'' - f_1''f_2')(t) \neq 0$ *almost everywhere on I. Then* Γ *is extremal.*

This theorem shows that if the curve Γ "does not behave too much like a straight line" then it is extremal. One might easily suppose that if the manifold Γ is "not too planar" then it will be extremal, but this naive point of view is far from reality; for any $n \geqslant 2$ there exist in $\mathbf{R}^n$ extremal lines and planes. Nevertheless, one may impose general conditions for any $n \geqslant 2$ under which the curve Γ is extremal, but it is necessary to require that the dimension of Γ be at least $n/2$. The condition $\dim \Gamma > n/2$ cannot be weakened if we want to guarantee extremality of Γ merely on the basis of sufficiently general conditions.

The method of trigonometric sums turns out to be very effective for the investigation of manifolds under general conditions. This method was first applied by Kubilius (cf. [16], [17]) (for the quadratic case of Mahler's conjecture) and by the author (cf. [10] and [18]) (for an analogue to Mahler's conjecture for quadratic polynomials with an arbitrary number of variables); but these applications involve special manifolds only.

For a simple formulation of the general results we have to assume that Γ is given in the space $\mathbf{R}^{m+n}$ and that $\Gamma = (t_1, \cdots, t_m; f_1, \cdots, f_n)$, where $t_1, \cdots t_m$ are independent parameters and each $f_j = f_j(t_1, \cdots, t_m)$, $1 \leqslant j \leqslant n$, is a function of these parameters such that, in particular, $\dim \Gamma = m$ and $t_i \in I_i$ $(1 \leqslant i \leqslant m)$, where the I_i are given intervals.

THEOREM 1. *Let m and n be integers, $1 \leqslant n \leqslant m$, and $f_j = \alpha_{j1}\varphi_1(t_1) + \cdots + \alpha_{jm}\varphi_m(t_m)$ $(1 \leqslant j \leqslant n)$, where the φ_i are real-valued functions, three times continuously differentiable on the intervals I_i, $\varphi_i'' \neq 0$ almost everywhere on I_i; the α_{ji} are rational numbers, and the matirx (α_{ji}) $(1 \leqslant i \leqslant n,\ 1 \leqslant j \leqslant m)$ is of rank n. Then the curve Γ is extremal.*

For $m = n = 1$ this theorem is in essence identical with Schmidt's theorem. It is not possible to allow the case where $n > m$ or where the rank of the matrix (α_{ji}) is less than n, inasmuch as the functions f_j would then be linearly dependent and it could be shown that some of the inequalities of the form (3) are consequences of the remaining ones (i.e. some α_i in (3) may be disregarded). In such a situation the manifold Γ cannot be extremal. This simple remark shown that if general conditions are imposed on Γ, the possibility arises that Γ may be a manifold of the type described in Theorem 1, and thus it would be necessary that $n \leqslant m$, i.e. that the dimension of Γ should not be more than half the dimension of the space, in order that Γ be extremal.

THEOREM 2. *Let m and n be integers, $1 \leqslant n \leqslant m$; let Ω be a domain in $\mathbf{R}^m$ and let $f_j = f_j(t_1, \cdots, t_m)$ $(1 \leqslant j \leqslant n)$ be real-valued functions defined on Ω and satisfying the following conditions*:

a) *The partial derivaitves $\partial^2 f_j/\partial t_i \partial t_k$ are continuous in Ω $(1 \leqslant j \leqslant n;\ 1 \leqslant i,\ k \leqslant m)$.*

b) $\det(\partial^2 f_j/\partial t_1 \partial t_k)_{j,k=1,\dots,n} \neq 0$ *almost everywhere in Ω.*

c) *Aby integral linear combination*

$$\varphi(t_k) = c_1 \frac{\partial^2 f_1}{\partial t_1 \partial t_k} + \dots + c_n \frac{\partial^2 f_n}{\partial t_1 \partial t_k},$$

regarded as a funtion of the variable t_k $(1 \leqslant k \leqslant m)$ with fixed values for the remaining parameters, has the property that any interval on which it is defined splits into a bounded number of subintervals, not depending on $c_1, \cdots, c_n$, on which the function $\varphi(t_k)$ is monotonic.

Then Γ is extremal.

We see that extremality of the manifold Γ here depends on general analytic conditions. Condition c), whose verbal formulation is rather lengthy, is evidently satisfied by "ordinary" functions such as polynomials, exponential functions, logarithms, sine functions, etc., and also by their linear combinations. This condition may be replaced by a more "constructive" condition involving a higher order of smoothness of the function f_j and a stipulation that a certain system of Wronskians of the partial derivatives $\partial f_j/\partial t_1$ $(1 \leqslant j \leqslant n)$ be nonvanishing almost everywhere in Ω.

The inequality $n < m$ may perhaps be weakened to $n \leqslant m$, but not further because of the reasons specified in the comments to Theorem 1.

THEOREM 3. *Let* s, m *and* n *be integers,* $s \geqslant 3$, $m \geqslant 2^{s-1}$, *and*

$$f_1 = \alpha_{j1} t_1^s + \cdots + \alpha_{jm} t_m^s + \varphi_j(t_1, \cdots, t_m) \qquad (1 \leqslant j \leqslant n),$$

where the $\varphi_j(t_1, \cdots, t_m)$ *are polynomials of degrees not exceeding* $s-1$ *with real coefficients*; $\alpha_{ji} \neq 0$ *is a real number*; *furthermore suppose that*

$$w\left(\frac{\alpha_{2i}}{\alpha_{1i}}, \ldots, \frac{\alpha_{ni}}{\alpha_{1i}}\right) \leqslant m+n \qquad (1 \leqslant i \leqslant m). \tag{5}$$

Then Γ *is extremal.*

In this theorem there is no restriction on the relation between m and n, so that, for any m, we may construct within the space $\mathbf{R}^{m+n}$ with sufficiently large n extremal manifolds of dimension m by means of polynomials. We see that the arithmetical condition (5), imposed upon the coefficients of the "principal term" of the polynomial f_j, guarantees extremality of Γ. Since the assumption and the assertion are of a similar nature, we may construct extremal manifolds by induction. For example, we may take

$$f_j = t^j (t_1^s + \ldots + t_m^s) + \varphi_j (t_1, \ldots, t_m)$$

and we may assert that $\Gamma = (t, t_1, \cdots, t_m; f_1, \cdots, f_n)$ is extremal in $\mathbf{R}^{m+n+1}$ inasmuch as $w(t, t^2, \cdots, t^{n-1}) = n-1 < m+n$ for almost all $t \in \mathbf{R}$.

If we suppose that the numbers α_{ji} are rational then it is necessary to make the stipulations $n \leqslant m$ and $\operatorname{rank}(\alpha_{ji}) = n$.

The condition $m \geqslant 2^{s-1}$ is made in order to use in the proof an estimate of a multiple Weyl sum by a method also due to Weyl. This inequality may be weakened essentially, replacing it by an inequality of the form $m \geqslant s^2 \ln s$, if we apply the method of I. M. Vinogradov (cf. [19] and [20]) for estimating Weyl sums.

In concluding we wish to discuss one possible approach to applying the method of trigonometric sums for solving problems of the type under consideration. This approach was used by the author for the analysis of quadratic polynomials in many variables, while it turns out to be effective also in many other cases (for example, it is used in the proof of Theorem 3).

It is sufficient to analyze a system of inequalities of the form (3) in the case where the manifold $\Gamma = (t_1, \cdots, t_m; f_1, \cdots, f_n)$ is of the supposed form with

$$\max(\|t_1 q\|, \ldots, \|t_m q\|, \|f_1 q\|, \ldots, \|f_n q\|) < q^{-v}. \tag{6}$$

We have to show that under certain restrictions on $f_1, \cdots, f_n$ for almost all $(t_1, \cdots, t_m)$ the inequality (6) is satisfied by at most finitely many integers $q > 0$, provided that $v > 1/(m+n)$. It follows from (6) that there exist integers $a_1, \cdots, a_m$ for which

$$\max(|t_1 q - a_1|, \ldots, |t_m q - a_m|) < q^{-v}$$

and for fixed $a_1, \cdots, a_m$ the measure of the set of such $(t_1, \cdots, t_m)$ is equal to $(2q^{-1-v})^m$. However, we do not admit an arbitrary m-tuple $(a_1, \cdots, a_m)$ since it follows from (6) that we must have

$$\max \left\| q f_j \left(\frac{a_1}{q}, \ldots, \frac{a_m}{q} \right) \right\| \ll q^{-v_0}, \quad v_0 = 1/(m+n). \tag{7}$$

In order to prove the extremality of Γ we have to define the true order of the number $N(q)$ of solutions of the inequality (7) in terms of integers $a_1, \cdots, a_m$ with each a_i running through an interval whose length is of the order q. We can show that $N(q)$ satisfies an estimation of the form

$$\ll q^{-nv_0} \sum_{\mathbf{c}} \left| \sum_{\mathbf{a}} e^{2\pi i q F\left(\frac{\mathbf{a}}{q}\right)} \right|, \tag{8}$$

where

$$F\left(\frac{\mathbf{a}}{q}\right) = c_1 f_1 \left(\frac{a_1}{q}, \ldots, \frac{a_m}{q} \right) + \ldots + c_n f_n \left(\frac{a_1}{q}, \ldots, \frac{a_m}{q} \right),$$

the integral vector $\mathbf{c} = (c_1, \cdots, c_n)$ satisfies the inequality $\max |c_i| < q^{v_0}$ $(1 \leqslant i \leqslant n)$, and $\mathbf{a} = (a_1, \cdots, a_n)$ runs through its range of values. Furthermore, in order to obtain a lower bound for the exponential sum we apply known methods, summing the results with respect to all $\mathbf{c}$. For the extremality of the manifold Γ we have to show the "best possible" estimate

$$N(q) \ll q^{m - nv_0 + \varepsilon}. \tag{9}$$

Finally, in the analysis of the relation (7) it is possible to use the "glass method" of I. M. Vinogradov [20], but in a number of cases with finite summations (8) it is more convenient than in the case of infinite series to refer to the "glass method" in order to obtain for the number $N(q)$ an asymptotic formula with a principal term of the form σq^{m-nv_0} also under the same conditions under which the estimate (9) is proved by means of (8).

In the proofs of Theorems 1 and 2 a more refined reduction to a trigonometric sum is necessary.

Recently, E. I. Kovalevskaja has used this approach in the investigation of approximations of zero by linear combinations of integral polynomials.

Let $w_0(t_1, \cdots, t_m)$ *be the supremum of the set of those* $w > 0$ *for which there exist infinitely many systems of integral polynomials* $P_1(x), \cdots, P_m(x)$ *of degrees* $k_1, \cdots, k_m$, *respectively, without constant terms, satisfying the inequality*

$$\|\lambda_1 P(t_1) + \ldots + \lambda_m P_m(t_m)\| < h^{-w}, \quad h = \max_{1 \leqslant j \leqslant m} h(P_j),$$

where $\lambda_1 \neq 0, \cdots, \lambda_m \neq 0$ *are fixed numbers. Then for almost all* $(t_1, \cdots, t_m) \in \mathbf{R}^m$ *the equation*

$$w_0(t_1, \ldots, t_m) = k_1 + \ldots + k_m$$

holds, provided that $m \gg k^2 \ln k$, *where* $k = \max k_j$ $(1 \leqslant j \leqslant m)$.

The restriction on the relation between m and k is imposed by the corresponding reduction in estimates of Weyl sums by the method of I. M. Vinogradov (cf. [19], [20]).

The proofs of Theorems 1–3 have been published in the paper [21].

BIBLIOGRAPHY

1. A. Khintchine [Hinčin], *Zwei Bemerkungen zu einer Arbeit des Herrn Perron,* Math. Z. **22** (1925), 274–284.

2. J. W. S. Cassels, *An introduction to Diophantine approximation,* Cambridge Tracts in Math. and Math. Phys., no. 45, Cambridge Univ. Press, New York, 1957; Russian transl., IL, Moscow, 1961. MR **19**, 396; **22** #10976.

3. A. Khintchine [Hinčin], *Zur metrischen Theorie der Diophantischen Approximationen,* Math. Z. **24** (1926), 706–714.

4. W. Schmidt, *A metrical theorem in diophantine approximation,* Canad. J. Math. **12** (1960), 619–631. MR **22** #9482.

5. ———, *Metrical theorems on fractional parts of sequences,* Trans. Amer. Math. Soc. **110** (1964), 493–518. MR **28** #3018.

6. P. X. Gallagher, *Metric simultaneous diophantine approximation,* J. London Math. Soc. **37** (1962), 387–390. MR **28** #1167.

7. ———, *Metric simultaneous diophantine approximation.* II, Mathematika **12** (1965), 123–127. MR **32** #5593.

8. V. G. Sprindžuk, *A contribution to the metric theory of linear Diophantine approximations,* Dokl. Akad. Nauk SSSR **176** (1967), 43–45 = Soviet Math. Dokl. **8** (1967), 1051–1054. MR **36** #3731.

9. ———, *On the metric theory of "nonlinear" diophantine approximations,* Dokl. Akad. Nauk BSSR **13** (1969), 298–301. (Russian) MR **40** #4210.

10. ———, *Mahler's problem in metric number theory,* Izdat. "Nauka i Tehnika", Minsk, 1967; English transl., Transl. Math. Monographs, vol. 25, Amer. Math. Soc., Providence, R. I., 1969. MR **39** #6832; #6833.

11. W. M. Schmidt, *Metrische Sätze über simultane Approximation abhändiger Grössen,* Monatsh. Math. **68** (1964), 154–166. MR **30** #1980.

12. K. Mahler, *Über das Mass der Menge aller S-Zahlen,* Math. Ann. **106** (1932), 131–139.

13. V. G. Sprindžuk, *On a conjecture of Mahler,* Dokl. Akad. Nauk SSSR **154** (1964), 783–786 = Soviet Math. Dokl. **5** (1964), 183–187. MR **28** #2090a.

14. V. G. Sprindžuk, *Further remarks on a conjecture of Mahler,* Dokl. Akad. Nauk SSSR **155** (1964), 54–56 = Soviet Math. Dokl. **5** (1964), 361–363. MR **28** #2090b.

15. ———, *Proof of Mahler's conjecture on the measure of the set of S-numbers,* Izv. Akad. Nauk SSSR Ser. Mat. **29** (1965), 379–436; English transl., Amer. Math. Soc. Transl. (2) **51** (1966), 215–273. MR **31** #4762.

16. J. P. Kubilius, *On the application of I. M. Vinogradov's method to the solution of a problem of the metric theory of numbers,* Dokl. Akad. Nauk SSSR **67** (1949), 783–786. (Russian) MR **11**, 82.

17. ———, *Probabilistic methods in the theory of numbers,* Gos. Izdat. Politič. i Naučn. Lit. Litovsk. SSR, Vilna, 1959; rev. ed., 1962; English transl., Transl. Math. Monographs, vol. 11, Amer. Math. Soc., Providence, R. I., 1964. MR **23** #A134; **26** #3691; **28** #3956.

18. V. G. Sprindžuk, *On theorems of Khinchin and Kubilius,* Litovsk. Mat. Sb. **2** (1962), no. 1, 147–152. (Russian) MR **28** #2088.

19. I. M. Vinogradov, *The method of trigonometrical sums on the theory of numbers,* Trudy Mat. Inst. Steklov. **23** (1947); rev. ed., "Nauka", Moscow, 1971; English transl., Interscience, New York, 1954. MR **10**, 599; **15**, 941.

20. ———, *Selected works,* Izdat. Akad. Nauk SSSR, Moscow, 1952. (Russian) MR **14**, 610.

21. V. G. Sprindžuk, *The method of trigonometric sums in the metric theory of diophantine approximations of dependent quantities,* Trudy Mat. Inst. Steklov. **128** (1972), 212–228 = Proc. Steklov Inst. Math. **128 (1972), 251–270.**

Translated by B. VOLKMANN

Trudy Mat. Inst. Steklov.
132 (1973)

Proc. Steklov Inst. Math.
132 (1973)

REDUCIBILITY OF QUADRINOMIALS

A. ŠINCEL' [A. SCHINZEL]

ABSTRACT. Conditions are studied for reducibility of quadrinomials in the field of rational numbers.

Bibliography: 4 items.

Unless otherwise stated, reducibility will mean here reducibility over the field of rational numbers. We let

$$f(y_1, \ldots, y_k) \overset{\text{can}}{=} \text{const} \prod_{\sigma=1}^{s} f_\sigma(y_1, \ldots, y_k)^{e_\sigma}$$

signify that the polynomials $f_\sigma(y_1, \cdots, y_k)$ are irreducible and relatively prime in pairs. If $\Phi(y_1, \cdots, y_k)$ is a rational function of the form

$$y_1^{\alpha_1} \ldots y_k^{\alpha_k} f(y_1, \ldots, y_k),$$

where $\alpha_1, \cdots, \alpha_k$ are integers and $(f(y_1, \cdots, y_k), y_1, \cdots, y_k) = 1$, then we put

$$J\Phi(y_1, \ldots, y_k) = f(y_1, \ldots, y_k), \qquad L\Phi(y_1, \ldots, y_k) = \text{const}\, \Pi' f_\sigma(y_1, \ldots, y_k)^{e_\sigma},$$

where the product Π' is taken over all polynomials f_σ such that

$$Jf_\sigma(y_1^{-1}, \ldots, y_k^{-1}) \neq \pm f_\sigma(y_1, \ldots, y_k)$$

(in other words, $L\Phi(y_1, \cdots, y_k)$ is the result of removing from Φ its irreducible reciprocal factors and powers of y_i).

The basic theorem of [1] gives the following corollary.

COROLLARY. *For any polynomial with integer coefficients* $a_0 + \Sigma_1^k a_j y^{n_j} \neq 0$ *there exists an integral matrix* $N = [\nu_{ij}]$, $i \leq r$, $j \leq k$, *of rank* r, *and an integral vector* $[v_1, \cdots, v_r]$, *such that*

(i) $\max |\nu_{ij}| \leq c_r(k, A)$,

(ii) $[n_1, \ldots, n_k] = [v_1, \ldots, v_r] N$,

(iii) $L\left(a_0 + \sum_{j=1}^{k} a_j \prod_{i=1}^{r} y_i^{\nu_{ij}}\right) \overset{\text{can}}{=} \text{const} \prod_{\sigma=1}^{s} F_\sigma(y_1, \ldots, y_r)^{e_\sigma}$

and thus

AMS (MOS) subject classifications (1970). Primary 12D05.

$$L\left(a_0+\sum_{j=1}^{k} a_j x^{n_j}\right) \overset{\text{can}}{=} \text{const} \prod_{\sigma=1}^{s} LF_\sigma(x^{v_1}, \ldots, x^{v_r})^{e_\sigma}.$$

Here $A = a_0^2 + \cdots + a_k^2$, *and*

$$c_r(k, A) = \begin{cases} \exp(9k \cdot 2^{A-5}), & \textit{if} \quad r=k, \\ \exp(5 \cdot 2^{A-4} + 2A \log 2), & \textit{if} \quad r+k=3, \\ \exp_{(k-r)(k+r-3)}(k \cdot 2^{A+2} \log 2) & \textit{otherwise}. \end{cases}$$

If all $a_j \neq 0$ and $r = k \geqslant 2$, then the polynomial

$$J\left(a_0 + \sum_{j=1}^{k} a_j \prod_{i=1}^{r} y_i^{v_{ij}}\right)$$

is irreducible (see [1], Lemma 14) and thus $L(a_0 + \Sigma_1^k a_j x^{n_j})$ is also irreducible. The case $k = 2$ (trinomials) is considered in [1] (Theorem 5). Here we consider the case $k = 3$ (quadrinomials), which has also been studied in a joint paper by the author and M. Fried [2]. We have

THEOREM 1. *The quadrinomial*

$$Q(y_1, y_2) = J\left(a_0 + \sum_{i=1}^{3} a_i y_1^{v_{1i}} y_2^{v_{2i}}\right),$$

where $a_i \neq 0$ $(0 \leqslant i \leqslant 3)$, $[v_{1i}, v_{2i}]$ *are pairwise distinct and different from* $[0, 0]$, *with the rank of the matrix* $[v_{ij}]$ *equal to* 2, *is reducible in a field* K *of characteristic zero if and only if either it can be decomposed into two parts with greatest common divisor* D *being binomial, or it can be represented in one of the forms*

$$\begin{gathered} k(U^3+V^3+W^3-3UVW) \\ = k(U+V+W)(U^2+V^2+W^2-UV-UW-VW), \\ k(U^2-4TUVW-T^2V^2-4T^2W^4) \\ = k(U-TV^2-2TVW-2TW^2)(U+TV^2-2TVW+2TW^2), \qquad (1) \\ k(U^2+2UV+V^2-W^2) = k(U+V+W)(U+V-W), \end{gathered}$$

where $k \in K$ *and* T, U, V, W *are monomials in* $K[y_1, y_2]$. *In the first case, either* QD^{-1} *is irreducible in* K *and not reciprocal, or it is a binomial. In the second case, the factors on the right are irreducible in* K *and not reciprocal, except only in case* $\zeta_3 \in K$, *when we have the decomposition*

$$U^2+V^2+W^2-UV-UW-VW = (U+\zeta_3 V+\zeta_3^2 W)(U+\zeta_3^2 V+\zeta_3 W).$$

Hence, using the Corollary, we have the following.

THEOREM 2. *Let* a, b, c, d *be integers different from zero, and* $m > n > p$ *positive integers. Suppose that* $q(x) = ax^m + bx^n + cx^p + d$ *is not the product of two binomials. Then* $Lq(x)$ *is reducible if and only if either* $q(x)$ *can be decomposed into two parts with a common factor which is not reciprocal, or* $q(x)$ *may be represented in one of the forms* (1), *where* $k \in Q$, T, U, V, W *are monomials in* $Q[x]$, *and the factors on the right in* (1) *are not reciprocal, or, finally,*

$$m = vm_1, \qquad n = vn_1, \qquad p = vp_1,$$

$$m_1 < c_1(3,\ a^2+b^2+c^2+d^2) = \exp_2(3 \cdot 2^{a^2+b^2+c^2+d^2+2} \log(a^2+b^2+c^2+d^2))$$

and $L(ax^{m_1} + bx^{n_1} + cx^{p_1} + d)$ *is reducible.*

By means of some additional argument, we may derive from this a generalization of results of Ljunggren [3], [4] on the irreducibility of quadrinomials $x^m \pm x^n \pm x^p \pm r$, where r is one or a prime number, to quadrinomials of the form $x^n \pm x^m + cx^p + d$, where c and d are integers with $|c| \leqslant |d|$.

A detailed account is given in [2].

BIBLIOGRAPHY

1. A. Schinzel, *Reducibility of lacunary polynomials.* I, Acta Arith. **16** (1969/70), 123–159. MR **40** #5583.

2. M. Fried and A. Schinzel, *Reducibility of quadrinomials*, Acta Arith. **21** (1972), 153–171.

3. W. Ljunggren, *On the irreducibility of certain trinomials and quadrinomials*, Math. Scand. **8** (1960), 65–70. MR **23** #A1627.

4. ————, *On the irreducibility of certain lacunary polynomials*, Norske Vid. Selsk. Forh. (Trondhein) **36** (1963), 159–164. MR **28** #5059.

Translated by J. E. and T. P. VAUGHAN

Trudy Mat. Inst. Steklov.
132 (1973)

Proc. Steklov Inst. Math.
132 (1973)

LATTICE POINTS IN MOREDIMENSIONAL ELLIPSOIDS*

B. NOVÁK

ABSTRACT. The author studies the dependence of the asymptotic behavior of the remainder in a particular formula on arithmetic properties of the coefficients of a quadratic form determining a multidimensional ellipsoid.

Bibliography: 14 items.

Let $n \geqslant 2$, let $Q(u)$ be a positive definite quadratic form in n variables, α_j, v_j, M_j $(M_j > 0)$ real numbers, $j = 1, 2, \ldots, n$. For $x \geqslant 0$ let (see [5])

$$A(x) = \sum_{\substack{Q(u) \leqslant x \\ u_j \equiv v_j (M_j)}} e^{2\pi i \sum_{j=1}^{n} \alpha_j u_j}.$$

The effect of the arithmetical properties of the coefficients of the form Q and the numbers α_j, v_j, M_j on the asymptotic behaviour of the «lattice remainder»

$$P(x) = A(x) - V(x)$$

($V(x)$ is the volume of the ellipsoid $Q(u) \leqslant x$ modified with respect to α_j, v_j, M_j) was studied mostly in two cases:

I. $v_j = \alpha_j = 0$, $M_j = 1$, $Q(u)$ *has the «almost diagonal» form* $Q(u) = a_1 Q_1(u_1, \ldots, u_{n_1}) + a_2 Q_2(u_{n_1+1}, \ldots, u_{n_1+n_2}) + \ldots + a_\sigma Q_\sigma(u_{n-n_\sigma}, \ldots, u_n)$, *where* $a_1, a_2, \ldots, a_\sigma$ *are positive numbers,* $n_1 + n_2 + \ldots + n_\sigma = n$, *and* $Q_1, Q_2, \ldots, Q_\sigma$ *are positive definite forms with integer coefficients* (Walfisz, Jarník, Diviš).

II. v_j, M_j *as well as the coefficients of the form* Q *are integers* (Walfisz, author).

Let us remind the reader that (D being the determinant of the form Q)

$$V(x) = \frac{\pi^{\frac{n}{2}} x^{\frac{n}{2}} e^{2\pi i \sum_{j=1}^{n} \alpha_j v_j}}{\sqrt{D} M_1 M_2, \ldots, M_n \Gamma\left(\frac{n}{2} + 1\right)},$$

provided all numbers $\alpha_1 M_1, \alpha_2 M_2, \ldots, \alpha_n M_n$ are integers, $V(x) \equiv 0$ in the

AMS (MOS) subject classifications (1970). Primary 10E05.

* *Editor's note.* This reprinting of the original paper incorporates suggestions made by the author.

other cases. If the trivial case $A(x) \equiv 0$ is in the sequel excluded, then the fundamental results by Landau yield (see [5])

$$P(x) = O\left(x^{\frac{n}{2} - \frac{n}{n+1}}\right), \qquad P(x) = \Omega\left(x^{\frac{n-1}{4}}\right).$$

The aim of the present communication is to discuss the best contemporaneous results in the case I and especially in the case II, to characterize briefly the corresponding methods and to present some open questions.

The starting point is the relation

$$\frac{1}{2}(P(x+0) + P(x-0)) = \frac{1}{2\pi i} \int_{(a)} \frac{e^{xs} F(s)}{s}\, ds, \tag{1}$$

where we integrate along the line $\operatorname{Re} s = a > 0$ (putting usually $a = 1/x$) and where we put

$$F(s) = \theta(s) - \frac{c}{s^{\frac{n}{2}}}, \qquad \theta(s) = \sum_{u_j = v_j (M_j)} \exp\left(-sQ(u) + 2\pi i \sum_{j=1}^{n} \alpha_j u_j\right),$$

c denotes here (as well as in the sequel) a constant depending on Q, α_j, v_j and M_j. For the purpose of estimating integral (1) (neglecting the technical difficulties caused by its non-absolute convergence) it is suitable to divide the integration path by Farey fractions and to use the fact that for $s = 1/x + it$, $\left|t - \frac{2\pi h}{k}\right| \leqslant \frac{1}{k\sqrt{x}}$, $h \neq 0$, k prime integers, $0 < k \leqslant \sqrt{x}$, $F(s)$ may be replaced by the expression

$$c\,\frac{S_{h,k} \exp\left(-\dfrac{\pi^2 R_k}{k^2\left(s - 2\pi i \dfrac{h}{k}\right)}\right)}{k^n\left(s - 2\pi i \dfrac{h}{k}\right)^{\frac{n}{2}}} \tag{2}$$

with good results, provided that the generalized Gaussian sum $S_{h,k}$ is nonzero (here $R_k \asymp \max_j \langle \alpha_j M_j k\rangle^2$, $\langle y\rangle = \min_{p = \text{integer}} |y - p|$). If $S_{h,k} = 0$ then the following estimate holds in above mentioned range:

$$|F(s)| \leqslant \frac{x^{\frac{n}{2}} \exp\left(-\dfrac{cx}{k^2\left(1 + x\left|t - \dfrac{2\pi h}{k}\right|\right)}\right)}{k^{\frac{n}{2}}\left(1 + x\left|t - \dfrac{2\pi h}{k}\right|\right)^{\frac{n}{2}}}.$$

1. The outlined process was worked out in the case I by Jarník (see [2, 3]). He reached a number of results which have definitive character. So e. g. a very ingenious decomposition of the integration path together with some specially deduced refined theorems from the the theory of diophantine approximations yield, if Jarník's Ω-method is used (see [2, 3])

$$f = \limsup_{x \to +\infty} \frac{\lg |P(x)|}{\lg x} = \frac{n}{2} - 1 - \frac{1}{\gamma},$$

provided $n_j \geqslant 2(\gamma + 1)/\gamma$, $j = 1, 2, \ldots, n$ where $\gamma = \gamma(a_2/a_1, a_3/a_1, \ldots, a_\sigma/a_1)$ is the supremum of all $\beta > 0$ for which

$$\liminf_{q \to +\infty} q^\beta \left\langle q \frac{a_j}{a_1} \right\rangle < +\infty, \ j = 1, 2, \ldots, n.$$

For $\sigma = 2$ this result is due to Jarník (see [3]), for $\sigma > 2$ his reasoning was generalized by his student B. Diviš (see [1]).

2. In the case II the following estimate can be shown for $n > 4$:

$$P(x) \ll x^{\frac{n}{4} - \frac{1}{2}} \sum_{k \leqslant \sqrt{x}}{}' \lg 2k \min{}^{\frac{n}{4} - \frac{1}{2}} \left(\frac{x}{k^2}, \frac{1}{R_k} \right) \tag{3}$$

(see [6]; $\sum'$ indicates that if for the given k it is $S_{h,k} = 0$ for all h, $(h, k) = 1$, we may write 1 instead of R_k). Hence the problem of finding the O-estimates is transformed to the study of the above sum and hence to a very refined study of the simultaneous approximation of the numbers $\alpha_1 M_1$, $\alpha_2 M_2, \ldots, \alpha_n M_n$.

The Ω-methods used before fail to give results in the case II. It has appeared that a very efficient method is that based on the relation of the «inverse» with (1)

$$F(s) = s \int_0^\infty e^{-xs} P(x)\, dx,$$

which again uses the approximation of $F(s)$ by (2) and the limit process for $s = s_n$, $\operatorname{Re} s_n \to +0$. Essential difficulties occur in the so called singular case ($S_{h,k} \equiv 0$). It turns out that these troubles are caused by certain properties of the numbers v_j; they disappear for $v_j = 0$. To illustrate this, let us introduce three results which can be formulated in a simple way (see [6, 11]):

If $v_j = 0$, $\gamma = \gamma(\alpha_1, \alpha_2, \ldots, \alpha_n)$, $n \geqslant 4$, *then*

$$\left(\frac{n}{4} - \frac{1}{2}\right) \frac{2\gamma + 1}{\gamma + 1} \leqslant f \leqslant \left(\frac{n}{2} - \frac{1}{2}\right) \frac{2\gamma + 1}{\gamma + 1} + \frac{1}{2(\gamma + 1)}; \tag{4}$$

if moreover $\alpha_1 = \alpha_2 = \ldots = \alpha_n = \alpha$, $\gamma = \gamma(\alpha)$, $n > 6$ *then*

$$f = \left(\frac{n}{4} - \frac{1}{2}\right) \frac{2\gamma + 1}{\gamma + 1}. \tag{5}$$

For any $n > 4$, $\gamma \geqslant \frac{1}{n}$ there exist Q, α_j, v_j, M_j so that $f \leqslant \frac{n}{4}$ and $\gamma = \gamma(\alpha_1, \alpha_2, \ldots, \alpha_n)$. Further, Walfisz proved in [14] that for $n > 4$ in the singular case and for rational α_j it is even

$$P(x) = O\left(x^{\frac{n}{4} - \frac{1}{10}}\right).$$

The following three problems are the main ones which are still open:

a detailed study of the singular case, finding of a definitive estimate of the sum in (3) if $\alpha_1, \alpha_2, \ldots, \alpha_n$ are not equal (cf. [4, 5]) and finally an improvement of the relation (3) e. g. by means of Kloosterman's sums with regard to Linnik's conjecture.

3. It was already Landau's general method which started from the study of the functions

$$P_0(x) = P(x), \qquad P_1(x) = \int_0^x P(t)\,dt_1, \ldots$$

Recently, more general functions ($\rho > 0$)

$$P_0(x) = P(x), \qquad P_\rho(x) = \frac{1}{\Gamma(\rho)} \int_0^x P(t)(x-t)^{\rho-1}\,dt = \frac{1}{2\pi i} \int_{(a)} \frac{e^{xs} F(s)}{S^{\rho+1}}\,ds \tag{6}$$

have been studied by Chandrasekharan with Narasimhan (the problems being much more general) and by Jarník (in our case). It was shown that the value

$$f_\rho = \limsup_{x \to +\infty} \frac{\lg |P_\rho(x)|}{\lg x}$$

is always equal to $(n-1)/4 + \rho/2$ provided $\rho > (n-1)/2$ (see [4]) and it is to be expected that f_ρ will change continuously with increasing ρ up to certain value $\rho = \rho_0 \leqslant (n-1)/2$ while being $(n-1)/4 + \rho/2$ for $\rho \geqslant \rho_0$. So e. g. in the case II for $\alpha_j = 0$, we have $f_\rho = n/2 - 1$ for $0 \leqslant \rho < n/2 - 2$ (see [4]).

It is of interest that the behaviour of the cases I and II is in a sense inverse. So we have e. g.

$$f_\rho = \left(\frac{n}{4} - \frac{1}{2}\right) \frac{2\gamma + 1}{\gamma + 1} + \frac{\rho}{2(\gamma + 1)}$$

in the case II provided $\alpha_1 = \alpha_2 = \ldots = \alpha_n = \alpha$, $\gamma = \gamma(\alpha)$, $v_j = 0$, $0 \leqslant \rho < < n/2 - 3$ and

$$f_\rho = \frac{n}{2} - 1 - \frac{\rho + 1}{\gamma}$$

in case I provided $u_j \geqslant 2(\gamma+1)\gamma+1/\gamma$, $\rho < n/2 - 1$, $\gamma = \gamma(a_2/a_1, a_3/a_1, \ldots, a_\sigma/a_1)$ (see [11]).

It turns out that determination of the value $P_\rho(x)$ and of the value f_ρ in its neighbourhood corresponds to the classical problems of the circle and the sphere and will be immensely difficult.

4. A certain information on the function $P_\rho(x)$ is obtained from the study of mean values. E. g. it can be shown (for $\rho = 0$ see [9], for $\rho > 0$ not published) that

$$\int_{\mathfrak{M}} |P_\rho(x)|^2\, d\alpha_1 d\alpha_2 \ldots d\alpha_n = c x^{\frac{n}{2}+2\rho} + O\left(x^{\frac{n}{2}+2\rho}\right)$$

$(c > 0)$ and

$$\int\limits_{\mathfrak{N}} |P_\rho(x)|^2 \, dv_1 dv_2 \ldots dv_n = \begin{cases} O\left(x^{\frac{n}{2}+\rho-\frac{1}{2}}\right), \\ \Omega\left(x^{\frac{n}{2}+\rho-\frac{1}{2}}\right) \end{cases}$$

where $\mathfrak{M} = \left\langle 0, \frac{1}{M_1}\right\rangle \times \ldots \times \left\langle 0, \frac{1}{M_n}\right\rangle$, $\mathfrak{N} = \langle 0, M_1\rangle \times \ldots \times \langle 0, M_n\rangle$.

Even more interesting are the results for the function

$$M_\rho(x) = \int\limits_0^x |P_\rho(t)|^2 \, dt.$$

The starting point is the formula $(a, v > 0)$

$$M_\rho(x) = -\frac{1}{4\pi^2} \int\limits_{(a)} \int\limits_{(v)} \frac{e^{x(s+s')} F(s) \overline{(F\,\overline{s'})}}{s^{\rho+1} (s')^{\rho+1} (s+s')} ds ds'$$

which was used (for $\rho = 0$ and in the case I) for the first time by Jarník. If the considerations are restricted to the case II, it is possible to establish the estimate

$$M_\rho(x) \ll x^{\frac{n}{2}-\frac{1}{2}} \sum_{k \leqslant \sqrt{x}}' k^{2\rho+1} \min{}^{\frac{n}{2}-\frac{1}{2}} \left(\frac{x}{k^2}, \frac{1}{R_k}\right)$$

similarly as it was outlined for the function $P(x)$. In the singular case and for $\rho > (n-3)/2$ we shall have

$$M_\rho(x) \asymp x^{\frac{n}{2}+\rho+\frac{1}{2}}.$$

Considering only rational $\alpha_1, \alpha_2, \ldots, \alpha_n$, we can even separate the main term in the form $cx^{\frac{n}{2}+\rho+\frac{1}{2}}$ $(c > 0)$ in the singular case und for $\rho > \frac{n}{2} - 1$ (not published; for $\rho = 0$ see [13]), in the form cx^{n-1} for $0 \leqslant \rho < (n-3)/2$ and in the form $cx^{n-1} \lg x$ (1) for $0 \leqslant \rho = (n-3)/2$ in the nonsingular case (see [7, 8]).

Again the relation

$$F(s) = s^{\rho+1} \int\limits_0^\infty e^{-xs} P_\rho(x) \, dx$$

may be used to find Ω-estimates if we write

$$\left|\frac{F(s)}{s^{\rho+1}}\right| \leqslant \left|\int\limits_0^T \ldots dx\right| + \left|\int\limits_T^\infty \ldots dx\right| \ll \sqrt{\frac{M_\rho(T)}{\operatorname{Re} s}} + \int\limits_T^\infty e^{-x \operatorname{Re} s} |P_\rho(x)| \, dx$$

and choose suitably T in dependance on $s = s_n$, $\operatorname{Re} s_n \to +0$ (see [11]).

Hence e. g. for $0 \leqslant \rho < (n-3)/2$, $\alpha_1 = \alpha_2 = \ldots = \alpha_n = \alpha$, $\gamma = \gamma(\alpha)$,

$v_j = 0$ it holds

$$\limsup_{x \to +\infty} \frac{\lg M_\rho(x)}{\lg x} = \max\left(\frac{n-1}{2}\frac{2\gamma+1}{\gamma-1} + \frac{2\rho+1}{2(\gamma+1)},\ \frac{n}{2}+\rho+\frac{1}{2}\right).$$

Consequently, it may be expected that the above mentioned value of ρ_0 will be equal to $n/2 - 3/2 - 1/2\gamma$.

5. The function $P_\rho(x)$ is given by the form Q and the parameters α_j, v_j, M_j (in this order). Let $\tilde{P}_\rho(x)$ be the same function for the form conjugated to Q with parameters v_j, $-\alpha_j$, $\frac{1}{M_j}$ (in this order). The following relation holds for these functions for $\rho > n/2$, $x > 0$.

$$P_\rho(x) - \frac{a_0 x^\rho}{\Gamma(\rho+1)} = \frac{x^{\frac{n}{4}+\frac{\rho+1}{2}} \exp 2\pi i \sum_{j=1}^{n} \alpha_j v_j}{\pi^{\rho-1}\sqrt{D}\, M_1 M_2 \ldots M_n} \int_0^\infty (\tilde{P}(t) - \tilde{a}_0) \frac{I_{n/2+\rho+1}(2\pi\sqrt{tx})}{t^{\frac{n}{4}+\frac{\rho+1}{2}}}\, dt,$$

$a_0 = P_0(0)$, $\tilde{a}_0 = \tilde{P}_0(0)$, which is an analogue of Landau's formula in terms of Bessel functions. Using the results for the case II we can establish e. g. the following results:

Let $n > 5$, $\alpha_j = 0$, M_j integers, let Q have integer coefficients. If at least one number from $v_1, v_2, \ldots, v_n$ is irrational, then

$$P(x) = o\left(x^{\frac{n}{2}-\frac{n}{n+1}}\right).$$

For almost all systems $v_1, v_2, \ldots, v_n$ (in the sense of Lebesgue measure) it is

$$P(x) = O\left(x^{\frac{n}{3}+\varepsilon}\right)$$

for any $\varepsilon > 0$ (see [10]).

Note added in proof. In the author's paper *Mittelwertsätze der Gitterpunktlehre.* III (Czechoslovak Math. J. **23 (98)** (1973), 467–482) the main term of the function $M_\rho(x)$ for $\rho > r/2 - 1$ is separated. Further, in the paper *On a certain sum in number theory.* II (Trans. Amer. Math. Soc. 195 (1974), 357–364; but see also Comment. Math. Univ. Carolinae **13** (1972), 763–775) it is shown that many of the results concerning O-estimates in the case II hold without assumption $\alpha_1 = \alpha_2 = \cdots = \alpha_r = \alpha$.

BIBLIOGRAPHY

1. B. Diviš, *Über Gitterpunkte in mehrdimensionalen Ellipsoiden.* I, Czechoslovak Math. J. **20** (95) (1970), 130–139. MR **41** #165.

2. V. Jarník, *Über Gitterpunkte in mehrdimensionalen Ellipsoiden*, Math. Ann. **100** (1928), 699–721.

3. V. Jarník, *Über Gitterpunkte in mehrdimensionalen Ellipsoiden*, Tôhoku Math. J. **30** (1929), 354–371.

4. ———, *Bemerkungen zu Landauschen Methoden in der Gitterpunktlehre*, Number Theory and Analysis (Papers in Honor of Edmund Landau), Plenum Press, New York, 1969, pp. 137–156. MR **42** #4494.

5. E. Landau, *Ausgewählte Abhandlungen zur Gitterpunktlehre*, VEB Deutscher Verlag der Wissenschaften, Berlin, 1962. MR **27** #112.

6a. B. Novák, *Verallgemeinerung eines Peterssonschen Satzes und Gitterpunkte mit Gewichten*, Acta Arith. **13** (1967/68), 423–454. MR **37** #5151.

6b. ———, *On lattice points with weight in high-dimensional ellipsoids*, Acta Arith. **14** (1969), 371–397. MR **37** #6243.

7. ———, *Mean value theorems in the theory of lattice points with weight*. II, Comment. Math. Univ. Carolinae **11** (1970), 53–81. MR **41** #6807.

8. ———, *Mittelwertsätze der Gitterpunktlehre*. II, Časopis Pěst. Mat. **96** (1971), 245–261, 300. MR **45** #191.

9. ———, *Mean value theorems in the theory of lattice points with weight,* Comment. Math. Univ. Carolinae **8** (1967), 711–733. MR **38** #121.

10. ———, *A remark on the theory of lattice points in ellipsoids*. II, Comment. Math. Univ. Carolinae **9** (1968), 547–561. MR **39** #5501.

11. ———, *Über eine Methode der Ω-Abschätzungen*, Czechoslovak Math. J. **21** (96) (1971), 257–279.

12. ———, *On certain sum in number theory*, Comment. Math. Univ. Carolinae **12** (1971), 669–685. MR **46** #7169.

13. A. Walfisz, *Über die Koeffizientensummen einiger Modulformen*, Math. Ann. **108** (1933), 75–90.

14. ———, *Convergence abscissae of certain Dirichlet series*, Akad. Nauk Gruzin. SSR. Trudy Tbiliss. Mat. Inst. Razmadze **22** (1956), 33–75. (Russian) MR **19**, 943.

Trudy Mat. Inst. Steklov.
132 (1973)

Proc. Steklov Inst. Math.
132 (1973)

THE GEOMETRY OF LINEAR ALGEBRAIC GROUPS

V. E. VOSKRESENSKIĬ

ABSTRACT. Birational properties of algebraic groups are studied.
Bibliography: 12 items.

§1. Birational characteristics of linear groups

1. Let G be a connected linear algebraic group defined over a field k, and let $k(G)$ be the field of rational functions on the group G. If k is a perfect field, G as a group variety is unirational over k, i.e. the field $k(G)$ is a subfield of a purely transcendental extension of k. This was proved by Chevalley in 1954 for groups defined over a field of characteristic zero [1], and then was generalized by Rosenlicht to the case of an arbitrary perfect field of definition [2]. If k is algebraically closed, then $k(G)$ is purely transcendental over k, but for nonalgebraically closed fields Chevalley and Rosenlicht gave examples of groups G which are not rational over k. However, in the years immediately following this, no one seems to have become engaged in a more detailed study of the birational characteristics of algebraic groups. In all the literature known to the present author no work on this theme occurs. In this lecture I wish to present the results which I have obtained during the years 1967–1971 in studying birational properties of linear algebraic groups. My methods are based, on the one hand, on Hironaka's fundamental result concerning the possibility of resolving singularities of rational maps of algebraic varieties, and, on the other, on the theory of birational invariants developed for rational surfaces in the work of Manin and Šafarevič [3]. Since Hironaka's theorems have only been proved in full generality in characteristic zero, I shall restrict my consideration to groups G which are defined over a field k of characteristic zero, although many results are probably also true in the case of an arbitrary perfect field of definition.

2. Let $\bar{k}$ be an algebraic closure of k, $\mathfrak{F}=\mathfrak{F}(\bar{k}/k)$ its Galois group. If X is an algebraic variety over k, then $\bar{X} = X \otimes_k \bar{k}$ is a variety over $\bar{k}$, and the group $\mathfrak{F}$ naturally acts on the various objects determined by the variety $\bar{X}$. In what follows we shall to a large part be interested in the $\mathfrak{F}$-module $\operatorname{Pic} \bar{X}$. Following Manin, we shall say that a $\mathfrak{F}$-module which is isomorphic to a finite direct sum of modules of the form $\mathbf{Z}[\mathfrak{F}] \otimes_{\mathbf{Z}[U]} \mathbf{Z}$, where U is a subgroup of finite index in $\mathfrak{F}$, is *$\mathfrak{F}$-trivial.*

AMS (MOS) subject classifications (1970). Primary 20G30, 12F20, 14E05.

Two $\mathfrak{F}$-modules A and B will be called *similar* if there exist $\mathfrak{F}$-trivial modules M_1 and M_2 such that we have an isomorphism of direct sums $A + M_1 \approx B + M_2$. We have the following necessary condition for birational equivalence:

THEOREM 1 [3], [4]. *Let X and Y be nonsingular projective varieties over a field k of characteristic zero. If X and Y are birationally equivalent over k, then the $\mathfrak{F}$-modules* Pic $\overline{X}$ *and* Pic $\overline{Y}$ *are similar.*

The proof is based on the possibility of resolving the singularities of rational maps.

COROLLARY. *The similitude class of the $\mathfrak{F}$-module* Pic $\overline{X}$ *is a birational invariant; in particular, if X is a nonsingular projective variety which is rational over k, then* Pic $\overline{X}$ *is similar to a $\mathfrak{F}$-trivial one, and*

$$H^1(K, \operatorname{Pic} \overline{X}) = 0$$

for any field K, $k \subseteq K \subseteq \overline{k}$.

3. A nonsingular projective variety V over a field k will be called a *projective model* of the group G if the underlying variety of G is mapped biregularly onto an open subset of V over k. The inclusion morphism $G \subset V$ defines an exact local cohomology sequence of the sheaf $O^*_{\overline{V}}$, which gives an exact sequence of $\mathfrak{F}$-modules

$$0 \to \hat{G} \to \hat{M} \to \operatorname{Pic} \overline{V} \to \operatorname{Pic} \overline{G} \to 0, \tag{1}$$

where $\hat{G}$ is the $\mathfrak{F}$-module of rational characters of G and $\hat{M}$ is a $\mathfrak{F}$-trivial module. If $G = T$ is a torus, then Pic $\overline{T} = 0$, and with regard for Theorem 1 we will obtain a necessary condition for a torus to be rational over its field of definition.

THEOREM 2 [4], [5]. *In order that a torus T be rational over a field k, it is necessary that its character module $\hat{T}$ be able to be inserted in an exact sequence*

$$0 \to \hat{T} \to \hat{M} \to \hat{N} \to 0, \tag{2}$$

where $\hat{M}$ and $\hat{N}$ are $\mathfrak{F}$-trivial modules.

While it is unclear whether or not this condition is sufficient, in applications it turns out to be very subtle. For example, Theorem 2 allows us to construct an example of a torus which splits over a cyclic extension of its field of definition, but is not rational over that field of definition.

We write the exact sequence of k-tori dual to sequence (2):

$$1 \to N \xrightarrow{\alpha} M \xrightarrow{\beta} T \to 1. \tag{3}$$

Since the module $\hat{N}$ is $\mathfrak{Z}$-trivial, the torus N is a product of tori of the form $R_{L/k}(G_m)$, from which it follows that the mapping β possesses a rational k-section. Therefore the torus M is birationally equivalent over k to the product $N \times T$. The tori M and N are rational over k. We use the terminology of algebraic topology and introduce the following definition.

DEFINITION. An affine variety X defined over a field k is said to be *stably rational over* k if for some natural number n the direct product $X \times \mathbf{A}^n$ is birationally equivalent over k to the affine space $\mathbf{A}^m$, where $m = n + \dim X$.

Thus it follows from sequence (2) that the torus T is stably rational over k. The converse is also obviously true.

THEOREM 3. *The following conditions are equivalent*:

1) *The torus* T *is stably rational over the field* k.
2) *The* $\mathfrak{Z}$-*module* $\operatorname{Pic} \overline{V}(T)$ *is similar to a* $\mathfrak{Z}$-*trivial module.*
3) *The module* $\hat{T}$ *is included in the exact sequence*

$$0 \to \hat{T} \to \hat{M} \to \hat{N} \to 0,$$

where $\hat{M}$ *and* $\hat{N}$ *are* $\mathfrak{Z}$-*trivial modules.*

CONJECTURE. *A torus that is stably rational over* k *is rational over* k.

There is a more general statement that any stably rational variety is rational. This statement is quite likely to be true, but it is probably very difficult to prove. Our conjecture is based on the conviction that we have not yet completely extracted all the information lurking in the sequence (2). The following two special cases serve as a corroboration of this thought.

THEOREM 4 [5], [6]. *Let* T *be a torus defined over a field* k *and splitting over a cyclic extension of prime degree* p *of* k. *If* T *is stably rational over* k, *then it is rational over* k.

Theorem 4 is trivial for those primes p for which the pth cyclotomic field has class number one. However, if the field $\mathbf{Q}(\zeta_p)$ has class number greater than one, then there exist tori which are not rational over the field of definition; for example, a torus T whose character module is isomorphic to a projective module, but not to a free ideal of the group ring of the cyclic group of order p. Later on it will turn out that this example has a direct relation to the problem of the rationality of the field of invariants of a cyclic group of transformations.

THEOREM 5. *Let* $N = G_m^e$ *in the sequence* (3). *Then the torus* T *is rational over its field of definition.*

PROOF. The group $M(k)$ is the group of invertible elements of its enveloping algebra over the field k. The additive group of this direct sum is the k-linear space $\mathbf{A}^d$, where $d = \dim M$. The homomorphism $\alpha\colon G_m^e \to M$ is a faithful representation of the group G_m^e by linear operators of the space $\mathbf{A}^d$. This representation is completely reducible and splits over k into a direct sum of one-dimensional representations $\mathbf{A}^d = V_1 + \cdots + V_d$, with $\alpha(x)v = (x_1v_1, \cdots, x_dv_d)$, where $x \in G_m^e$, $v = (v_1, \cdots, v_d) \in \mathbf{A}^d$, $v_i \in V_i$ and $x \to x_i$ is the representation $G_m^e \to \operatorname{Aut}(V_i) = G_m$. Since M is open in $\mathbf{A}^d$, the epimorphism $\beta\colon M \to T$ extends to a rational map $\beta\colon \mathbf{A}^d \to T$, and we see that the variety T is birationally isomorphic over k to the torus G_m^d/G_m^e, which is rational over k. This proves Theorem 5.

As a consequence we construct an example of a torus T which is rational over k, and whose character module is projective but not free. We use the example of a projective nonfree $\mathfrak{g}$-module $\hat{T}$ constructed by Swan [7], for which there is an isomorphism of direct sums $\hat{T} + \mathbf{Z} = \mathbf{Z}[\mathfrak{g}] + \mathbf{Z}$, where $\mathfrak{g}$ is the quaternion group. By Theorem 5 the torus T is rational.

4. We have already said that the group $H^1(k, \operatorname{Pic} \overline{V}(G))$ is a birational invariant of the group G. We have the following important fact.

THEOREM 6. *If the group G is defined over a field k and is split over a cyclic extension of k, then* $H^1(k, \operatorname{Pic} \overline{V}(G)) = 0$.

This theorem was proved for tori in [4], [5], while a proof for arbitrary connected G will be published in [8]; this is true under the condition that k is an algebraic number field. Theorem 6 explains the well-known fact that the arithmetic properties of a group with cyclic splitting field are good. This will be seen from the following theorem.

THEOREM 7 [4], [5]. *Let T be a torus defined over an algebraic number field; then we have an exact sequence*

$$0 \to A(T) \to H^1(k, \operatorname{Pic} \overline{V}(T)) \to Ш(T) \to 0.$$

(*Here* $Ш(G) = \operatorname{Ker}[H^1(k, G) \to \Pi_v H^1(k_v, G)]$, *and* $A(G) = (\Pi_v G(k_v))/\overline{G(k)}$ *is the set showing the measure of deviation from the weak approximation theorem in G; $\overline{G(k)}$ is the closure of $G(k)$ in the direct product topology, and k_v is the v-adic completion of the field k.*)

Theorem 7 yields a number of interesting arithmetic consequences.

THEOREM 8 [5]. *Let $T = R_{L/k}^{(1)}(G_m)$, where L is a normal extension of the field k. Then* $H^1(k, \operatorname{Pic} \overline{V}(T)) = H^3(L/K, \mathbf{Z})$, $k \subseteq K \subseteq L$.

Let $\omega_1, \cdots, \omega_n$ be a basis of the field L over k and let $x = \Sigma_1^n k_i\omega_i$ be a "general element" of the field. The norm $N_{L/k}(x) = F(x_1, \cdots, x_n)$ is a homogeneous polynomial of degree n of n variables with coefficients in k. The variety

$R^{(1)}_{L/k}(G_m)$ is biregularly equivalent to the hypersurface

$$F(k_1, \ldots, x_n) = 1 \tag{4}$$

in the affine space $\mathbf{A}^n$. Hypersurfaces of the form $F(x_1, \cdots, x_n) = a$ will be called *normed*. They are principal homogeneous spaces of the torus $R^{(1)}_{L/k}(G_m)$.

Theorem 8 shows that the normed hypersurfaces (4) are as a rule not rational over k.

We now consider a series of normed hypersurfaces defined over a field K which is not normal over $\mathbf{Q}$.

Let $K = \mathbf{Q}(\sqrt[q]{a})$, $a \in \mathbf{Q}$; let q be a prime number and let the polynomial $x^q - a$ be irreducible over $\mathbf{Q}$; hence $(K:\mathbf{Q}) = q$. Consider the torus $T = R^{(1)}_{K/\mathbf{Q}}(G_m)$.

THEOREM 9. $H^1(\mathbf{Q}, \operatorname{Pic} \overline{V}(T)) = 0$.

PROOF. Let L be the minimal normal closure of K over $\mathbf{Q}$, and let $\mathfrak{g}$ be the Galois group of the extension $L/\mathbf{Q}$. The group $\mathfrak{g}$ is metacyclic, of order $q(q-1)$. Hence any Sylow subgroup $\mathfrak{g}_p$ of $\mathfrak{g}$ is cyclic. The field L is the splitting field for the torus T. Let V be a projective model of T over $\mathbf{Q}$. Then

$$H^1(\mathbf{Q}, \operatorname{Pic} \overline{V}) = H^1(\mathfrak{g}, \operatorname{Pic} \overline{V}^U),$$

where U is the subgroup of $\mathfrak{F}$ consisting of elements leaving elements of L in place. Since $H^1(\mathfrak{g}_p, \operatorname{Pic} \overline{V}^U) = 0$ for all $p | q(q-1)$ by Theorem 6, it follows that $H^1(\mathbf{Q}, \operatorname{Pic} \overline{V}) = 0$. Theorem 7 then shows that $A(T) = Ш(T) = 0$, i.e. the weak approximation theorem holds for the torus T, and the Hasse-Minkowski principle holds for normed hypersurfaces. For $q = 2, 3$ this fact was known before.

5. We pass to semisimple groups. A number of the results obtained for tori also carry over to semisimple groups. It is true that a part of the information is only obtained in the relative case, i.e. modulo simply connected groups. Here there are many unsolved questions.

Since for a connected semisimple group G the group of rational characters $\hat{G} = 0$, it follows that the sequence (1) takes the form

$$0 \to \hat{M} \to \operatorname{Pic} \overline{V} \to \operatorname{Pic} \overline{G} \to 0.$$

The group $\operatorname{Pic} \hat{G}$ is computed in [9] and is isomorphic (as a $\mathfrak{F}$-module) to the group $\hat{\pi} = \operatorname{Hom}(\pi, \bar{k}^*)$, where π is the fundamental group of G. If G is simply connected, then $\operatorname{Pic} \hat{G} = 0$, and hence the $\mathfrak{F}$-module $\operatorname{Pic} \overline{V}$ is $\mathfrak{F}$-trivial. To us it seems that this fact should contain nontrivial arithmetic information about the simply connected group G, although we do not know how to extract it.

As in the case of tori, for a connected semisimple group G there is an analogue of Theorem 7, connecting the group $H^1(k, \operatorname{Pic} \overline{V}(G))$ with the arithmetic of the group G.

THEOREM 10 [4]. *Let G be a connected semisimple algebraic group over an algebraic number field k. Then the sequence*

$$0 \to A \to H^1(k, \operatorname{Pic} \bar{V}(G)) \to B \to 0,$$

is exact, where

$$A = \operatorname{Coker}\left[H^1(k, \pi) \to \prod_{v \in S_0} H^1(k_v, \pi)\right],$$

and

$$B = \operatorname{Ker}\left[H^2(k, \pi) \to \prod_{v} H^2(k_v, \pi)\right].$$

(*Here S_0 is a finite set of valuations of the field k.*)

If L is a normal splitting field of the group G of finite degree over k, then we can throw out of S_0 all the places with cyclic decomposition group in L. In particular, it is possible to *not* include the real valuations in S_0. If $S \cap S_0 = \varnothing$, then the mapping

$$H^1(k, \pi) \to \prod_{v \in S} H^1(k_v, \pi)$$

is an epimorphism. According to Kneser [10], it follows from this that weak approximation holds in G at the places of S. In particular, if the group G splits over a cyclic extension, then the weak approximation theorem holds in G for any S. This was known earlier for tori. Furthermore, if G is a torus or a semisimple group that is rational over k, then Ш$(G) = 0$ [4].

§2. The problem of the rationality of a field of invariants

6. Let $\mathbf{Q}$ be the field of rational numbers, $K = \mathbf{Q}(x_1, \cdots, x_n)$ a purely transcendental extension of $\mathbf{Q}$, and G the subgroup of the symmetric group S_n, acting naturally on the field K, which permutes the elements of the set $\{x_1, \cdots, x_n\}$. We denote by $L = K^G \subset K$ the field of invariants of G.

Years ago E. Noether conjectured that the field K^G was purely transcendental over $\mathbf{Q}$ for any subgroup $G \subset S_n$. If $G = S_n$, this is a classical theorem about symmetric functions. The general case turned out to be very difficult, and only recently Swan [11] constructed the first counterexamples to Noether's problem. At approximately the same time, while occupied with the problem of rationality of algebraic groups, I succeeded in constructing examples of tori with a cyclic splitting field which are not rational over their field of definition [4]. Becoming interested by Swan's results, I discovered that the counterexamples which Swan had constructed for a different reason and mine essentially coincide and that the field of functions K^G for a cyclic group G can be treated as the field of rational functions on some torus. This allows us to apply the already known rules for the rationality of tori to Noether's problem. We shall show that the results of the preceding section allow us to completely solve the problem of the pure transcendence of the field K^G for a cyclic group G of prime order p. We prove a criterion for the rationality of such fields which complements Swan's results. This criterion reduces the question of the rationality of the field K^G to a problem in algebraic number theory and allows us to obtain counterexamples to Noether's problem, and also gives the possibility of making a list of the purely transcendental fields of invariants, including all the fields known before, as well as some new ones. We also examine some examples of fields K^G, where G is a cyclic group of not necessarily prime order.

We proceed to sketch this theory briefly.

7. Let G denote the group generated by a cyclic permutation of the unknowns $\{x_1, \cdots, x_n\}$. Since the group G and the field K^G are completely determined by the natural number n, we shall denote the field K^G by L_n. Let ϵ be a primitive nth root of 1, and let $P_n = \mathbf{Q}(\epsilon)$ be the nth cyclotomic field. Consider the linear forms

$$y_i = \sum_{k=1}^{n} x_k \varepsilon^{ki}, \qquad i = 0, 1, \ldots, n-1.$$

Then $P_n(x_1, \cdots, x_n) = P_n(y_0, \cdots, y_{n-1})$, and the field of invariants $P_n(x_1, \cdots, x_n)^G$ is purely transcendental over P_n with the system of generators $z_0, z_1, \cdots, z_{n-1}$, where $z_0 = y_0$, $z_1 = y_1^n$ and $z_m = y_m y_1^{-m}$ for $2 \leqslant m \leqslant n - 1$. Let $\mathfrak{g}$ be the Galois group of the field P_n over $\mathbf{Q}$, isomorphic to the group of residue classes modulo n. The group $\mathfrak{g}$ naturally acts on the field $P_n(x_1, \cdots, x_n)$, leaving the elements $x_1, \cdots, x_n$ fixed. We have the relation

$$L_n = \mathbf{Q}(x_1, \ldots, x_n)^G = P_n(z_0, \ldots, z_{n-1})^{\mathfrak{g}}.$$

From the above fact it is easy to verify that the free multiplicative subgroup of the field $P_n(z_0, \cdots, z_{n-1})$ generated by the elements $\{z_0, \cdots, z_{n-1}\}$ is stable relative to operators of the group $\mathfrak{g}$, and hence is a $\mathfrak{g}$-module. We denote this module by $\hat{T}_n$. We have

THEOREM 11 [6]. *The field of invariants L_n is the field of rational functions $\mathbf{Q}(T_n)$, where T_n is a torus defined over $\mathbf{Q}$ and split over P_n, with character module $\hat{T}_n$.*

The torus T_n is a very convenient model of the field L_n to study the structure of this field. In the work of Swan [11] and Masuda [12] the investigation proceeds on the level of $\mathfrak{g}$-modules. The use of tori allows one to reveal finer relations and to obtain stronger results.

We consider the $\mathfrak{g}$-module $\hat{T}_n$ in more detail. Let $\hat{I}_n$ be the subgroup of $\hat{T}_n$ generated by the elements z_m, with $(m, n) = 1$. Since every element $\sigma \in \mathfrak{g}$

takes the primitive root ϵ into ϵ^k with $(k, n) = 1$, it is easy to show, for example, by the formulas of [6], that $\hat{I}_n$ is a $\mathfrak{g}$-module. Consider the quotient module $\hat{T}_n/\hat{I}_n = \hat{M}_n$. The same formulas also show that the $\mathfrak{g}$-module $\hat{M}_n$ is $\mathfrak{g}$-trivial. The exact sequence of $\mathfrak{g}$-modules

$$0 \to \hat{I}_n \to \hat{T}_n \to \hat{M}_n \to 0$$

defines a dual sequence of tori over $\mathbf{Q}$:

$$1 \to M_n \to T_n \to I_n \to 1.$$

It is obvious that there exists a section $s\colon I_n \to T_n$ rational over $\mathbf{Q}$, so that the torus T_n is birationally equivalent over $\mathbf{Q}$ to the product of the tori I_n and M_n, i.e. $T_n \approx M_n \times I_n$. Hence the field L_n is purely transcendental over $\mathbf{Q}$ if the torus I_n is rational over $\mathbf{Q}$. The module $\hat{I}_n$ is a submodule of finite index of the free $\mathfrak{g}$-module generated by the elements y_m, $(m, n) = 1$, and isomorphic to the group ring $\mathbf{Z}[\mathfrak{g}]$. Denoting this isomorphism by ψ, it is always possible to assume that $\psi(z_1) = 1$. We shall find a representation of the module $\hat{I}_n$ as an ideal of the ring $\mathbf{Z}[\mathfrak{g}]$. Let $\epsilon^{\sigma_i} = \epsilon^{m_i}$, $\sigma_i \in \mathfrak{g}$, $(m_i, n) = 1$, $1 < m_i < n$. Then $\sigma_i y_1 = y_{m_i}$, and we have

$$\psi(z_1) = \psi(y_1^n) = n, \quad \psi(z_m) = \psi(y_m y_1^{-m}) = \psi(y_m) - m = \psi(\sigma y_1) - m = \sigma - m.$$

Thus, if $\sigma_1, \cdots, \sigma_t$ are generators of $\mathfrak{g}$, then the ideal $\hat{I}_n$ has the form

$$\hat{I}_n = (n, \sigma_1 - m_1, \ldots, \sigma_t - m_t). \tag{5}$$

8. We now consider the field L_p, where p is a prime number. Then the group $\mathfrak{g}$ is cyclic of order $p - 1$ with generator σ, and the ideal $\hat{I}_p = (p, \sigma - a)$, where a is a primitive root modulo p. The ideal $\hat{I}_p$ has index p in $\mathbf{Z}[\mathfrak{g}]$ and is therefore a projective $\mathfrak{g}$-module. Let $F = \mathbf{Q}(\zeta)$ be the cyclotomic field of $(p - 1)$th roots of 1, ζ a primitive root, $\zeta^{p-1} = 1$. We let $\mathbf{Z}[\zeta] = R$ denote the ring of integers of F, and let $\varphi\colon \mathbf{Z}[\mathfrak{g}] \to R$ be the natural epimorphism $f(\sigma) \to f(\zeta)$. Under the homomorphism φ the ideal $\hat{I}_p = (p, \sigma - a)$ goes to the ideal $\mathfrak{p} = (p, \zeta - a)$ of the ring R. Since $p \equiv 1 \pmod{p-1}$, the prime number p is the norm of the ideal $\mathfrak{p}$: $p = \mathrm{Nm}\,\mathfrak{p}$. If A is an arbitrary $\mathfrak{g}$-module, then the groups $A^\varphi = \mathrm{Hom}_R(R, A)$ and $A_\varphi = R \otimes_{\mathfrak{g}} A$ can be considered as R-modules. If A is projective, then the R-modules A^φ and A_φ are isomorphic. If M is a $\mathfrak{g}$-trivial module, then the module M^φ is R-free. Furthermore, if

$$0 \to A \to B \to M \to 0$$

is an exact sequence of $\mathfrak{g}$-modules, where M is $\mathfrak{g}$-trivial, then

$$0 \to A^\varphi \to B^\varphi \to M^\varphi \to 0$$

is an exact sequence of R-modules [11].

Now suppose that the field L_p is purely transcendental over $\mathbf{Q}$. Then the torus I_p is stably rational over $\mathbf{Q}$, and by Theorem 3 the $\mathfrak{g}$-module $\hat{I}_p$ is inserted in an exact sequence

$$0 \to \hat{I}_p \to \hat{M} \to \hat{M}_1 \leftarrow 0,$$

where $\hat{M}$ and $\hat{M}_1$ are $\mathfrak{g}$-trivial modules. Then the sequence of R-modules

$$0 \to \hat{I}_p^\varphi \to \hat{M}^\varphi \to \hat{M}_1^\varphi \to 0$$

splits, and we have a direct sum $\hat{M}^\varphi = \hat{I}_p^\varphi + \hat{M}_1^\varphi$, where $\hat{M}^\varphi$ and $\hat{M}_1^\varphi$ are free R-modules. But over a ring of algebraic integers such an equality is only possible in the case where $\hat{I}_p^\varphi$ is R-free. Since $\hat{I}_p^\varphi \approx \hat{I}_\varphi \approx \mathfrak{p}$, the ideal $\mathfrak{p}$ is principal and we have the equality $p = |N_{F/\mathbf{Q}}(\alpha)|$, $\alpha \in R$. This is also the necessary condition for the rationality of the field L_p which was obtained by Swan [11]. It turns out that this condition is also sufficient. A detailed proof will be published in [8]. Thus we have the following criterion for the rationality of L_p.

THEOREM 12. *In order that the field L_p be purely transcendental over $\mathbf{Q}$ it is necessary and sufficient that the number p or $-p$ be the norm relative to $\mathbf{Q}$ of an integral element of the cyclotomic field $\mathbf{Q}(\zeta)$, $\zeta^{p-1} = 1$.*

This theorem immediately gives a whole series of fields L_p that are not rational over $\mathbf{Q}$, for example, for $p = 47, 167, 359, 383, 479$.

We apply this theorem to establish that certain fields L_p are purely transcendental. Since the mth cyclotomic fields have class number one for $m < 23$, for $p = 2, 3, 5, 7, 11, 13, 17, 19, 23, 31, 43$ the fields L_p are purely transcendental over $\mathbf{Q}$. The cases $p = 2, 3, 5$ were known long ago, while the rationality of the fields L_7 and L_{11} was proved by Masuda [12]. The cases $p = 13, 17, 19, 23, 31, 43$ are apparently new.

9. In conclusion we consider some fields L_n with n a composite number. It would be very interesting to have a positive answer to the following question.

Let $(m, n) = 1$ and suppose the fields L_m and L_n to be purely transcendental over $\mathbf{Q}$. Is the field L_{mn} also rational over $\mathbf{Q}$?

In [6] it is proved that the fields L_n with $n = 4, 6, 9, 10, 14, 18, 22$ are purely transcendental over $\mathbf{Q}$. Unfortunately, as a result of computational errors it was asserted that the field L_8 is rational over $\mathbf{Q}$. We shall show that this is not so. This will also exhibit the nonrational field L_n with smallest n, since all the L_n with $n < 8$ are purely transcendental over $\mathbf{Q}$.

The field P_8 has a Galois group $\mathfrak{g}$ isomorphic to $\mathbf{Z}_2 \times \mathbf{Z}_2$, with generators σ and τ. The ideal

$$\hat{I}_8 = (8, \sigma - 3, \tau - 5).$$

Let $\mathbf{Z}[\mathfrak{g}] \to \mathbf{Z}$ be the augmentation homomorphism. Under this homomorphism the ideal $\hat{I}_8$ is mapped onto the subgroup $2\mathbf{Z} \subset \mathbf{Z}$ with kernel $\hat{M}_8$, where $\hat{M}_8$ has the following $\mathbf{Z}$-basis:

$$m_1 = 4\sigma - 4, \qquad m_2 = 2\sigma + \tau - 3, \qquad m_3 = \sigma + \sigma\tau - 2.$$

We have an exact sequence of $\mathfrak{g}$-modules

$$0 \to \hat{M}_8 \to \hat{I}_8 \to \mathbf{Z} \to 0,$$

which defines an exact sequence of tori over $\mathbf{Q}$:

$$1 \to G_m \to I_8 \xrightarrow{u} M_8 \to 1.$$

It is clear that the mapping u has a rational section over $\mathbf{Q}$, so the torus I_8 is birationally equivalent over $\mathbf{Q}$ to a direct product of tori $I_8 \approx G_m \times M_8$. We explain how the torus M_8 is presented. Let $\hat{N}$ be the quotient-module $\mathbf{Z}[\mathfrak{g}]/\mathbf{Z}$ and let $\psi\colon \mathbf{Z}[\mathfrak{g}] \to N$ be the natural epimorphism. We write $\psi(1) = n_1$, $\psi(\sigma) = n_2$, $\psi(\tau) = n_3$, $\psi(\sigma\tau) = n_4$. It is clear that $n_1 + n_2 + n_3 + n_4 = 0$. Hence n_1, n_2, n_3 form a $\mathbf{Z}$-basis of the module $\hat{N}$.

We shall prove that the $\mathfrak{g}$-modules $\hat{M}_8$ and $\hat{N}$ are isomorphic. We have

$$\begin{aligned} \sigma n_1 &= n_2, & \tau n_1 &= n_3, \\ \sigma n_2 &= n_1, & \tau n_2 &= -n_1 - n_2 - n_3, \\ \sigma n_3 &= -n_1 - n_2 - n_3, & \tau n_3 &= n_1. \end{aligned}$$

In $\hat{M}_8$ we choose a new $\mathbf{Z}$-basis:

$$\begin{aligned} \bar{m}_1 &= -m_2 + 2m_3, \\ \bar{m}_2 &= -m_1 + 2m_2 - m_3, \\ \bar{m}_3 &= m_1 - m_2, \end{aligned} \qquad \begin{vmatrix} 0 & -1 & 2 \\ -1 & 2 & -1 \\ 1 & -1 & 0 \end{vmatrix} = -1.$$

Then

$$\begin{aligned} \sigma \bar{m}_1 &= \bar{m}_2, & \tau \bar{m}_1 &= \bar{m}_3, \\ \sigma \bar{m}_2 &= \bar{m}_1, & \tau \bar{m}_2 &= -\bar{m}_1 - \bar{m}_2 - \bar{m}_3, \\ \sigma \bar{m}_3 &= -\bar{m}_1 - \bar{m}_2 - \bar{m}_3, & \tau \bar{m}_3 &= \bar{m}_1. \end{aligned}$$

This also proves the isomorphism $\hat{M}_8 \approx \hat{N}$. Since $\hat{N} = \mathbf{Z}[\mathfrak{g}]/\mathbf{Z}$, the torus

$$N = R^{(1)}_{P_8/\mathbf{Q}}(G_m) \approx M_8$$

is a normed hypersurface of dimension 3. Theorem 8 shows that

$$H^1(\mathbf{Q}, \operatorname{Pic} \bar{V}(N)) = H^3(\mathfrak{g}, \mathbf{Z}) = \mathbf{Z}_2$$

for any projective model V of the torus N. From Theorem 3 it then follows that the torus M_8 cannot be stably rational over $\mathbf{Q}$; hence the field L_8 is not purely transcendental over $\mathbf{Q}$. However, nothing else is to be expected of the field P_8, or could be done.

Since in [6] the fields L_8 and L_{12} are considered simultaneously, it will also be necessary to consider the field L_{12}. Here $\mathfrak{g} = \mathbf{Z}_2 \times \mathbf{Z}_2$. The ideal $\hat{I}_{12} = (12, \sigma - 5, \tau - 7)$. Let $\hat{M}$ be the kernel of the map $\hat{I}_{12} \to \mathbf{Z} \cdot \hat{M} = \{m_1, m_2, m_3\}$, where $m_1 = (\sigma + 1)(1 - \tau)$, $m_2 = (2\tau + 1)(1 - \sigma)$ and $m_3 = (\tau + 2)(1 - \sigma)$. The formulas

$$\sigma m_1 = m_1, \qquad \tau m_1 = -m_1,$$

$$\sigma m_2 = -m_2, \qquad \tau m_2 = m_3,$$

$$\sigma m_3 = -m_3, \qquad \tau m_3 = m_2$$

show that the $\mathfrak{g}$-module $\hat{M}$ splits into a direct sum $\hat{M}_1 + \hat{M}_2$, where $\hat{M}_1$ is a module of $\mathbf{Z}$-rank 1, and $\hat{M}_2$ is a module of rank 2. Hence $M = M_1 \times M_2$, and since tori of dimension 1 or 2 are rational over $\mathbf{Q}$, the torus M is rational over $\mathbf{Q}$. Then the torus I_{12} is rational and hence the field L_{12} is purely transcendental over $\mathbf{Q}$. Summarizing: the fields L_n are purely transcendental over $\mathbf{Q}$ for $n = 2, 3, 4, 5, 6, 7, 9, 10, 11, 12, 13, 14, 17, 18, 19, 22, 23, 31, 43$.

BIBLIOGRAPHY

1. C. Chevalley, *On algebraic group varieties*, J. Math. Soc. Japan **6** (1954), 303–324. MR **16**, 672.

2. M. Rosenlicht, *Some rationality questions on algebraic groups*, Ann. Mat. Pura Appl. (4) **43** (1957), 25–50. MR **19**, 767.

3. Ju. I. Manin, *Rational surfaces over perfect fields*, Inst. Hautes Études Sci. Publ. Math. No. 30 (1966), 55–97; English transl., Amer. Math. Soc. Transl. (2) **84** (1969), 137–186. MR **37** #1373.

4. V. E. Voskresenskiĭ, *On the birational equivalence of linear algebraic groups*, Dokl. Akad. Nauk SSSR **188** (1969), 978–981 = Soviet Math. Dokl. **10** (1969), 1212–1215. MR **40** #5623.

5. ———, *Birational properties of linear algebraic groups*, Izv. Akad. Nauk SSSR Ser. Mat. **34** (1970), 3–19 = Math. USSR Izv. **4** (1970), 1–18. MR **41** #6861.

6. ———, *On the question of the structure of the subfield of invariants of a cyclic group of automorphisms of the field $Q(x_1, \cdots, x_n)$*, Izv. Akad. Nauk SSSR Ser. Mat. **34** (1970), 366–375 = Math. USSR Izv. **4** (1970), 371–380. MR **43** #192.

7. R. G. Swan, *Induced representations and projective modules*, Ann. of Math. (2) **81** (1960), 552–578. MR **25** #2131.

8. V. E. Voskresenskiĭ, *Rationality of certain algebraic tori*, Izv. Akad. Nauk SSSR Ser. Mat. **35** (1971), 1037–1046 = Math. USSR Izv. **5** (1971), 1049–1056. MR **44** #4007.

9. ———, *Picard groups of linear algebraic groups*, Studies in Number Theory, no. 3, Izdat. Saratov. Univ., Saratov, 1969, pp. 7–16. (Russian) MR **42** #5999.

10. M. Kneser, *Schwache Approximation in algebraischen Gruppen*, Colloq. sur la Théorie des Groupes Algébriques (Bruxelles, 1962), Librairie Universitaire, Louvain; Gauthier-Villars, Paris, 1962, pp. 41–52.

11. R. G. Swan, *Invariant rational functions and a problem of Steenrod*, Invent. Math. **7** (1969), 148–158. MR **39** #5532.

12. K. Masuda, *Application of the theory of the group of classes of projective modules to the existence problem of independent parameters of invariant*, J. Math. Soc. Japan **20** (1968), 223–232. MR **36** #6393.

Translated by J. S. JOEL

Trudy Mat. Inst. Steklov.
132 (1973)

Proc. Steklov Inst. Math.
132 (1973)

THE ARITHMETIC THEORY OF LINEAR ALGEBRAIC GROUPS AND NUMBER THEORY

V. P. PLATONOV

ABSTRACT. The author presents new results in the arithmetic theory of linear algebraic groups.

Bibliography: 19 items.

Introduction

The arithmetic theory of linear algebraic groups was established as an independent field of mathematics within the last decade. One of its characteristic features is a deep connection with many areas of mathematics, in particular with number theory. Almost all of the fundamental results in the arithmetic theory of linear algebraic groups obtained to the present time, such as the general structure and reduction theorems for arithmetic groups, the solution of the approximation problem, the computation of Tamagawa numbers and their applications, the congruence problem, and so on, are closely connected with classical ideas and results in number theory.

In this report we present new results in the arithmetic theory of linear algebraic groups which have been obtained recently, and some of which have not yet been published.(1) They are also naturally connected with certain questions in number theory.

In this sequel we only consider linear algebraic groups. As usual, Q, Q_p and R are the rational number field, the p-adic numbers and the real numbers, respectively, and $\mathbf{Z}$ and $\mathbf{Z}_p$ are the ring of integers and the ring of p-adic integers.

§1. The problem of the genus in arithmetic groups

Let G be a connected Q-defined algebraic group. We say that the elements $a, b \in G_{\mathbf{Z}}$ belong to the same G-genus if they are conjugate in the groups G_Q and $G_{\mathbf{Z}_p}$. We note that the G_Q-conjugacy of a and b implies that they are automatically conjugate in almost all of the groups $G_{\mathbf{Z}_p}$.

We let $[a]_G$ designate the set of elements in $G_{\mathbf{Z}}$ which belong to the G-genus of a; more briefly, $[a]_G$ denotes the G-genus a. $[a]_G$ is divided into classes relative to $G_{\mathbf{Z}}$-equivalence. The set of classes in $[a]_G$ is denoted by $M([a]_G)$. For every $a \in G_{\mathbf{Z}}$

AMS (MOS) subject classifications. Primary 10C30, 20G30, 20G40; Secondary 20E15.

(1) A significant part of the results in this report have already been published (see [1], [2]).

we may define

$$f_G(a) = \operatorname{ord}(M[a]_G)).$$

The (qualitative) genus problem consists in studying the function $f_G(a)$, and, in particular, obtaining estimates for f_G.

It turns out that always $f_G(a) < \infty$, and we have the following fundamental theorem.

THEOREM 1. *If G is a semisimple Q-isotropy group, i.e.* $\operatorname{rank}_Q G > 0$, *then* $\sup_{a \in H} f_G(a) = \infty$ *for every arithmetic subgroup* $H \subset G_Z$. *In other words, the number of classes in a genus is finite, but it is not totally bounded in the set.*

The proof of Theorem 1 is based on the approximation theorem [3], [4], uses the classification of semisimple algebraic groups [5], and relies on the following result.

THEOREM 2. *Let Ω^a be the centralizer of a in G; let $\nu(\Omega^a)$ be the set of dual classes of the adèle group Ω^0_A which lie in $G_Q G_{A(\infty)}$. Then there exists a unique surjection $\varphi \colon \nu(\Omega^a) \to M([a]_G)$.*

Theorem 2 reduces the proof of Theorem 1 to estimating the number of dual classes in adèle groups of certain reductive groups. It is well known that, in general, estimating the number of dual classes of reductive algebraic groups is a very difficult problem, which as very special cases includes the classical problems of the number of classes in a genus of quadratic forms (Gauss) and the class number of number fields. In our case the required estimates can be obtained on account of the following two central lemmas.

LEMMA 1. *If* $\sup_{a \in R_Z} f_R(a) = \infty$ *for every Q-decomposable three-dimensional subgroup $R \subset G$, then* $\sup_{a \in G_Z} f_G(a) = \infty$.

The proof of this lemma is very technical and lengthy. It uses Theorem 2 and the classification of simple algebraic groups. For the important case of a decomposable group G, there is almost a complete proof in [1].

LEMMA 2. *If $G = SL(2, C)$ then* $\sup_{a \in G_Z} f_G(a) = \infty$.

The proof of Lemma 2 using Theorem 2 quickly reduces to the following number-theoretic result, which was obtained relatively recently by Brauer and others (see [6]).

For a quadratic field $K \supset Q$ of discriminant D, we let $h(K)$ denote the order of the group of ideal classes, and we let $t(D)$ denote the number of prime divisors of D. Then for any $\epsilon > 0$ there exists an infinite sequence of real quadratic fields K_i with discriminants D_i such that $h(K_i) > D_i^{1/2-\epsilon}$. From this it easily follows that $\lim_{i\to\infty}(h(K_i)/2^{t(D_i)}) = \infty$. It is this last result which is used to prove Lemma 2.

REMARK 1. It seems probable that the condition $\operatorname{rank}_Q G > 0$ in Theorem 1 is not essential, although it remains unclear how to prove this (of course, it must be assumed that if $\operatorname{rank}_Q G = 0$ then the group of real points G_R is noncompact).(2)

(2) *Added in proof.* G. V. Matveev has recently proved this for the orthogonal and the unitary groups using analogous considerations.

REMARK 2. At the beginning of the section we indicated that the fundamental theorem gives a qualitative solution of the genus problem. It would be extremely interesting to obtain more precise results on the classes in a genus of the type of the analytic results of Siegel-Weil for the representations of quadratic forms by forms of a given genus which express the global characteristics (weight) in terms of local ones (p-adic densities).

Nothing is known yet in this regard.

§2. The isomorphism problem for arithmetic groups

The classification of arithmetic subgroups of semisimple algebraic groups to within abstract isomorphism is of interest from several points of view. It turns out that, as in many other arithmetic questions, the classification can be reduced to the local case. The purpose of this section is to present the fundamental result of this reduction.

From a conceptual standpoint the fundamental result was to a certain extent inspired by the article [7], while the strong approximation theorem [3], [4] was most important for its proof.

Let G and G' be semisimple Q-defined groups (the case of an arbitrary algebraic number field k reduces to this case by restricting the ground field). Arithmetic subgroups $H \subset G_Q$ and $H' \subset G'_Q$ are said to be locally equivalent if the corresponding p-adic closures $\overline{H}^{(p)}$ and $\overline{H}'^{(p)}$ are isomorphic and the groups G_R and G'_R are locally isomorphic.

These are natural conditions, since they are necessary for arithmetic groups to be isomorphic. If $\operatorname{rank}_Q G > 1$ this follows from a theorem of Raghunathan [8], and if $\operatorname{rank}_Q G = 0$ it was recently proved (even under more general assumptions concerning H) in [9]. The case $\operatorname{rank}_Q G = 1$ is certain not to be an exception (see [10]).

We let $\mathcal{G}(H)$ designate the set of isomorphism classes of arithmetic subgroups of different semisimple Q-groups which are locally equivalent to H. Let $\alpha(H) = \operatorname{Card}(\mathcal{G}(H))$.

Our main purpose is to describe $\mathcal{G}(H)$ and to compute $\alpha(H)$.

In the classical case of integral lattices (see [7]), every lattice H is uniquely determined by its localizations $\overline{H}^{(p)}$ for all p. It is natural to require the analogous condition in the case of the arithmetic group H. Namely, let $H = \bigcap_p \overline{H}^{(p)}$. This implies that H must be a congruence-subgroup of G_Q. The converse is not true in general, although the essential part of the fundamental theorem—the finiteness of $\alpha(H)$—is true for any congruence subgroup of G_Q. In particular, if the congruence theorem holds for G_Z, then $\alpha(H) < \infty$ for every arithmetic subgroup $H \subset G_Q$. We note that the above condition is automatically fulfilled for maximal arithmetic subgroups.

From now on we shall assume all the semisimple groups under consideration to be simply connected. It can be shown that this is actually not a loss of generality, but on the other hand the results become more transparent with this assumption.

We may also assume that G is a simple group and that G_Z is an infinite subgroup, since otherwise we obviously always have $\alpha(H) = 1$.

Let Aut (G) be the group of rational automorphisms of G. Since H is dense in G, we may choose a Q-realization of G such that Aut (G) is given as a linear group in a basis of the lattices $Z(H)$ which is contained in H.

We let $F =$ Aut (G). Let Ш(F) be the Šaferevič-Tate-Borel-Serre([3]) group of F, i.e.

$$Ш(F) = \operatorname{Ker}\left[H^1(Q, F) \xrightarrow{\tau} H^1(R, F) \times \prod_p H^1(Q_p, F)\right],$$

where τ is the canonical morphism.

A theorem of Borel and Serre (see [11], [12]) asserts that Ш(F) is always finite. We let Ш$_F$ denote the order of Ш(F).

Further, let F_A be the adèle group of F. By the Borel finiteness theorem [19], for any open subgroup $\Delta \subset F_A$ the set of dual classes $\Delta \times F_Q$ is finite. We let $\gamma(F, \Delta)$ denote the number of them; if $\Delta = F_{A(\infty)}$, we simply write $\gamma(F)$.

If $\overline{H}_{A(f)}$ is the closure of H in the group of finite adèles $G_{A(f)}$ and if $\overline{H}_{A(\infty)} = G_R \times \overline{H}_{A(f)}$, then the approximation theorem implies

LEMMA 3. *The subgroup* $T(\overline{H}_{A(\infty)}) = \{\sigma \in F_A \mid \sigma(\overline{H}_{A(\infty)}) = \overline{H}_{A(\infty)}\}$ *is open in* F_A.

We can now formulate the fundamental theorem.

THEOREM 3. *Suppose that* $\theta(H)$ *is the set of dual classes* $T(\overline{H}_{A(\infty)} x F_Q)$ *of the adèle group* F_A; $\Phi(H) = \{(\theta, \omega)\}$, *where* $\theta \in \Theta(H)$, $\omega \in Ш(F)$. *Then there exists a unique injection* $G(H) \to \Phi(H)$; *in particular,* $\alpha(H) \leqslant \gamma(F, T(\overline{H}_{A(\infty)}))Ш_F$. *But if* H *is a maximal arithmetic subgroup of* G_Q, *then* $\alpha(H) \leqslant \gamma(F)Ш_F$.

COROLLARY. *If* H *is a maximal arithmetic subgroup of the group* G *of the type* E_8, F_4 *or* G_2 *(only those forms* E_8 *are taken for which the Hasse principle has been proved), then* $\alpha(H) = 1$, *i.e. the groups locally equivalent to* H *are isomorphic.*

In fact, in this case $F = G$ and Ш$(F) = 0$ (concerning the Hasse principle for algebraic groups, see [12], [13]); in addition, by the approximation theorem, all the $\overline{H}^{(p)}$ are maximal compact subgroups in G_{Q_p} (see [14]), so that $\gamma(F, T(\overline{H}_{A(\infty)})) = \gamma(F) = \gamma(G)$. But by the approximation theorem $\gamma(G) = 1$.

---

(3) This long name for the group Ш(F) is not the most generally accepted (and convenient), but, in the author's opinion, it reflects the actual origin of the group. The group Ш(F) is called the Šafarevič-Tate group in the theory of abelian varieties, and this is what is generally accepted.

REMARK. The fundamental theorem takes on an especially complete form for a maximal arithmetic subgroup H. In fact, in this central case it is possible to obtain a significantly stronger result—an essentially complete classification up to conjugacy in G_Q (see [14]).

Theorem 3 implies the following finiteness theorem (the condition $H = \bigcap_p \overline{H}^{(p)}$ is no longer required).

THEOREM 4. *Let* H *be a congruence subgroup in* G_Q. *Then* $\alpha(H) < \infty$. *In particular, if the congruence kernel of the group* G_Z *is finite, then* $\alpha(H) < \infty$ *for any arithmetic subgroup* H.

The following assertion shows that the finiteness condition for the congruence kernel of G_Z cannot be omitted.

THEOREM 5. *If the congruence kernel of* G_Z *is infinite, then there exists an infinite set of arithmetic subgroups* H_i, $i \in I$, *such that* $\overline{H}_i^{(p)} = \overline{H}_j^{(p)}$ *for all* p $(i, j \in I)$.

We conclude this section with some additional remarks on the proof of the fundamental theorem.

The appearance of the group $\text{Ш}(F)$ is natural: it follows from local equivalence that the corresponding p-adic groups are isomorphic, since abstract isomorphism is analytic isomorphism in the p-adic case. This implies that the corresponding semisimple algebraic groups must be rationally Q_p-isomorphic for all p, since the groups under consideration are simply connected.

The dual classes arise from the modification of the corresponding reasoning for lattices in [7] using the strong approximation theorem.

§3. The congruence problem for nonsemisimple and nonarithmetic groups

Let Γ be an arbitrary subgroup of $GL(n, \mathbf{Z})$. The classical congruence problem is stated as follows: Is every subgroup of Γ of finite index a congruence subgroup? In other words: let $\hat{\Gamma}$ and $\overline{\Gamma}$ be the completions of Γ in the topology of subgroups of finite index and in the congruence-topology, respectively, and let $\varphi: \hat{\Gamma} \to \overline{\Gamma}$ be the canonical morphism; is it true that the congruence kernel $c(\Gamma) = \text{Ker}\, \varphi = 1$? In a broader sense, the congruence problem is understood to mean the computation of the congruence kernel $c(\Gamma)$. The congruence problem is the most natural and has the most depth in the case of an arithmetic group Γ. Essential progress has been made in recent years in investigating the congruence problem for arithmetic subgroups of semisimple groups (see the survey by Serre [15] and the memoir by Bass, Milnor and Serre [16]).

In the case of a group Γ which is not isomorphic to an arithmetic group, not a single result was known concerning the congruence problem. The basic result of this section is the following theorem.

THEOREM 6. *If* Γ *is a solvable subgroup of* $GL(n, \mathbf{Z})$, *then every subgroup of* Γ *of finite index is a congruence subgroup.*

The class of integral solvable groups is essentially broader than the class of solvable arithmetic groups (for which the congruence problem has also not been solved), so that, to the best of my knowledge, the above theorem gives the first examples of a positive solution to the congruence problem for groups which are not isomorphic to arithmetic groups.

The proof of Theorem 6, which is in [2], is based on the following general assertion, which reduces the congruence problem for arithmetic subgroups of arbitrary algebraic groups to the case of semisimple groups.

THEOREM 7. *Let* G *be an algebraic* Q*-group, and let* F *be any maximal semisimple* Q*-subgroup of* G. *Then* $c(G_{\mathbf{Z}}) = c(F_{\mathbf{Z}})$.

Then, in turn, to prove Theorem 7, along with general facts in [17] we have essential need of the following result, which was actually obtained by Chevalley [18]: if T is an algebraic Q-torus, then $c(T_{\mathbf{Z}}) = 1$.

The problem naturally arises of generalizing Theorems 6 and 7 to the case of arithmetic rings (see [16]). Namely, let k be an algebraic number field, and let V be the set of all (nonequivalent) nonarchimedean norms of k. For an arbitrary finite set $S \subset V$ we let $O(S)$ denote the subring of k consisting of the elements which are integral for all $v \in V - S$. The rings $O(S)$ are customarily called arithmetic rings.

Theorem 7 generalizes without any changes to the case of arithmetic rings; $c(G_{O(S)}) = c(F_{O(S)})$ for every k-defined algebraic group G and a maximal k-defined semisimple subgroup F. We note that, by Chevalley's theorem, $c(T_{O(S)}) = 1$ for every k-defined algebraic torus.

Nevertheless, it turns out that Theorem 6 is not true in general for arithmetic rings. In this case the congruence problem does not always have a positive solution, although it is possible to compute the congruence kernel precisely and to describe the class of subgroups of finite index which are not congruence subgroups. Namely, we have the following theorem, which was proved jointly by the author and A. A. Šaromet.

THEOREM 8. *Let* Γ *be a solvable subgroup of* $GL(n, O(S))$; *let* S *determine a finite set of prime ideals* $(p_1), \cdots, (p_k)$ *of the ring* $\mathbf{Z}$. Let Γ_u *be the unipotent part of the group* Γ; *let* $P_i(\Gamma_u)$ *be the Sylow pro-*p_i*-subgroups of the compact group* $\hat{\Gamma}_u$. *Then the congruence kernel* $c(\Gamma) \approx \Pi_{i=1}^k P_i(\Gamma_u)$.

In particular, $c(\Gamma) = 1$ *if and only if* Γ_u *is an S-arithmetic group.*

REMARK 1. The true reason for what is at first glance an unexpected difference in the results for solvable groups over $\mathbf{Z}$ and $O(S)$ is explained by the following difference in the structure of these groups: if $\Gamma \subset GL(n, \mathbf{Z})$ and H is an arbitrary subgroup of Γ, then $\hat{H}$ is a closed subgroup of $\hat{\Gamma}$; but if $\Gamma \subset GL(n, O(S))$, then even in the simplest cases $\hat{H}$ may not be a subgroup of $\hat{\Gamma}$; i.e. the induced topology of $\hat{\Gamma}$ in H is weaker than the topology of $\hat{H}$. This distinction can be expressed in abstract terms as follows. Every solvable integral group is semicyclic, but a solvable group over $O(S)$ does not come close to having this property.

REMARK 2. Theorem 7 can be used to answer the well-known question of extending an isomorphism of discrete uniform subgroups of solvable Lie groups to an analytic isomorphism of the corresponding Lie groups. Such an extension does not always exist in the general case, and Theorem 7 is useful for describing the "obstruction" to the extension.

BIBLIOGRAPHY

1. V. P. Platonov, *On the genus problem in arithmetical groups,* Dokl. Akad. Nauk SSSR **200** (1971), 793–796 = Soviet Math. Dokl. **12** (1971), 1503–1507. MR **45** #6824.

2. ———, *The convergence problem for solvable integral groups,* Dokl. Akad. Nauk BSSR **15** (1971), 869–872. (Russian) MR **44** #4009.

3. ———, *The problem of strong approximation and the Kneser-Tits conjecture for algebraic groups,* Izv. Akad. Nauk SSSR Ser. Mat. **33** (1969), 1211–1219 = Math. USSR Izv. **3** (1969), 1139–1148. MR **41** #3485.

4. ———, *Addendum to the paper "The problem of strong approximation and the Kneser-Tits conjecture for algebraic groups",* Izv. Akad. Nauk SSSR Ser. Mat. **34** (1970), 775–777 = Math. USSR Izv. **4** (1970), 784–786. MR **42** #7670.

5. J. Tits, *Classification of algebraic semisimple groups,* Proc. Sympos. Pure Math, vol. 9, Amer. Math. Soc., Providence, R. I., 1966, pp. 33–62. MR **37** #309.

6. N. C. Ankeny, R. Brauer and S. Chowla, *A note on the class-numbers of algebraic number fields,* Amer. J. Math. **78** (1956), 51–61. MR **18**, 565.

7. V. P. Platonov, *Adèle groups and integral representations,* Izv. Akad. Nauk SSSR Ser. Mat. **33** (1969), 155–162 = Math. USSR Izv. **3** (1969), 147–154. MR **41** #183a.

8. M. S. Raghunathan, *Cohomology of arithmetic subgroups of algebraic groups.* I, Ann. of Math. (2) **86** (1967), 409–424. MR **37** #2898.

9. G. D. Mostow, *The rigidity of locally symmetric spaces,* Proc. Internat. Congress Math. (Nice, 1970), vol. 2, Gauthier-Villars, Paris, 1971, pp. 187–198.

10. H. Garland and M. S. Raghunathan, *Fundamental domains for lattices in (R–) rank 1 semisimple Lie groups,* Ann. of Math. (2) **92** (1970), 279–326. MR **42** #1943.

11. A. Borel and J.-P. Serre, *Théorèmes de finitude en cohomologie galoisienne,* Comment. Math. Helv. **39** (1964), 111–164. MR **31** #5870.

12. J.-P. Serre, *Cohomologie galoisienne,* 2nd ed., Lecture Notes in Math., vol. 5, Springer-Verlag, Berlin and New York, 1964; Russian transl., "Mir", Moscow, 1968. MR **31** #4785; **40** #5704.

13. G. Harder, *Über die Galoiskohomologie halbeinfacher Matrizengruppen.* I, II, Math. Z. **90** (1965), 404–428; ibid. **92** (1967), 396–415. MR **32** #7677; **34** #2777.

14. V. P. Platonov, *On the maximality problem for arithmetic groups,* Dokl. Akad. Nauk SSSR **200** (1971), 530–533 = Soviet Math. Dokl. **12** (1971), 1431–1435. MR **45** #1922.

15. J.-P. Serre, *Groupes de congruence,* Séminaire Bourbaki 1966/67, Exposé 330, Benjamin, New York, 1968. MR **37** #10.

16. H. Bass, J. Milnor and J. -P. Serre, *Solution of the congruence subgroup problem for* SL_n $(n \geqslant 3)$ *and* Sp_{2n} $(n \geqslant 2)$, Inst. Hautes Études Sci. Publ. Math. No. **33** (1967), 59–137. MR **39** #5574.

17. A. Borel and Harish-Chandra, *Arithmetic subgroups of algebraic groups,* Ann. of Math. (2) **75** (1962), 485–535. MR **26** #5081.

18. C. Chevalley, *Deux théorèmes d'arithmétique,* J. Math. Soc. Japan **3** (1951), 36–44. MR **13**, 440.

Translated by N. KOBLITZ

Trudy Mat. Inst. Steklov.
132 (1973)

Proc. Steklov Inst. Math.
132 (1973)

ON ARITHMETIC PROPERTIES OF VALUES OF ANALYTIC FUNCTIONS

A. B. ŠIDLOVSKIĬ

ABSTRACT. This article contains a survey of the many results and investigations of various authors pertaining to the theory of transcendence and algebraic independence of values of analytic functions.

Bibliography: 75 items.

§1. The Hermite-Lindemann method

Let **Z** be the ring of integers, **Q**, **C** and **A** the fields of rational numbers, complex numbers and all algebraic numbers, respectively. If **K** is a field or ring, let **K**$[z]$ and **K**(z) denote the ring of polynomials and the field of rational functions over **K**.

Hermite [1] created the first analytic method useful for establishing the transcendence and algebraic independence of values of analytic functions, and it was by this means that he proved the transcendence of e.

By this method Lindemann [2] proved the transcendence of the values of the exponential function e^z when $z \in \mathbf{A}$, $z \neq 0$, and, as a consequence, the transcendence of π and the value of $\ln z$ when $z \in \mathbf{A}$, $z \neq 1$. Lindemann also indicated that the following more general theorem, proved by Weierstrass in [3], could be established analogously.

LINDEMANN-WEIERSTRASS THEOREM. *If* $\alpha_1, \cdots, \alpha_n$ *are distinct numbers in* **A** *and* $c_1, \cdots, c_n \in \mathbf{A}$ *are not all zero, then* $c_1e^{\alpha_1} + \cdots + c_ne^{\alpha_n} \neq 0$.

This theorem is equivalent to the following: *if* $\beta_1, \cdots, \beta_m \in \mathbf{A}$ *are linearly independent over* **Q**, *then the numbers* $e^{\beta_1}, \cdots, e^{\beta_m}$ *are algebraically independent.*

These theorems show that the Hermite-Lindemann method completely settles the question of the transcendence and algebraic independence of the values of the exponential function e^z at algebraic points.

Many well-known mathematicians have improved and simplified the Hermite-Lindemann method without changing its essence. But important new results have not been obtained. There appeared many papers in which the transcendence of values of certain functions was established by reducing the proof to the Lindemann-Weierstrass

AMS (MOS) subject classifications (1970). Primary 10F35, 10-02; Secondary 10F30.

Theorem by various transformations. A relevant bibliography can be found in [4].

The Hermite-Lindemann method can also be used to establish numerical characteristics of transcendence and algebraic independence of values of the functions under consideration at algebraic points. Such characteristics are defined by the lower bounds of the measure of linear independence, transcendence and mutual transcendence, which are the minima of linear forms and polynomials with integral coefficients in the numbers under consideration as functions of their heights and degrees.

By the measure of linear independence of the numbers $\alpha_1, \cdots, \alpha_m$ we mean the function

$$L(\alpha_1, \ldots, \alpha_m; H) = \min |a_1\alpha_1 + \ldots + a_n\alpha_n|, \tag{1}$$

where $a_k \in \mathbf{Z}$, $|a_k| \leqslant H$, $k = 1, \cdots, m$, and $a_1^2 + \cdots + a_m^2 > 0$.

Similarly, by the measure of mutual transcendence of the numbers $\alpha_1, \cdots, \alpha_n$ we mean the function

$$\Phi(\alpha_1, \ldots, \alpha_m; n_1, \ldots, n_m; H) = \min |P(\alpha_1, \ldots, \alpha_m)|, \tag{2}$$

where $P(x_1, \cdots, x_m) \not\equiv 0$ is a polynomial in $x_1, \cdots, x_m$ with coefficients in $\mathbf{Z}$, of height not greater than H, and of degrees with respect to $x_1, \cdots, x_m$ not greater than $n_1, \cdots, n_m$, respectively, and the minimum is taken over all polynomials satisfying these conditions. When $m = 1$ the function

$$\Phi(\alpha, n, H) = \min |P(\alpha)| \tag{3}$$

is called the measure of transcendence of the number α.

As a measure of mutual transcendence we often use the function $\Phi(\alpha_1, \cdots, \alpha_m; n; H)$, which is defined like the function (2), except that n is the degree of $P(x_1, \cdots, x_m)$ with respect to the set $x_1, \cdots, x_m$.

By means of Dirichlet's principle we obtain upper estimates for the functions (1), (2), and (3) valid for any sets of numbers α (see, for example, [4]).

The first measure estimate for the number e was obtained by Borel [5]. He showed that

$$\Phi(e, n, H) > H^{-c\ln\ln H}, \qquad c = c(n) > 0,$$

for bounded n and increasing H.

Pópken [6], [7] proved that

$$\Phi(e, n, H) > H^{-n-\frac{\lambda}{\ln\ln H}}, \qquad \lambda = \lambda(n),$$

and Mahler [8] refined this result and proved that

$$\Phi(e,\ n,\ H) > H^{-n-\frac{cn^2\ln(n+1)}{\ln\ln H}}, \tag{4}$$

where $c > 0$ is an absolute constant.

By this method estimates were obtained for $\Phi(\ln\alpha, n, H)$ and $\Phi(e^{\alpha_1}, \cdots, e^{\alpha_m}; n; H)$ with algebraic $\alpha, \alpha_1, \cdots, \alpha_m$ (see [4]).

The Hermite-Lindemann method is based, essentially, on two properties of the exponential function, the addition theorem and the fact that it is a solution of the very simple linear differential equation $y' = y$. Naturally, there arose the problem of extending the results of Hermite and Lindemann to other functions satisfying more general linear differential equations.

Legendre, Stridsberg, Maier and others investigated the irrationality of values of certain functions which are solutions of second order linear differential equations with polynomial coefficients (see [4]). But for a long time nobody succeeded in proving the transcendence of values of similar functions.

§2. Siegel's method

In 1929 C. L. Siegel [9] published a paper in which he set forth a method for proving the transcendence and algebraic independence of values at algebraic points of a class of entire functions. This method is a direct generalization of the classical Hermite-Lindemann method. In it the author does not require that the functions considered satisfy the addition theorem or any other functional equation, but assumes that they satisfy linear differential equations with coefficients in $\mathbf{C}(z)$.

The inspiration for this method, as Siegel himself says, was the paper of Maier [10] referred to above, which studied the irrationality of the values of certain functions. This paper suggested to him the idea of constructing approximate linear forms with coefficients in $\mathbf{C}(z)$ from the functions under consideration. Essential to Siegel's method is a generalization of an idea of Thue [11] from the theory of approximation of algebraic numbers by rational fractions.

Suppose $\mathbf{K}$ is an algebraic field of degree h over $\mathbf{Q}$, $\alpha \in \mathbf{K}$, and $\alpha_1, \cdots$ α_h are the conjugates of α with respect to $\mathbf{K}$. We will use the notation

$$\overline{|\alpha|} = \max_{1 \leqslant k \leqslant h} |\alpha_k|.$$

Siegel calls the entire function

$$f(z) = \sum_{n=0}^{\infty} c_n \frac{z^n}{n!}$$

an E-function if it satisfies the following conditions:

1) $c_n \in \mathbf{K}$, $n = 0, 1, 2, \cdots$.

2) For any $\epsilon > 0$, $\overline{|c_n|} = O(n^{\epsilon n})$ as $n \to \infty$.

3) There exists a sequence of natural numbers $\{q_n\}$ such that each $q_n c_k$, $k = 0, 1, \cdots, n$, is an algebraic integer, and for any $\epsilon > 0$ we have $q_n = O(n^{\epsilon n})$ as $n \to \infty$.

The class of E-functions forms a ring of functions which is closed under the operations of differentiation, integration from 0 to z, and replacement of the argument z by γz, where $\gamma \in \mathbf{A}$.

Examples of E-functions are an algebraic constant, a polynomial in z with coefficients in A, e^z, $\cos z$, $\sin z$, the Bessel function $J_0(z)$, and others.

A method is known for constructing E-functions satisfying linear differential equations with coefficients in $\mathbf{C}[z]$ (see [9], [12]).

Put

$$[\alpha, 0] = 1, \qquad [\alpha, n] = \alpha(\alpha+1)\dots(\alpha+n-1), \qquad n = 1, 2, \dots.$$

If $a_1, \cdots, a_l, b_1, \cdots, b_m \in \mathbf{Q}$, $b_k \neq 0, -1, -2, \cdots, k = 1, \cdots, m$, and $m - l = t > 0$, we define

$$c_n = \frac{[a_1, n] \dots [a_l, n]}{[b_1, n] \dots [b_m, n]}, \qquad n = 0, 1, 2, \dots,$$

and

$$\varphi(z) = \sum_{n=0}^{\infty} c_n z^{tn}. \tag{5}$$

The function $\varphi(z)$ is a solution of a linear differential equation of order m with coefficients in $\mathbf{C}[z]$. Siegel called the E-functions thus defined hypergeometric E-functions.

The main results obtained by Siegel [9] pertain to the functions

$$K_\lambda(z) = \sum_{n=0}^{\infty} \frac{(-1)^n}{n!(\lambda+1)\dots(\lambda+n)} \left(\frac{z}{2}\right)^{2n}, \qquad \lambda \neq -1, -2, \dots. \tag{6}$$

These functions satisfy the differential equation

$$y'' + \frac{2\lambda+1}{z} y' + y = 0.$$

The function $K_\lambda(z)$ differs from the Bessel function $J_\lambda(z)$ only by a factor of $(z/2)^\lambda/\Gamma(\lambda+1)$, and $K_0(z) = J_0(z)$.

THEOREM 1. *If* $\lambda \in \mathbf{Q}$, $\lambda \neq \pm 1/2, -1, \pm 3/2, -2, \cdots, \alpha \in \mathbf{A}$ *and* $\alpha \neq 0$, *then the numbers* $K_\lambda(\alpha)$ *and* $K'_\lambda(\alpha)$ *are algebraically independent.*

Theorems on the transcendence of function values possess duality.

Lindemann's theorem on the transcendence of e^z when $z \in \mathbf{A}, z \neq 0$, immediately implies the transcendence of $\ln z$ when $z \in \mathbf{A}$, $z \neq 1$. Thus Theorem 1 implies that all algebraic points of the functions $K_\lambda(z)$ and $K'_\lambda(z)$ which are different from zero are transcendental. In particular, all zeros of these functions, hence also the zeros of $J_\lambda(z)$ and $J'_\lambda(z)$, which are different from zero are transcendental.

Theorem 1 also implies that the following continued fraction is transcendental:

$$1+\cfrac{1}{2+\cfrac{1}{3+\ddots}}$$

Siegel proved a theorem more general than Theorem 1.

THEOREM 2. *If* $\lambda_1, \cdots, \lambda_m \in \mathbf{Q}$, $\lambda_k \neq \pm 1/2, -1, \pm 3/2, -2, \cdots, k = 1, \cdots,$ *m, none of the numbers* $\lambda_k - \lambda_l$, $1 \leqslant k < l \leqslant m$, *are integers, and* $\alpha_1, \cdots, \alpha_m$ *are nonzero numbers in* $\mathbf{A}$ *whose squares are distinct, then the* $2mn$ *numbers* $K_{\lambda_k}(\alpha_i)$ *and* $K'_{\lambda_k}(\alpha_i)$, $k = 1, \cdots, m$, $i = 1, \cdots, n$, *are algebraically independent.*

Without giving the proof, Siegel asserted that the following theorem, which generalizes both Theorem 1 and the Lindemann-Weierstrass Theorem, could be proved analogously.

THEOREM 3. *Suppose* $\xi \in \mathbf{A}$, $\xi \neq 0$, *and* $\alpha_1, \cdots, \alpha_n$ *are distinct numbers in* $\mathbf{A}$, *and also that* $P_k(x, y) \not\equiv 0$, $k = 1, \cdots, m$, *are polynomials with algebraic coefficients. Then*

$$P_1[J_0(\xi),\ J'_0(\xi)]\,e^{\alpha_1}+\ldots+P_n[J_0(\xi),\ J'_0(\xi)]\,e^{\alpha_n}\neq 0.$$

This theorem implies that for $\xi \in \mathbf{A}$, $\xi \neq 0$, and $\alpha \in \mathbf{A}, \alpha \neq 0$, the numbers $J_0(\xi)$, $J'_0(\xi)$ and e^α are algebraically independent.

Siegel's method makes it possible to obtain corresponding numerical estimates for the measures of linear independence and mutual transcendence of numbers. Siegel proved that

$$\Phi\,(J_0(\xi),\ J'_0(\xi);\ n;\ H) > cH^{-123n^2h^3}, \tag{7}$$

where $\xi \in \mathbf{A}$, $\xi \neq 0$, h is the degree of ξ, and $c > 0$ is a constant depending only on ξ and h.

He then states, without proofs, that it can be shown that when $r \in \mathbf{Q}, r \neq 0$, we have

$$L(J_0(r),\ J_0'(r),\ 1;\ H) > c_1 H^{2-\varepsilon}, \tag{8}$$

$$\Phi(J_0(r),\ J_0'(r);\ n;\ H) > c_2 H^{1-\frac{(n+1)(n+2)}{2}-\varepsilon} \tag{9}$$

for any $\epsilon > 0$, where $c_1 > 0$ is a constant depending only on r and ϵ, and $c_2 > 0$ is a constant depending only on r, n and ϵ. The latter estimates are close to the natural upper bounds obtained by means of Dirichlet's principle.

In 1946 Siegel lectured at Princeton on transcendental numbers. These lectures were published as a monograph [12]. In it he set forth his method in the form of a general theorem on the algebraic independence of values at algebraic points of a set of E-functions satisfying a system of first order homogeneous linear differential equations with coefficients in $\mathbf{C}(z)$. This monograph does not contain results beyond those appearing in the article [9].

Siegel's general theorem easily implies the Lindemann-Weierstrass Theorem and, with the aid of rather complicated additional investigations, all of the results established by Siegel in 1929.

The meaning of Siegel's general theorem is as follows. Suppose the E-functions $f_1(z), \cdots, f_m(z)$ comprise a solution of the system

$$y_k' = \sum_{i=1}^{m} Q_{k,i} y_i, \qquad k = 1, \ldots, m, \tag{10}$$

where all $Q_{k,i} \in \mathbf{C}(z)$. Henceforth for the system (10) we will denote by $T_0(z)$ a polynomial which is a least common multiple of all the functions $Q_{k,i}$.

Assume that $\alpha \in \mathbf{A}$, $\alpha\, T_0(\alpha) \neq 0$, and the set of products of powers

$$\begin{gathered} f_1^{k_1}(z) \ldots f_m^{k_m}(z),\ k_i \geqslant 0,\ i = 1, \ldots, m, \\ 0 \leqslant k_1 + \ldots + k_m \leqslant N, \end{gathered} \tag{11}$$

for each $N = 1, 2, \cdots$ satisfies a certain analytic normality condition (we omit the definition because of its length). Then the numbers $f_1(\alpha), \cdots, f_m(\alpha)$ are algebraically independent.

The proof of Siegel's general theorem is a very convenient blueprint of the method. However, in essence it only reduces the proof of its arithmetic assertion to the verification of an analytic normality condition. But the latter is very complicated, even in the simplest cases. Siegel succeeded in verifying the normality condition and applying his general theorem only to sets of concrete functions which mainly satisfy second order homogeneous linear differential equations.

Attempts to verify the normality condition for a set of functions satisfying linear differential equations of order greater than 2 introduce insurmountable difficulties. Siegel understood this and mentioned the difficulty of this problem.

Thus his general theorem, despite its apparent generality, has had few applications to concrete functions.

§3. Development and generalization of the method

Until almost the middle of the 50's there appeared no paper connected with Siegel's method beside the paper and monograph [9], [12] of Siegel himself mentioned above.

In 1954 there was published in the article [13] a theorem analogous to Siegel's general theorem, but proved under less restrictive assumptions. Moreover, the possibilities for its application extended to the case of a set of E-functions $f_1(z), \cdots, f_m(z)$ satisfying a system of first order nonhomogeneous linear differential equations

$$y_k' = Q_{k,0} + \sum_{i=1}^{m} Q_{k,i} y_i, \qquad k = 1, \ldots, m, \tag{12}$$

where all $Q_{k,i} \in \mathbf{C}(z)$. A detailed proof is given in [14].

Henceforth in considering the system (12) we will denote by $T(z)$ a polynomial which is a least common multiple of all the functions $Q_{k,i}$.

In the hypothesis of this theorem Siegel's normality condition for the products of powers (11) when $N = 1, 2, \cdots$ is replaced by an irreducibility condition on the same set of functions. The latter condition is also sufficient for the assertion of the theorem, but can be stated very simply and imposes fewer restrictions on the functions under investigation. The verification of this condition for concrete functions is also complicated, but is possible in a number of cases where Siegel's condition cannot be verified.

This theorem provided the means to prove the transcendence and algebraic independence of values of certain E-functions satisfying third and fourth order linear differential equations, and also nonhomogeneous linear equations [13], [14]. We will state the main results.

Consider the functions

$$\varphi_\lambda(z) = \sum_{n=0}^{\infty} \frac{z^n}{(\lambda+1)\ldots(\lambda+n)}, \qquad \lambda \neq -1, -2, \ldots, \tag{13}$$

satisfying the equation

$$y' + \left(\frac{\lambda}{z} - 1\right) y = \frac{\lambda}{z}, \tag{14}$$

and also the functions

$$K_{\lambda,\mu}(z)=\sum_{n=0}^{\infty}\frac{(-1)^n}{(\lambda+1)\dots(\lambda+n)(\mu+1)\dots(\mu+n)}\left(\frac{z}{2}\right)^{2n}, \tag{15}$$

$$\lambda,\ \mu\neq -1,\ -2,\ \dots,$$

and

$$A_\lambda(z)=\int_0^z \frac{K_\lambda(t)-1}{t}\,dt, \tag{16}$$

where the function $K_\lambda(z)$ is defined in (6). The functions (15) and (16) satisfy nonhomogeneous linear differential equations of orders 2 and 3, respectively.

THEOREM 1. *If* $\lambda \in \mathbf{Q}$, *and* $\alpha_1, \cdots, \alpha_n \in \mathbf{A}$ *are linearly independent over* $\mathbf{Q}$, *then the numbers* $\varphi_\lambda(\alpha_1), \cdots, \varphi_\lambda(\alpha_n)$ *are algebraically independent.*

When $\lambda = 0$ this theorem becomes the Lindemann-Weierstrass Theorem.

THEOREM 2. *If* $\lambda, \mu \in \mathbf{Q}$ *and* $\lambda - \mu \neq \pm 1/2, \pm 3/2, \cdots$, *then for any* $\alpha \in \mathbf{A}$, $\alpha \neq 0$, *the numbers* $K_{\lambda,\mu}(\alpha)$ *and* $K'_{\lambda,\mu}(\alpha)$ *are algebraically independent.*

THEOREM 3. *If* $\lambda \in \mathbf{Q}$, $\lambda \neq -1, \pm 1/2, -2, \pm 3/2, \cdots$, *then for any* $\alpha \in \mathbf{A}$, $\alpha \neq 0$, *the numbers* $A_\lambda(\alpha)$, $K_\lambda(\alpha)$ *and* $K'_\lambda(\alpha)$ *are algebraically independent.*

Suppose $f(z)$ is an entire function having a zero of order k at $z = 0$, and $\beta \neq -k, -(k+1), \cdots$. Then the symbol

$$\int^z t^{\beta-1} f(t)\,dt$$

will henceforth denote that primitive function of the integrand which is obtained after term-by-term multiplication of the terms of the power series for $f(z)$ by $z^{\beta-1}$ and term-by-term integration with zero as constant.

We introduce the notation

$$\omega_{\lambda,0}(z)=e^z,$$

$$\omega_k(z)=\omega_{0,k}(z)=1+\int_0^z\frac{\omega_{k-1}(t)-1}{t}\,dt=1+\sum_{n=1}^{\infty}\frac{z^n}{n!\,n^k},\qquad k=1,\dots,r. \tag{17}$$

$$\omega_{\lambda,k}(z)=z^{-\lambda}\int^z t^{\lambda-1}\omega_{\lambda,k-1}(t)\,dt=\sum_{n=0}^{\infty}\frac{z^n}{n!\,(n+\lambda)^k},\qquad k=1,\dots,r.$$

$$H_0(z)=K_\lambda(z),$$

$$H_k(z)=z^{-\beta_k}e^z\int^z t^{\beta_k-1}e^{-t}H_{k-1}(t)\,dt,\qquad \beta_k\neq 0,\ -1,\ -2,\ \dots,$$

$$k=1,\dots,r. \tag{18}$$

The functions (17) and (18) satisfy linear differential equations of orders $k+1$ and $k+2$, respectively.

THEOREM 4. *The numbers* e^{α} *and* $\omega_1(\alpha)$ *are algebraically independent for any* $\alpha \in \mathbf{A}$, $\alpha \neq 0$.

THEOREM 5. *If* $\lambda \in \mathbf{Q}$ *is not an integer, and* $\alpha \in \mathbf{A}$, $\alpha \neq 0$, *then at least one of the two numbers* $\omega_{\lambda,1}(\alpha)$ *and* $\omega_{\lambda,2}(\alpha)$ *is transcendental.*

THEOREM 6. *If* $\lambda \in \mathbf{Q}$ *is not equal to half an odd integer, and* $\beta_1 = \beta_2 = \beta$, $\beta \in \mathbf{Q}$, *then for any* $\alpha \in \mathbf{A}$, $\alpha \neq 0$, *the numbers* $H_2(\alpha)$, $H_1(\alpha)$, $K_\lambda(\alpha)$ *and* $K'_\lambda(\alpha)$ *are algebraically independent.*

In the article [15] there appeared a theorem on the algebraic independence of values of E-functions whose conditions were both necessary and sufficient. A detailed proof of this theorem is given in [16]. It represents a further development and generalization of the functional part of Siegel's method.

FIRST FUNDAMENTAL THEOREM. *Suppose the set of E-functions*

$$f_1(z), \ldots, f_m(z), \tag{19}$$

$m \geqslant 1$, *is a solution of the system of differential equations* (12), *where all* $Q_{k,i} \in \mathbf{C}(z)$, *and* $\alpha \in \mathbf{A}$, $\alpha T(\alpha) \neq 0$. *Then in order that the numbers*

$$f_1(\alpha), \ldots, f_m(\alpha) \tag{20}$$

be algebraically independent it is necessary and sufficient that the functions (19) *be algebraically independent over* $\mathbf{C}(z)$.

COROLLARY. *If the E-functions* (19) *are algebraically independent over* $\mathbf{C}(z)$ *and satisfy the conditions of the First Fundamental Theorem, then the following assertions are true*:

1) *Each of the numbers* (20) *is transcendental.*

2) *By the duality of the theorems on the transcendence of function values the zeros and, in general, all algebraic points of the functions* (19) *different from zero and the zeros of* $T(z)$ *are transcendental.*

Suppose the E-function $f(z)$ is a solution of the linear differential equation

$$P_m y^{(m)} + \ldots + P_1 y' + P_0 y = Q, \tag{21}$$

where all coefficients belong to $\mathbf{C}[z]$. Let

$$f_1(z) = f(z), \qquad f_k(z) = f^{(k-1)}(z), \qquad k = 2, \ldots, m.$$

The set of these E-functions comprises a solution of the system of differential equations

$$y'_k = y_{k+1}, \qquad k=1, \ldots, m-1,$$
$$y'_m = -\frac{P_{m-1}}{P_m} y_m - \cdots - \frac{P_0}{P_m} y_1 + \frac{Q}{P_m}.$$

Hence the First Fundamental Theorem is applicable to it, and we have the following theorem.

THEOREM 7. *Suppose the E-function* $f(z)$ *is a solution of the mth order linear differential equation* (21), *where all coefficients belong to* $\mathbf{C}[z]$, *and* $\alpha \in \mathbf{A}$, $\alpha P_m(\alpha) \neq 0$. *Then in order that the numbers* $f(\alpha), f'(\alpha), \cdots, f^{(m-1)}(\alpha)$ *be algebraically independent it is necessary and sufficient that* $f(z)$ *not satisfy any algebraic differential equation over the field* $\mathbf{C}(z)$ *of order less than* m.

When $m = 1$ we obtain from this theorem

THEOREM 8. *Suppose the E-function* $f(z)$ *is transcendental and is a solution of the first order linear differential equation*

$$P_1 y' + P_0 y = Q, \tag{22}$$

where $P_0, P_1, Q \in \mathbf{C}(z)$, *and* $\alpha \in \mathbf{A}$, $\alpha P_1(\alpha) \neq 0$. *Then* $f(\alpha)$ *is transcendental.*

In particular, the E-function e^z satisfies the equation $y' = y$. Hence e^α is transcendental for any $\alpha \in \mathbf{A}$, $\alpha \neq 0$.

The function $\varphi_\lambda(z)$ (13) for $\lambda \in \mathbf{Q}$ is an E-function and satisfies equation (14). Hence the number $\varphi_\lambda(\alpha)$ is transcendental for any $\lambda \in \mathbf{Q}$, $\lambda \neq -1, -2, \cdots$, and any $\alpha \in \mathbf{A}$, $\alpha \neq 0$.

If $\alpha_1, \cdots, \alpha_m \in \mathbf{A}$ are linearly independent over $\mathbf{Q}$, it is easy to verify that the functions $e^{\alpha_1 z}, \cdots, e^{\alpha_m z}$ are algebraically independent over $\mathbf{C}(z)$. They comprise a solution of the system $y'_k = \alpha_k y_k$, $k = 1, \cdots, m$. Hence by the First Fundamental Theorem the numbers $e^{\alpha_1}, \cdots, e^{\alpha_m}$ are algebraically independent. This means that the Lindemann-Weierstrass Theorem is included in the First Fundamental Theorem.

The natural condition of algebraic independence over $\mathbf{C}(z)$ of the set of functions under consideration is much simpler than the conditions of normality and irreducibility. There are methods which allow us to establish the algebraic independence of certain classes of analytic functions. We can develop these methods and create new ones. Hence by means of the First Fundamental Theorem we can establish the transcendence and algebraic independence of values of E-functions satisfying linear differential equations of any orders.

The proofs of any of Siegel's results stated in §2 and of Theorems 1–6 of this presnet section can be greatly simplified using the First Fundamental Theorem, and Theorems 1–6 can be generalized.

Analogously to the First Fundamental Theorem can be proved

THEOREM 9. *Suppose the set of E-functions* (19) *is a solution of the system of homogeneous linear differential equations* (10), *where all* $Q_{k,\,i} \in \mathbf{C}(z)$, *and* $\alpha \in \mathbf{A}$, $\alpha T_0(\alpha) \neq 0$. *Then in order that the numbers* (20) *not be related by a homogeneous algebraic equation with coefficients in* $\mathbf{A}$ *it is necessary and sufficient that the functions* (19) *not be related by a homogeneous algebraic equation with coefficients in* $\mathbf{C}[z]$.

When $m > 1$ we obtain from this theorem a number of corollaries pertaining to a set of E-functions (19) not related by a homogeneous algebraic equation.

1. *None of the numbers* (20) *is equal to zero; hence all zeros of each of the functions* (19) *are transcendental, except possibly for zeros coinciding with the number* 0 *and the zeros of* $T_0(z)$.

2. *The* $m-1$ *numbers*

$$\frac{f_k(\alpha)}{f_s(\alpha)}, \qquad k=1, \ldots, m, \quad k \neq s,$$

for $s = 1, \cdots, m$, *are algebraically independent.*

3. *At least* $m-1$ *of the numbers* (20) *are transcendental.*

4. *If one of the numbers* (20) *is algebraic, the remaining* $m-1$ *are algebraically independent.*

As a very example of the way in which the theorem is applied, consider the functions

$$y_1 = \sin z, \qquad y_2 = \cos z.$$

They are a solution of the system

$$y_1' = y_2, \qquad y_2' = -y_1,$$

transcendental and algebraically dependent, since $y_1^2 + y_2^2 = 1$. A homogeneous algebraic equation between y_1 and y_2 over $\mathbf{C}[z]$ is impossible, since otherwise the function $y_1/y_2 = \tan z$ would be algebraic. Hence, by Theorem 9, when $\alpha \in \mathbf{A}$, $\alpha \neq 0$, the number $(\tan \alpha)$ is transcendental. But then from the realtion $\sin^2\alpha + \cos^2\alpha = 1$ we obtain that $\sin \alpha$ and $\cos \alpha$ are also transcendental.

It is easy to see that the First Fundamental Theorem is included in Theorem 9.

In 1955 [15] there were published a number of applications of the First Fundamental Theorem to concrete functions. A detailed proof of these results is given in [17]. We will state the main ones.

The following theorem completely settles the question of the algebraic independence of the values of the set of functions $\varphi_\lambda(z)$ (13) for distinct values of λ and at distinct algebraic points.

THEOREM 10. *Suppose* $\lambda_0 \in \mathbf{Q}$, $\lambda_0 \geqslant 0$; $\lambda_1, \cdots, \lambda_m$ *are fractions in* $\mathbf{Q}$ *such that the differences* $\lambda_{j_1} - \lambda_{j_2}$, $1 \leqslant j_1 < j_2 \leqslant m$, *are not integers*; $\beta_1, \cdots, \beta_m \in \mathbf{A}$ *are linearly independent over* $\mathbf{Q}$; $\alpha_1, \cdots, \alpha_n \in \mathbf{A}$ *are distinct and different from zero. Then the* $(m+1)n$ *numbers* $\varphi_{\lambda_0}(\alpha_i)$ *and* $\varphi_{\lambda_j}(\alpha_i)$, $i = 1, \cdots, n$, $j = 1, \cdots, m$, *are algebraically independent.*

Next consider the functions

$$\varphi_{s,\lambda}(z) = \sum_{n=0}^{\infty} \frac{z^n}{(\lambda+1)\ldots(\lambda+n)(\mu_1+n)\ldots(\mu_s+n)},$$
$$\lambda, \mu_i \neq -1, -2, \ldots, \qquad i = 1, \ldots, s, \qquad s = 0, 1, \ldots, r. \tag{23}$$

We can prove a general theorem on the algebraic independence of the set of such functions with distinct $\lambda \in \mathbf{Q}$ and distinct sets of numbers $\mu_1, \cdots, \mu_r \in \mathbf{Q}$ at distinct algebraic points.

In particular, this theorem includes the following assertion about the functions (17): suppose λ is a fraction in $\mathbf{Q}$ or zero, and $\alpha \in \mathbf{A}$, $\alpha \neq 0$. Then the $m+1$ numbers $e^\alpha, \omega_{\lambda,1}(\alpha), \cdots, \omega_{\lambda,r}(\alpha)$ are algebraically independent.

From this we obtain theorems on the values of the "incomplete" Euler-Poisson integral and Euler gamma function

$$\Phi(z) = \int_0^z e^{-t^2} dt, \qquad F(z) = \int_0^z t^{p-1} e^{-t} dt.$$

1. *The numbers* $\Phi(\alpha)$ *and* e^{α^2} *are algebraically independent for any* $\alpha \in \mathbf{A}$, $\alpha \neq 0$.

2. *The numbers* $F(\alpha)$ *and* e^α *are algebraically independent for any fractional* $p \in \mathbf{Q}$, $p > 0$, *and any* $\alpha \in \mathbf{A}$, $\alpha \neq 0$.

Analogously, for any $\lambda > 0$, $\lambda \in \mathbf{Q}$, and any $z \in \mathbf{A}$, $z \neq 0$, the integral $\int_0^z e^{-t^\lambda} dt$ assumes a transcendental value. If, moreover, $\lambda \neq 1/k$, where $k \in \mathbf{Z}$, then the value of the integral and the number e^{z^λ} are algebraically independent.

We can then prove an analogous theorem pertaining to the values of the set of functions

$$K_{0,\lambda}(z) = K_\lambda(z),$$

$$K_{l,\lambda}(z) = \sum_{n=0}^{\infty} \frac{(-1)^n}{n!(\lambda+1)\ldots(\lambda+n)(\mu_1+n)\ldots(\mu_l+n)} \left(\frac{z}{2}\right)^{2n},$$

$$\lambda \neq -1, -2, \ldots; \qquad \mu_i \neq -1, 2, \ldots; \qquad i = 1, \ldots, l; \qquad l = 1, \ldots, r.$$

This theorem includes Theorem 3.

The question of the algebraic independence at algebraic points of the values of the functions (18) is completely settled.

THEOREM 11. *Suppose* $\lambda \in \mathbf{Q}$ *is not equal to half an odd integer*; $\beta_1, \cdots, \beta_r \in \mathbf{Q}$ *and the differences* $\beta_{l_1} - \beta_{l_2}$, $1 \leqslant l_1 < l_2 \leqslant r$, *are not nonzero integers*; $\alpha \in \mathbf{A}$, $\alpha \neq 0$. *Then the* $r + 3$ *numbers* e^{α}, $K_\lambda(\alpha)$, $K'_\lambda(\alpha)$, $H_{l,\lambda}(\alpha)$, $l = 1, \cdots, r$, *are algebraically independent.*

In [18] it is shown that an analogue of the Lindemann-Weierstrass Theorem is true for any E-function which is a solution of a first order linear differential equation, and even in a stronger form.

THEOREM 12. *Suppose the E-function* $f(z)$ *is transcendental and satisfies the differential equation* (22), *where* $P_0, P_1, Q \in \mathbf{C}(z)$, *and* $\alpha_1, \cdots, \alpha_m \in \mathbf{A}$ *are linearly independent over the field* $\mathbf{Q}$ *and* $\Pi_1^m \alpha_i P_1(\alpha_i) \neq 0$. *Then the numbers* $f(\alpha_1), \cdots f(\alpha_m)$ *are algebraically independent. If no solution of* (22) *belongs to* $\mathbf{C}(z)$, *then* $f(\alpha_1), \cdots, f(\alpha_m)$ *are algebraically independent for any distinct* $\alpha_1, \cdots, \alpha_m \in \mathbf{A}$ *such that* $\Pi_1^m \alpha_i P_1(\alpha_i) \neq 0$.

The theorem is true, in particular, for the functions (13) with rational λ.

Proving the algebraic independence of functions over $\mathbf{C}(z)$ in all of the above theorems is accomplished by algebra-analytic methods, making extensive use of the differential properties of the functions. These methods represent a further development of the method employed by Siegel in [9].

A number of applications of the fundamental theorem to concrete functions are given in articles [19] – [21]. Here the proof of the algebraic independence of functions is accomplished by an arithmetic method based on comparison of coefficients of the power series of the functions under consideration. This method is also a generalization of a method employed by Siegel in [9].

In [20] there is a modern proof of the algebraic independence of the values of the functions (17), and the following assertion is also proved.

Let

$$\psi_k(z) = \sum_{n=0}^{\infty} \frac{1}{(n!)^k} \left(\frac{z}{k}\right)^{kn}, \quad k \geqslant 1. \tag{24}$$

THEOREM 13. *For any* $\alpha \in \mathbf{A}$, $\alpha \neq 0$, *the* $r(r+1)/2$ *numbers* $\psi_k^{(l)}(\alpha)$, $l = 0, 1, \cdots, k-1$; $k = 1, \cdots, r$, *are algebraically independent.*

The function (24) is a solution of a kth order linear differential equation with coefficients in $\mathbf{C}[z]$.

In [21] this last result is generalized. Let

$$A_{\lambda, k}(z) = \sum_{n=0}^{\infty} \frac{1}{[(\lambda+1)\dots(\lambda+n)]^k}\left(\frac{z}{k}\right)^{kn}, \quad \lambda \neq -1, -2, \dots; \quad k \geqslant 1. \quad (25)$$

THEOREM 14. *Suppose* $\lambda \in \mathbf{Q}$, *and* $\alpha \in \mathbf{A}$, $\alpha \neq 0$. *Then the numbers* $A_{\lambda, k}(\alpha), A'_{\lambda, k}(\alpha), \cdots, A^{(k-1)}_{\lambda, k}(\alpha)$ *are algebraically independent.*

One also finds there a proof of a theorem on the algebraic independence of values of a set of functions (25) with distinct values of λ and k and their successive derivatives, with certain restrictions on λ and k.

§4. Further development and generalization of the method in the case of a set of functions related by algebraic equations over the field of rational functions

Naturally, there arose the problem of extending the arithmetic results mentioned in §3 to a set of E-functions related by algebraic equations over the field $\mathbf{C}(z)$.

Note that Theorem 9 of §3 already admits the possibility that the functions under consideration are related by a nonhomogeneous algebraic equation.

In [22], [23] there are obtained a number of theorems on the transcendence and algebraic independence of values of a set of E-functions related by one algebraic equation, and in [24], [25] there are general theorems of this type for a set of E-functions related by any number of algebraic equations over the field $\mathbf{C}(z)$. A detailed proof of the latter theorems is given in [26]. We state below the main ones. Both the functional and arithmetic part of the method have been altered and generalized.

SECOND FUNDAMENTAL THEOREM. *Suppose the set of E-functions* (19), $m \geqslant 1$, *comprises a solution of the system of linear differential equations* (12), *where all* $Q_{k,i} \in \mathbf{C}(z)$, $l \in \mathbf{Z}$, $0 \leqslant l \leqslant m$, *and* $\alpha \in \mathbf{A}$, $\alpha T(\alpha) \neq 0$. *Then in order that among the numbers* (20) *the greatest number of algebraically independent ones be equal to* l *it is necessary and sufficient that among the functions* (19) *the greatest number of algebraically independent ones over* $\mathbf{C}(z)$ *also be equal to* l.

When $l = m$ this theorem becomes the First Fundamental Theorem. It is easy to reformulate the Second Fundamental Theorem for the case of an E-function $f(z)$ satisfying the linear differential equation (21). Analogously to Theorem 9 we can state and prove a homogeneous analogue of the Second Fundamental Theorem.

There is a certain indefiniteness in the Second Fundamental Theorem, since it does not answer the question of the algebraic independence of values of l concrete algebraically independent functions occurring in the set (19). But with the aid of various additional investigations this indefiniteness can be made more precise.

We say that a statement is valid for almost all $\alpha \in \mathbf{A}$ if it is valid for all $\alpha \in \mathbf{A}$ with the possible exception of finitely many.

THEOREM 1. *Suppose the set of E-functions* (19), $m \geqslant 1$, *comprises a solution of the system of differential equations* (12), *where all* $Q_{k,i} \in \mathbf{C}(z)$, $l \in \mathbf{Z}$, $1 \leqslant l \leqslant m$, *and the functions* $f_1(z), \cdots, f_l(z)$ *are algebraically independent over* $\mathbf{C}(z)$. *Then, for almost all* $\alpha \in \mathbf{A}$, l *of the numbers* $f_1(\alpha), \cdots, f_m(\alpha)$ *are algebraically independent.*

This theorem implies

THEOREM 2. *Suppose the E-function* $f(z)$ *is transcendental and is a solution of the linear differential equation* (21), *all of whose coefficients belong to* $\mathbf{C}[z]$. *Then for almost all* $z \in \mathbf{A}$, *the values of* $f(z)$ *and all of its successive derivatives are transcendental.*

These theorems solve the problems of the transcendence and algebraic independence of values of the functions under consideration in general. But they do not specifically indicate those exceptional points at which they are not valid. In a number of cases we can prove general theorems in which either there are no additional exceptional points or such points are specified.

THEOREM 3. *Suppose the set of E-functions* (19), $m \geqslant 2$, *is a solution of the system of differential equations* (12), *where all* $Q_{k,i} \in \mathbf{C}(z)$. *Suppose also that among these functions the greatest number of algebraically independent ones over* $\mathbf{C}(z)$ *is* l, $1 \leqslant l \leqslant m-1$, *the functions* $f_1(z), \cdots, f_l(z)$ *are algebraically independent, and the functions* $f_{l+1}(z), \cdots, f_m(z)$ *are related to* $f_1(z), \cdots, f_l(z)$ *by the algebraic equations*

$$P_\nu = P_\nu[f_1(z), \ldots, f_l(z), f_\nu(z)] = 0, \qquad \nu = l+1, \ldots, m, \tag{26}$$

where P_ν *is an irreducible polynomial in* $l+1$ *variables, including* $f_\nu(z)$, *with coefficients in* $\mathbf{C}$. *Furthermore, suppose* $\alpha \in \mathbf{A}$, $\alpha T(\alpha) \neq 0$. *Then the* l *numbers* $f_1(\alpha), \cdots, f_l(\alpha)$ *are algebraically independent.*

If the coefficients of the P_ν *belong to* $\mathbf{C}[z]$, *and for each* $\nu = l+1, \cdots, m$ *the term* $A_\nu(z) f_\nu^{s_\nu}(z)$, $A_\nu(z) \in \mathbf{C}[z]$, $A_\nu(z) \not\equiv 0$, *in the set of homogeneous terms of* P_ν *contains the highest power* s_ν *with respect to* $f_1(z), \cdots, f_\nu(z)$, *and also* $\Pi_{\nu=l+1}^{m} A_\nu(\alpha) \neq 0$, *then the numbers* $f_1(\alpha), \cdots, f_l(\alpha)$ *are algebraically independent.*

In the general case, it follows from Lemma 9 of [26] that the additional exceptional points at which the transcendence and algebraic independence of the function values are not claimed are determined from equations (26). For each concrete set of functions we can, in general, having found the equations (26), determine all exceptional algebraic points.

As a very simple example of the way in which the above theorems are applied we can again use the functions $\sin z$ and $\cos z$.

In the paper [21] the following sets of functions are considered:

$$\gamma_{\lambda,0}(z) = \cos z,$$

$$\gamma_{0,k}(z) = 1 + \int_0^z \frac{\gamma_{k-1}(t) - 1}{t}\,dt = 1 + \sum_{n=1}^{\infty} \frac{(-1)^n z^{2n}}{(2n)!\,(2n)^k}, \qquad k = 1, \ldots, m,$$

$$\gamma_{\lambda,k}(z) = z^{-\lambda}\int_0^z t^{\lambda-1}\gamma_{\lambda,k-1}(t)\,dt = \sum_{n=0}^{\infty} \frac{(-1)^n z^{2n}}{(2n)!\,(2n+\lambda)^k}, \; k = 1, \ldots, m, \quad \lambda \neq 0, -1, -2, \ldots,$$

$$\delta_{\lambda,0}(z) = \sin z,$$

$$\delta_{\lambda,k}(z) = z^{-\lambda}\int_0^z t^{\lambda-1}\delta_{\lambda,k-1}(t)\,dt = \sum_{n=0}^{\infty} \frac{(-1)^n z^{2n-1}}{(2n-1)!\,(2n-1+\lambda)^k},$$

$$k = 1, \ldots, m, \quad \lambda \neq -1, -2, \ldots.$$

The functions $\gamma_{\lambda,k}(z)$ and $\delta_{\lambda,k}(z)$ are solutions of linear differential equations of order $k+2$ with polynomial coefficients. Concerning their values we can prove

THEOREM 4. *Suppose* λ *is a fractional number in* $\mathbf{Q}$ *or zero, and* $\alpha \in \mathbf{A}$, $\alpha \neq 0$. *Then the following assertions are true.*

1) *The* $m+1$ *numbers* $\gamma_{\lambda,k}(\alpha)$, $k = 0, 1, \cdots, m$, *are algebraically independent.*

2) *The* $m+1$ *numbers* $\delta_{\lambda,k}(\alpha)$, $k = 0, 1, \cdots, m$, *are algebraically independent.*

Consider the "incomplete" integral sine and integral cosine

$$\operatorname{si} z = \int_0^z \frac{\sin t}{t}\,dt, \qquad \operatorname{ci} z = \int_0^z \frac{1 - \cos t}{t}\,dt$$

and the "incomplete" Fresnel integrals

$$F_1(z) = \int_0^z \sin t^2\,dt, \qquad F_2(z) = \int_0^z \cos t^2\,dt.$$

Theorem 4 implies that for any $\alpha \in \mathbf{A}$, $\alpha \neq 0$, the following are true:

1) *The numbers* $\operatorname{si}\alpha$ and $\sin\alpha$, *and also the numbers* $\operatorname{ci}\alpha$ *and* $\cos\alpha$, *are algebraically independent.*

2) *The numbers* $F_1(\alpha)$ *and* $\sin\alpha^2$, *and also the numbers* $F_2(\alpha)$ *and* $\cos\alpha^2$, *are algebraically independent.*

Similarly, the integrals $\int_0^z \sin t^\lambda\, dt$ and $\int_0^z \cos t^\lambda dt$ for any $\lambda \in \mathbf{Q}$, $\lambda > 0$, and any $z \in \mathbf{A}$, $z \neq 0$, assume transcendental values. If, in addition, $\lambda \neq 1/k$, where $k \in \mathbf{Z}$, then the value of each integral and the corresponding number $\sin z^\lambda$ or $\cos z^\lambda$ are algebraically independent.

Theorem 2 of §2, established by Siegel, implies that if $\lambda_1, \lambda_2 \in \mathbf{Q}$ are different from $-1, \pm 1/2, -2, \pm 3/2, \cdots$, and $\lambda_1 \pm \lambda_2$ are not integers, then the numbers $K_{\lambda_1}(\alpha)$, $K'_{\lambda_1}(\alpha)$, $K_{\lambda_2}(\alpha)$ and $K'_{\lambda_2}(\alpha)$ are algebraically independent for any $\alpha \in \mathbf{A}$, $\alpha \neq 0$. If we take $\lambda_2 = -\lambda_1$, the analogous result is not true, since the corresponding four functions turn out to be algebraically dependent over $\mathbf{C}(z)$.

With the aid of the Second Fundamental Theorem the following result is proved in [26]:

THEOREM 5. *Suppose* $\lambda \in \mathbf{Q}$, $\lambda \neq 0, \pm 1/2, \pm 1, \pm 3/2, \cdots$, *and* $\alpha \in \mathbf{A}$, $\alpha \neq 0$. *Then any three of the four numbers* $K_\lambda(\alpha)$, $K'_\lambda(\alpha)$, $K_{-\lambda}(\alpha)$ *and* $K'_{-\lambda}(\alpha)$ *are algebraically independent.*

In [27], Theorem 2 of §2 is generalized with regard to the connections between the functions $K_\lambda(z)$, $K_{-\lambda}(z)$ and their derivatives.

THEOREM 6. *Suppose* $\lambda \in \mathbf{Z}$, $\lambda \geqslant 0$; $\lambda_1, \cdots, \lambda_m$ *are fractional numbers of* $\mathbf{Q}$, *none of which is half an odd integer, such that the* $\lambda_i - \lambda_j$, $1 \leqslant i < j \leqslant m$, *are not integers*; μ_0 *is half an odd integer*; $\mu_1, \cdots, \mu_m \in \mathbf{Q}$ *are such that the sums* $\mu_k + \lambda_k$ *are integers for all* $k = 1, \cdots, m$. *Also, suppose* $\beta_1, \cdots, \beta_n \in \mathbf{A}$ *are linearly independent over the field* $\mathbf{Q}$, *and* $\alpha_1, \cdots, \alpha_n$ *are nonzero numbers in* $\mathbf{A}$ *whose squares are distinct. Then the* $3n(m+1)$ *numbers* $K_{\lambda_0}(\alpha_i)$, $K'_{\lambda_0}(\alpha_i)$, $K_{\lambda_k}(\alpha_i)$, $K'_{\lambda_k}(\alpha_i)$, $K_{\mu_0}(\beta_i)$, $K_{\mu_k}(\beta_i)$, $i = 1, \cdots, n$; $k = 1, \cdots, m$, *are algebraically independent.*

§5. Refinement of the method's fundamental lemma

In Siegel's method the lemma on estimating the rank of a set of values of the functions under investigation is fundamental (the rank is the greatest number of numbers linearly independent with respect to the considered algebraic field). Its proof involves the construction of functional and numerical linear forms in the functions under investigation and their values, and the estimation of the values of these forms and their coefficients. The general theorems on the algebraic independence of the function values follow from the rank lemma very simply.

In [28], thanks to a better choice of the method's fundamental parameter, several estimates were improved in the auxiliary lemmas. Hence the estimate in the rank lemma was significantly improved. In particular, if as the ground field we take an imaginary quadratic field $\mathbf{J}$ over $\mathbf{Q}$ or the field $\mathbf{Q}$ itself, then under the hypothesis of the lemma we obtain the largest possible value for the rank.

E-functions whose power series coefficients belong to an algebraic field $\mathbf{K}$ will be called KE-functions.

In this section $\mathbf{K}$ will denote either the field $\mathbf{J}$ or the field $\mathbf{Q}$.

The following results are established in [28].

THEOREM 1. *Suppose the set of KE-functions* (19) *comprises a solution of the system* (12), *where all* $Q_{k,i} \in \mathbf{C}(z)$, *and* $\alpha \in \mathbf{K}$, $\alpha T(\alpha) \neq 0$. *Then the following are true.*

1) *In order that the numbers* (20) *and the number* 1 *be linearly independent over* $\mathbf{K}$ *it is necessary and sufficient that the functions* (19) *together with* $f(z) \equiv 1$ *be linearly independent over* $\mathbf{C}(z)$. *If the latter condition is satisfied, each of the numbers* (20) *lies outside* $\mathbf{K}$ *and so is irrational, and none of the functions has zeros in* $\mathbf{K}$ *different from zero and the zeros of* $T(z)$.

2) *In order that the numbers* (20) *not be related by an algebraic equation of degree* k, $k \geqslant 1$, *with coefficients in* $\mathbf{K}$ *it is necessary and sufficient that the functions* (19) *not be related by an algebraic equation of degree* k *with coefficients in* $\mathbf{C}(z)$.

3) *If the system* (12) *is homogeneous, then in order that the numbers* (20) *not be related by a homogeneous algebraic equation of degree* k, $k \geqslant 1$, *with coefficients in* $\mathbf{K}$ *it is necessary and sufficient that the function* (19) *not be related by a homogeneous algebraic equation of degree* k *with coefficients in* $\mathbf{C}(z)$. *In particular, if the latter condition is satisfied with* $k = 1$, *none of the functions has zeros in* $\mathbf{K}$ *different from zero and the zeros of* $T_0(z)$.

In the case of the field $\mathbf{J}$, the First Fundamental Theorem is the limiting case of Theorem 1.

Analogously, we can state and prove the theorem in the case where among the numbers (20) there are l, $l < m$, which are linearly independent and not related by an algebraic equation or a homogeneous algebraic equation of degree k.

Theorem 1 can easily be reformulated in the case of an E-function satisfying a linear differential equation with polynomial coefficients and the sequence of its successive derivatives. We state the simplest part.

THEOREM 2. *Suppose the KE-function* $f(z)$ *is a solution of the mth order linear differential equation* (21), $m \geqslant 1$, *all of whose coefficients belong to* $\mathbf{C}(z)$, *and does not satisfy a linear differential equation of the same type of order less than* m (*or, if* $m = 1$, *does not belong to* $\mathbf{C}[z]$), *and* α *is any number in* $\mathbf{K}$ *such that* $\alpha P_m(\alpha) \neq 0$. *Then the numbers* $f(\alpha), f'(\alpha), \cdots, f^{(m-1)}(\alpha)$ *are, together with the number* 1, *linearly independent over* $\mathbf{K}$; *hence each lies outside* $\mathbf{K}$ *and is therefore irrational.*

A refinement of the rank lemma allows us to improve estimates of the measures of linear independence and mutual transcendence. In the case of the field **J**, close to the natural bounds are obtained. We will discuss this further in §6.

§6. Estimates of measures

Using Siegel's paper [9] and the results of [16] and [26], it is very easy to obtain estimates for the mutual transcendence measure of values of E-functions at algebraic points. The author has discussed this in [26] and in a lecture at the Fourth All-Union Mathematical Congress in 1961 [29].

A general theorem of this type was published by S. Lang in 1962 [30].

THEOREM 1. *Suppose the KE-functions* (19) *are algebraically independent over* $\mathbf{C}(z)$ *and comprise a solution of the system of homogeneous linear differential equations* (10), *where all* $Q_{k,i} \in \mathbf{C}(z)$, *and* $\alpha \in \mathbf{K}$, $\alpha T_0(\alpha) \neq 0$. *Then*

$$\Phi(f_1(\alpha), \ldots, f_m(\alpha); n; H) > cH^{-\gamma n^m},$$

where $c > 0$ *and depends only on the functions* (19), m, α *and* n, *and* $\gamma > 0$ *depends only on* m *and the degree of the field* $\mathbf{K}$ *over* $\mathbf{Q}$.

The author does not know if this result is also true in the nonhomogeneous case, i.e. when the functions (19) satisfy the system (12).

Suppose again that the field $\mathbf{K}$ is either $\mathbf{J}$ or $\mathbf{Q}$. Using the results of [28], we established the following theorems in [31].

We assume that the coefficients of all systems of differential equations considered belong to $\mathbf{C}(z)$. Also, ϵ denotes any positive real number, the positive constants c will depend only on the functions (19), m, α and ϵ; and the positive constants σ, on the set of KE-functions under consideration and the numbers α, the number of functions, the powers n, n_l, $n_{k,l}$ occurring in the estimated measures, and the number ϵ. In this regard, different instances of c and σ denote different constants.

THEOREM 2. *Suppose the set of KE-functions* (19), $m \geqslant 2$, *is linearly independent over* $\mathbf{C}(z)$ *and comprises a solution of the system* (10), *and* $\alpha \in \mathbf{K}$, $\alpha T_0(\alpha) \neq 0$. *Then* $L(f_1(\alpha), \cdots, f_m(\alpha); H) > cH^{1-m-\epsilon}$.

THEOREM 3. *Suppose the set of KE-functions* (19), $m \geqslant 1$, *is, together with* 1, *linearly independent over* $\mathbf{C}(z)$ *and comprises a solution of the system* (12), *and* $\alpha \in \mathbf{K}$, $\alpha T(\alpha) \neq 0$. *Then* $L(1, f_1(\alpha), \cdots, f_m(\alpha); H) > cH^{1-m-\epsilon}$.

Theorems 2 and 3, and also Theorems 5 and 6 below, can easily be reformulated for the case of a KE-function satisfying equation (21) and the sequence of its successive derivatives. We state the theorem corresponding to Theorem 3.

THEOREM 4. *Suppose the KE-function* $f(z)$ *is a solution of equation* (21), $m \geqslant 1$, *and satisfies no differential equation of the same type of order less than* m *(or, if* $m = 1$, *is not a polynomial), and* $\alpha \in \mathbf{K}$, $\alpha P_m(\alpha) \neq 0$. *Then*

$$L(1,\ f(\alpha),\ f'(\alpha),\ \ldots,\ f^{(m-1)}(\alpha);\ H) > cH^{-m-\varepsilon}.$$

THEOREM 5. *Suppose the set of KE-functions* (19), $m \geqslant 1$, *is algebraically independent over* $\mathbf{C}(z)$ *and comprises a solution of the system* (12), *and* $\alpha \in \mathbf{K}$, $\alpha T(\alpha) \neq 0$. *Then*

$$\Phi(f_1(\alpha),\ \ldots,\ f_m(\alpha);\ n;\ H) > \sigma H^{1-\frac{(n+m)!}{n!\,m!}-\varepsilon}$$

Theorems 3 and 5 contain Siegel's results (8) and (9).

THEOREM 6. *Suppose each of the* r *sets of KE-functions*

$$f_{1,l}(z),\ \ldots,\ f_{m_l,l}(z), \qquad m_l \geqslant 1,\ \ l = 1,\ \ldots,\ r, \tag{27}$$

is a solution of the corresponding system of equations

$$y'_{k,l} = Q_{k,0,l} + \sum_{i=1}^{m_l} Q_{k,i,l}\, y_{i,l}, \qquad k = 1,\ \ldots,\ m_l,\ \ l = 1,\ \ldots,\ r. \tag{28}$$

Suppose also that the $m_1 + \cdots + m_r$ *functions* (27) *are algebraically independent over* $\mathbf{C}(z)$, *and* $\alpha \in \mathbf{K}$, $\alpha\prod_{l=1}^{r} T_l(\alpha) \neq 0$. *Then*

$$\Phi(f_{1,1}(\alpha), \ldots, f_{m_1,1}(\alpha), \ldots, f_{1,r}(\alpha), \ldots, f_{m_r,r}(\alpha);\ n_1, \ldots, n_r;\ H) > \sigma H^{1-N-\varepsilon},$$
$$N = \prod_{l=1}^{r} \frac{(n_l + m_l)!}{n_l!\, m_l!}, \tag{29}$$

where for each of the systems (28) $T_l(z)$ *has the same meaning as* $T(z)$ *has for the system* (12).

COROLLARY. *Under the hypothesis of Theorem* 6 *with* $m_l = 1$, $l = 1, \cdots$ r, *if each of the KE-functions* (19) *is a solution of the corresponding equation,* $\Phi_l(z)y' + Q_l(z)y = R_l(z)$, $l = 1, \cdots, r$, *and* $\alpha\prod_{l=1}^{r} P_l(\alpha) \neq 0$, *then*

$$\Phi(f_1(\alpha),\ \ldots,\ f_r(\alpha);\ n_1,\ \ldots,\ n_r) > \sigma H^{1-(n_1+1)\cdots(n_r+1)-\varepsilon}. \tag{30}$$

The measure estimates in Theorems 2–6 are very close to the natural upper bounds established with the aid of Dirichlet's principle. Refinements are possible only at the expense of replacing ϵ by a decreasing function of H. Note that (8) and (9) are contained in the above theorems.

In the proof of all the theorems on the transcendence and algebraic independence of values of concrete sets of E-functions stated above and contained in §7 one first proves the algebraic independence of the corresponding functions over $\mathbf{C}(z)$. Hence we can apply to these functions those of Theorems 1–6 which are appropriate and obtain the corresponding measure estimates. We give only a few examples.

Suppose the KE-function $f(z)$ is transcendental and satisfies the first order differential equation (22), and $\alpha \in \mathbf{K}$, $\alpha P_1(\alpha) \neq 0$. Then by Theorem 5

$$\Phi(f(\alpha), n, H) > \sigma H^{-n-\varepsilon};$$

in particular, $\Phi(e, n, H) > \sigma H^{-n-\epsilon}$.

If, moreover, $f(z)$ satisfies the hypothesis of Theorem 12 of §3, then by Theorems 5 and 6 we obtain the inequalities

$$\begin{gathered}\Phi(f(\alpha_1), \ldots, f(\alpha_m); n; H) > \sigma H^{1-\frac{(n+m)!}{n!\,m!}-\varepsilon}, \\ \Phi(f(\alpha_1), \ldots, f(\alpha_m); n_1, \ldots, n_m; H) > \sigma H^{1-(n_1+1)\ldots(n_m+1)-\varepsilon}.\end{gathered} \tag{31}$$

If $\xi_0, \xi_1, \cdots, \xi_m$ are distinct numbers in $\mathbf{K}$, then

$$L(e^{\xi_0}, e^{\xi_1}, \ldots, e^{\xi_m}; H) > cH^{-m-\varepsilon}.$$

Using the proofs of Theorems 10 and 13 of §3 concerning the functions (17) and (24), we obtain by Theorems 5 and 6 the estimates

$$\Phi(\omega_0(\alpha), \omega_1(\alpha), \ldots, \omega_r(\alpha); n; H) > H^{1-\frac{(n+r+1)!}{n!\,(r+1)!}-\varepsilon}, \tag{32}$$

$$\Phi(\psi_1(\alpha); \psi_2(\alpha), \psi_2'(\alpha); \ldots; \psi_r(\alpha), \psi_r'(\alpha), \ldots, \psi_r^{(r-1)}(\alpha); n; H) > \sigma H^{1-\frac{(n+t)!}{n!\,t!}-\varepsilon}, \quad t = \frac{r(r+1)}{2}, \tag{33}$$

$$\Phi(\psi_1(\alpha); \psi_2(\alpha), \psi_2'(\alpha); \ldots; \psi_r(\alpha), \psi_r'(\alpha), \ldots, \psi_r^{(r-1)}(\alpha); n_1, \ldots, n_r; H) > \sigma H^{1-N-\varepsilon}, \quad N = \prod_{l=1}^{r} \frac{(n_l + l)!}{n_l!\,l!}. \tag{34}$$

If in the definition of an E-function we replace the estimates $O(n^{\epsilon n})$ for the coefficients c_n, their conjugates and the least common multiple of the denominators of the first n coefficients c_n by c^n, where $c > 0$ is some constant (all known E-functions satisfy this condition), then after making the appropriate changes in the lemmas of the method Theorems 2–6 become more precise [32]. For example, in Theorems 2, 3 and 4 the ϵ in the exponent can be replaced by

$$\frac{\gamma m^3 \sqrt{\ln(n+1)}}{\sqrt{\ln\ln(H+2)}},$$

and in Theorem 5 by

$$\frac{\gamma\,(2en)^{3m}\sqrt{\ln\,(n+1)}}{\sqrt{\ln\ln\,(H+2)}},$$

where γ depends on the functions (19) and α, provided that $m = o[(\ln\ln H)^{1/6}]$ and $n^m = o[(\ln\ln H)^{1/6}]$ respectively.

In 1968 A. I. Galočkin [33], using the results of [28], proved the analogue of Theorem 6 for the case of an arbitrary algebraic field **K** of degree h over **Q**. Here the right-hand side of (29) is replaced by

$$\sigma H^{-(2er)^q h^{q+1} n_1^{m_1} \cdots n_r^{m_r}}.$$

When $r = 1$ this theorem refines Theorem 1 to the result

$$\Phi\,(f_1(\alpha),\ \ldots,\ f_m(\alpha);\ n;\ H) > \sigma H^{-(2e)^m h^{m+1} n^m}.$$

If $m_l = 1$, $l = 1, \cdots, r$, then when each of the functions (19) is a solution of a first order differential equation we obtain a generalization of (30):

$$\Phi\,(f_1(\alpha),\ \ldots,\ f_r(\alpha);\ n_1,\ \ldots,\ n_r;\ H) > \sigma H^{-2(2r)^r h^{r+1} n_1 \cdots n_r}.$$

In particular, when $r = 1$

$$\Phi\,(f\,(\alpha),\ n,\ H) > \sigma H^{-4h^2 n}.$$

The analogue of (31) assumes the form

$$\Phi\,(f\,(\alpha_1),\ \ldots,\ f\,(\alpha_m);\ n_1,\ \ldots,\ n_m;\ H) > \sigma H^{-2(2m)^m h^{m+1} n_1 \cdots n_m}.$$

The author lowers the constant 123 in the inequality (7) to 16. Note that in Schneider's book [34] it is lowered to 83. Then Galočkin gives measures for the sets of functions considered in Theorem 2 of §2 and Theorem 10 of §3 and indicates that one can prove analogues of inequalities (32), (33) and (34).

Galočkin [35] considered the hypergeometric functions of the form (5)

$$\varphi\,(z) = \sum_{n=0}^{\infty} \frac{1}{[(\lambda_1+1),\ n] \cdots [(\lambda_m+1),\ n]} \left(\frac{z}{m}\right)^{mn};$$
$$\lambda_i \in \mathbf{A};\quad \lambda_i \neq -1,\ -2,\ \ldots;\qquad i = 1,\ \ldots,\ m.$$

Let a be a natural number such that each $a\lambda_k$ $(k = 1, \cdots, m)$ is an integer, and suppose $\Lambda = \max(a, |\lambda_1|, \cdots, |\lambda_m|)$,

$$\varphi_1(z) = \varphi\,(z),\qquad \varphi_k(z) = z\frac{d}{dz}\varphi_{k-1}(z),\qquad k = 2,\ \ldots,\ m.$$

Two theorems concerning these functions are proved, in which the field **K** is again either **J** or **Q**.

THEOREM 7. *Suppose* $\lambda_1, \cdots, \lambda_m \in \mathbf{Q}$ *and* $\alpha^m \in \mathbf{K}$, $\alpha \neq 0$. *Then when* $H > H_1(m, \alpha, \Lambda)$

$$L(1, \varphi_1(\alpha), \ldots, \varphi_m(\alpha); H) > H^{-m-\frac{\gamma_1 m^2}{\ln\ln H}} \tag{35}$$

and for any integers $p_1, \cdots, p_k \in \mathbf{K}$ *and any natural number* $q > q_1(m, \alpha, \Lambda)$

$$\max_{1 \leqslant k \leqslant m} \left| \varphi_k(\alpha) - \frac{p_k}{q} \right| > q^{-1-\frac{1}{m}-\frac{\gamma_2 m^{-1}}{\ln\ln q}},$$

where γ_1 *and* γ_2 *are constants depending only on* α *and* Λ.

THEOREM 8. *Suppose* $\lambda_1, \cdots, \lambda_m \in \mathbf{A}$ *are such that all coefficients of the polynomial* $P(z) = (z + \lambda_1) \cdots (z + \lambda_m)$ *belong to* **K**, $P(0) = 0$, $m \geqslant 2$, $\alpha^m \in \mathbf{K}$ *and* $\alpha \neq 0$. *Then when* $H > H_2(m, \alpha, \Lambda)$

$$L(\varphi_1(\alpha), \ldots, \varphi_m(\alpha); H) > H^{1-m-\frac{\gamma_3 m^2}{\ln\ln H}},$$

and for any integers $p_1, \cdots, p_m \in \mathbf{K}$, *any natural number* $q > q_2(m, \alpha, \Lambda)$ *and any* t, $1 \leqslant t \leqslant m$,

$$\max_{\substack{1 \leqslant k \leqslant m \\ k \neq t}} \left| \frac{\varphi_k(\alpha)}{\varphi_t(\alpha)} - \frac{p_k}{q} \right| > q^{-1-\frac{1}{m-1}-\frac{\gamma_4 m^{-1}}{\ln\ln q}},$$

where γ_3 *and* γ_4 *are constants depending only on* α *and* Λ.

Mahler's estimate (4) is similar to (35). An analogous estimate for the functions $\varphi_\lambda(z)$ (13) was obtained by Fel′dman [36]. Estimates less precise than those in Theorems 7 and 8 follow from the more general results of Osgood [37].

As examples, consider the functions $K_\lambda(z)$ (6) and the Bessel functions $J_\lambda(z)$. For these functions we have for any $\alpha \in \mathbf{K}$, $\alpha \neq 0$, and $\lambda \in \mathbf{Q}$, $p, q \in \mathbf{Z}$, $q > 0$,

$$L(1, K_\lambda(\alpha), K'_\lambda(\alpha); H) > H^{-2-\frac{c_1(\alpha, \lambda)}{\ln\ln H}}$$

and

$$\left| \frac{J'_\lambda(\alpha)}{J_\lambda(\alpha)} - \frac{p}{q} \right| > q^{-2-\frac{c_2(\alpha, \lambda)}{\ln\ln q}}.$$

All estimates in Theorems 7 and 8 are valid for the functions $\psi_k(z)$ (24).

In [38] Galočkin obtained estimates more precise than previously known for $\Phi(e^\alpha, n, H)$ with $\alpha \in \mathbf{A}$, valid for increasing n, and for $\Phi(\xi, n, H)$, where ξ

is a root of the equation $P(z, e^z) = 0$ and $P(x, y)$ is a polynomial with rational coefficients.

§7. Other results on arithmetic properties of values of E-functions

This section is a survey of other papers in which arithmetic properties of values of E-functions at algebraic points are established with the aid of the method's general theorems discussed in §§3 – 5.

Siegel [9] indicated that using his method one could investigate analogously to the functions $K_\lambda(z)$ (6) the Kummer functions

$$A_{\lambda,\mu}(z) = \sum_{n=0}^{\infty} \frac{\mu(\mu+1)\ldots(\mu+n-1)}{n!\,\lambda(\lambda+1)\ldots(\lambda+n-1)} z^n, \qquad \lambda, \mu \neq 0, -1, -2, \ldots, \tag{36}$$

satisfying the differential equation

$$y'' + \left(\frac{\lambda}{z} - 1\right) y' = \frac{\mu}{z} y.$$

Yakabe [39], using only Siegel's book [12], showed that if $\lambda, \mu \in \mathbf{Q}$, then the zeros of $A_{\lambda,\mu}(z)$ are transcendental, and if $\mu \in \mathbf{Z}$, $\lambda - \mu \notin \mathbf{Z}$ and $\alpha \in \mathbf{A}$, $\alpha \neq 0$, then at least one of the numbers $A_{\lambda,\mu}(\alpha)$ or $A'_{\lambda,\mu}(\alpha)$ is transcendental. Then Oleĭnikov [40], using the First Fundamental Theorem, proved the following result.

THEOREM 1. *Suppose* $\lambda, \mu \in \mathbf{Q}$, $\mu \notin \mathbf{Z}$, $\mu - \lambda \neq 0, -1, -2, \cdots$, *and* $\alpha \in \mathbf{A}$, $\alpha \neq 0$. *Then the numbers* $A_{\lambda,\mu}(\alpha)$ *and* $A'_{\lambda,\mu}(\alpha)$ *are algebraically independent.*

In establishing the algebraic independence of the functions $A_{\lambda,\mu}(z)$ and $A'_{\lambda,\mu}(z)$ in the proof of this theorem the author used the method by which Siegel handled the analogous case of the functions $K_\lambda(z)$ and $K'_\lambda(z)$. It turned out that in this method the homogeneity of the differential equation satisfied by the function under investigation was essential. But in [41] the author extended the possible application of this method to the nonhomogeneous case. The proof of the algebraic independence of a function satisfying a nonhomogeneous differential equation and its derivative reduced to the analogous problem for solutions of the corresponding homogeneous equation. A similar transition was justified for a set of functions satisfying a nonhomogeneous system of differential equations, related only by an algebraic equation over the field $\mathbf{C}(z)$.

Thanks to this transition, the very complicated proof of Theorem 2 of §3 became very simple. With the aid of the proof of Theorem 1, the following theorem can be proved very easily. Consider the functions

$$A_{\lambda,\mu,\nu}(z) = \sum_{n=0}^{\infty} \frac{(\nu+1)\ldots(\nu+n)}{(\lambda+1)\ldots(\lambda+n)(\mu+1)\ldots(\mu+n)} z^n, \quad \lambda, \mu, \nu \neq -1, -2, \ldots, \tag{37}$$

satisfying a second order nonhomogeneous linear differential equation with coefficients in $\mathbf{C}[z]$.

THEOREM 2. *If* $\lambda, \mu, \nu \in \mathbf{Q}$, $\nu - \lambda \neq 0, 1, 2, \cdots; \nu - \mu \neq 0, 1, 2, \cdots$, *and* $\alpha \in \mathbf{A}$, $\alpha \neq 0$, *then the numbers* $A_{\lambda, \mu, \nu}(\alpha)$ *and* $A'_{\lambda, \mu, \nu}(\alpha)$ *are algebraically independent.*

Oleĭnikov [42]–[47] developed a new analytic method for proving the algebraic independence of functions and their successive derivatives, applicable to functions satisfying third order homogeneous linear differential equations. Using this method and the transition from a nonhomogeneous problem to a homogeneous one mentioned above, he proved the transcendence and algebraic independence of values at algebraic points of the main hypergeometric E-functions satisfying third order linear differential equations:

$$K_{\beta_1\beta_2\beta_3}(z) = \sum_{n=0}^{\infty} \frac{1}{[\beta_1, n]\,[\beta_2, n]\,[\beta_3, n]} \left(\frac{z}{3}\right)^{3n},$$

$$K_{\beta_1\beta_2\beta_3\alpha_1}(z) = \sum_{n=0}^{\infty} \frac{[\alpha_1, z]}{[\beta_1, n]\,[\beta_2, n]\,[\beta_3, n]} \left(\frac{z}{2}\right)^{2n},$$

$$K_{\beta_1\beta_2\beta_3\alpha_1\alpha_2}(z) = \sum_{n=0}^{\infty} \frac{[\alpha_1, n]\,[\alpha_2, n]}{[\beta_1, n]\,[\beta_2, n]\,[\beta_3, n]} z^n$$

and their first and second derivatives.

In [45] the conditions under which the values of the function

$$K_{\lambda, \mu, 0}(z) = \sum_{n=0}^{\infty} \frac{1}{n!\,(\lambda+1)\ldots(\lambda+n)\,(\mu+1)\ldots(\mu+n)} \left(\frac{z}{3}\right)^{3n}, \quad \lambda, \mu \neq -1, 2, \ldots,$$

and its derivatives are algebraically independent were made more precise.

THEOREM 3. *Suppose* $\lambda, \mu \in \mathbf{Q}$ *and* $\alpha \in \mathbf{A}$, $\alpha \neq 0$. *The numbers* $K_{\lambda, \mu, 0}(\alpha)$, $K'_{\lambda, \mu, 0}(\alpha)$ *and* $K''_{\lambda, \mu, 0}(\alpha)$ *are algebraically independent if and only if one of the following conditions is violated*:

$$\lambda = \pm \frac{1}{3} + n_1, \quad \mu = \mp \frac{1}{3} + n_2, \quad n_1, n_2 \in \mathbf{Z}.$$

In the series of papers [44] – [49] Oleĭnikov worked on establishing conditions for the differential irreducibility of algebraic differential equations. These problems are very closely related to the problem of proving the algebraic independence of a set of functions satisfying algebraic differential equations or that of finding algebraic equations relating such functions, the latter being necessary for the application of the general theorems on the algebraic independence of values of E-functions.

T. V. Peršikova [50] proved a general theorem on the transcendence of values of one class of E-functions. She considered an E-function $f(z)$ satisfying an mth order linear differential equation with coefficients in $\mathbf{C}(z)$. A function $\varphi(z)$ is obtained from $f(z)$ by multiplying its corresponding power series coefficient by γ_n, where $\gamma_n \in \mathbf{A}$ and the sequence $\{\gamma_n\}$ is a solution of a linear recurrence equation with coefficients in $\mathbf{A}$. It is established that the values of $f(z)$ for $z \in \mathbf{A}$, $z \neq 0$, are transcendental provided that $f(z)$ satisfies certain conditions and the largest multiplicity of a root of the characteristic equation of the recurrence equation does not exceed m.

This theorem generalizes a theorem of Redheffer [51] in which $f(z) = e^z$. Moreover, it is applicable to the functions $\varphi_\lambda(z)$ (13), $K_\lambda(z)$ (6) and $\varphi_{s,\lambda}(z)$ (23).

A. A. Šmelev [52] published a proof of a theorem on the algebraic independence of values of the set of functions $K_{\lambda,\mu}(z)$ (14) and their derivatives with distinct values of the parameters λ and μ and distinct values of the argument, analogous to Theorem 2 of §2 on the values of the set of functions $K_\lambda(z)$. In the functional part of the proof he used the method by which Theorem 2 of §3 was proved.

I. I. Belogrivov [53] considered the functions $K_\lambda(z)$ (6) and

$$u_\lambda(z) = \frac{\partial k_\lambda(z)}{\partial \lambda} = \sum_{n=1}^{\infty}\left[\frac{1}{\lambda+1} + \dots + \frac{1}{\lambda+n}\right] \times \frac{(-1)^n}{n!\,(\lambda+1)\dots(\lambda+n)}\left(\frac{z}{2}\right)^{2n}, \qquad \lambda \neq -1, -2, \dots.$$

Note that when $\lambda = 0$ the function $y = u_0(z) + J_0(z)\ln z$ is a solution of Bessel's differential equation and is linearly independent of the solution $J_0(z) = K_0(z)$; hence

$$z\,(u_0' J_0 - u_0 J_0') = 1 - J_0^2,$$

and the functions $u_\lambda(z)$ and $K_\lambda(z)$ are related by the equation

$$u_\lambda'' + \frac{2\lambda+1}{z} u_\lambda' + u_\lambda = \frac{-2}{z} K_\lambda'.$$

The following theorems are proved with the aid of the general theorems of §4.

THEOREM 4. *Suppose* $\alpha_i \in \mathbf{A}$, $\alpha_i \neq 0$, $i = 1, \cdots, n$, *and* $\alpha_i^2 \neq \alpha_j^2$ *when* $i \neq j$. *Then of the five sets of numbers*

$$\{J_0(\alpha_i)\},\ \{J_0'(\alpha_i)\},\ \{u_0(\alpha_i)\},\ \{u_0'(\alpha_i)\},\ \{u_0''(\alpha_i)\}, \qquad i = 1, \dots, n,$$

any three comprise an algebraically independent system.

THEOREM 5. *Suppose* $\lambda \in \mathbf{Q}$, $\lambda \neq s/2$, $s = 0, \pm 1, \pm 2, \cdots$, *and* $\alpha \in \mathbf{A}$, $\alpha \neq 0$. *Then of the six numbers* $K_\lambda(\alpha)$, $K'_\lambda(\alpha)$, $u_\lambda(\alpha)$, $u'_\lambda(\alpha)$, $u''_\lambda(\alpha)$, $u'''_\lambda(\alpha)$ *any four, except for the quadruple* $K'_\lambda(\alpha)$, $u_\lambda(\alpha)$, $u'_\lambda(\alpha)$, $u''_\lambda(\alpha)$, *are algebraically independent.*

Analogous theorems can be formulated for the functions $\varphi_\lambda(z)$ (13) and

$$v_\lambda(z) = \frac{\partial \varphi_\lambda(z)}{\partial \lambda} = \sum_{n=1}^{\infty} \left[\frac{1}{\lambda+1} + \cdots + \frac{1}{\lambda+n}\right] \frac{z^n}{(\lambda+1)\ldots(\lambda+n)},$$
$$\lambda \neq -1, -2, \ldots .$$

Note that the result concerning the algebraic independence of the set of numbers e^α, $\varphi_\lambda(\alpha)$, $v_\lambda(\alpha)$ for distinct values of λ and α is contained in [14].

Mahler [54] generalized Belogrivov's result. He considered the functions

$$u_{k,\lambda}(z) = \frac{1}{k!}\frac{\partial^k}{\partial \lambda^k} K_\lambda(z), \quad \lambda \neq -1, \quad -2, \ldots, \quad k = 0, 1, 2, \ldots,$$

when $\lambda = 0$, and proved the following theorem.

THEOREM 6. *Suppose* $\alpha \in \mathbf{A}$, $\alpha \neq 0$. *Then each of the four sets of numbers*

$$\{u_{0,0}(\alpha),\ u'_{0,0}(\alpha),\ u_{1,0}(\alpha),\ u_{2,0}(\alpha),\ u'_{2,0}(\alpha),\ u_{3,0}(\alpha)\},$$
$$\{u_{0,0}(\alpha),\ u'_{0,0}(\alpha),\ u'_{1,0}(\alpha),\ u_{2,0}(\alpha),\ u'_{2,0}(\alpha),\ u_{3,0}(\alpha)\},$$
$$\{u_{0,0}(\alpha),\ u'_{0,0}(\alpha),\ u_{1,0}(\alpha),\ u_{2,0}(\alpha),\ u'_{2,0}(\alpha),\ u'_{3,0}(\alpha)\},$$
$$\{u_{0,0}(\alpha),\ u'_{0,0}(\alpha),\ u'_{1,0}(\alpha),\ u_{2,0}(\alpha),\ u'_{2,0}(\alpha),\ u'_{3,0}(\alpha)\}$$

is algebraically independent. In particular, all of the numbers $u_{k,0}(\alpha)$ *and* $u'_{k,0}(\alpha)$, $k = 0, 1, 2, 3$, *are transcendental.*

Then Mahler considered the derivatives with respect to the parameter, up to those of third order, of the Bessel functions $J_\lambda(\alpha)$ at $\lambda = 0$, and proved a theorem, analogous to Theorem 4, concerning numbers related to the values of these functions and the numbers $\zeta(2)$, $\zeta(3)$ and γ, where γ is Euler's constant. In particular, this theorem implies the transcendence of the number

$$\frac{\pi}{2}\frac{Y(2)}{J_0(2)} - \gamma,$$

where $Y(z)$ is a Bessel function of the second kind.

Recently Belogrivov investigated the derivatives with respect to the parameter of the functions $K_\lambda(z)$ of any order and for any $\lambda \in \mathbf{Q}$. He first proved

THEOREM 7. *Suppose* $\lambda \in \mathbf{Q}$, $\lambda \neq s/2$, *where* $s \in \mathbf{Z}$, *and* $\alpha \in \mathbf{A}$, $\alpha \neq 0$. *Then for any* $m \geqslant 0$ *the* $2m + 2$ *numbers*

$$u_{0,\lambda}(\alpha),\ u'_{0,\lambda}(\alpha), \ldots, u_{m,\lambda}(\alpha),\ u'_{m,\lambda}(\alpha) \tag{38}$$

are algebraically independent.

Furthermore, under the hypothesis of Theorem 7, the author augments the set of numbers (38) by the $2m + 2$ numbers $u_{k,-\lambda}(\alpha)$ and $u'_{k,-\lambda}(\alpha)$, $k = 0, 1, \cdots, m$, and proves theorems on the algebraic independence of certain subsets of this set of numbers.

In addition, he considers the cases where $\lambda \in \mathbf{Z}$, $\lambda \geqslant 0$, and $\lambda = (2s + 1)/2$, $s \in \mathbf{Z}$, and indicates in (38) algebraically independent subsystems.

Belogrivov [55], [56], using the arithmetic method of proving the algebraic independence of functions and a further development of it, investigated certain classes of hypergeometric E-functions.

Concerning the functions

$$A_m(z) = \sum_{n=0}^{\infty} \frac{1}{[(\lambda+1)\ldots(\lambda+n)]^{m_1}[(\mu+1)\ldots(\mu+n)]^{m_2}}\left(\frac{z}{2}\right)^{mn},$$
$$m_1 + m_2 = m, \quad \lambda, \mu \neq -1, -2, \ldots, \tag{39}$$

he proved, in particular,

THEOREM 8. *Suppose* $\lambda, \mu \in \mathbf{Q}$ *are such that when one of them is an integer the other is not half an odd integer, and* $\alpha \in \mathbf{A}$, $\alpha \neq 0$. *Then the* m *numbers* $A_m(\alpha), A'_m(\alpha), \cdots, A_m^{(m-1)}(\alpha)$ *are algebraically independent.*

When $\mu = 0$, $m_1 = m_2 = 1$, this theorem includes Theorem 1 of §2, proved by Siegel in 1929. Theorem 8 also generalizes and refines Theorem 1 of § 3.

In the same paper a whole series of theorems are proved concerning the algebraic independence of values of the functions

$$A_{m,s}(z) = \sum_{n=0}^{\infty} \frac{1}{[\lambda_1+1, n]^{m_1}[\lambda_2+1, n]^{m_2}\ldots[\lambda_s+1, n]^{m_s}}\left(\frac{z}{m}\right)^{km},$$
$$m_1 + m_2 + \ldots + m_s = m, \quad \lambda_1, \ldots, \lambda_s \neq -1, -2, \ldots, \tag{40}$$

their successive derivatives and certain related functions under definite restrictions on the values of the parameters. We state the simplest result.

THEOREM 9. *Suppose* $\lambda_1, \cdots, \lambda_s \in \mathbf{Q}$ *are such that those fractional parts of these numbers which are different from zero are either all greater than* $1/2$ *or all less than* $1/2$, *and* $\alpha \in \mathbf{A}$, $\alpha \neq 0$. *Then the* m *numbers* $A_{m,s}(\alpha)$, $A'_{m,s}(\alpha), \cdots, A_{m,s}^{(m-1)}(\alpha)$ *are algebraically independent.*

In [57] Belogrivov significantly simplified in some places the method of proving the algebraic independence of a set of functions satisfying second order homogeneous linear differential equations and their derivatives used by Siegel to prove Theorem 2 of §2. He considered the Kummer functions $A_{\lambda,\ \mu}(z)$ (36) and the functions $B_{\nu,\ \mu}(z) = e^{-z/2}A_{\nu,\ \mu}(z)$ and proved the following theorems.

THEOREM 10. *Suppose* $\nu_j, \mu_j \in \mathbf{Q}$, $\nu_j \neq -k_1, \mu_j, \mu_j - \nu_j \neq k_2$, $j = 1, \cdots, m$; $2\mu_j - \nu_j - (2\mu_k - \nu_k) \pm (\nu_j \pm \nu_k) \neq 2k_3$, $1 \leqslant j < k \leqslant m$, *where* $k_1, k_2, k_3 \in \mathbf{Z}$, $k_1 \geqslant 0$, *and* $\alpha_i \in \mathbf{A}$, $\alpha_i \neq 0$, $\alpha_i \neq \alpha_s$ *when* $i \neq s$, $i, s = 1, \cdots, n$. *Then the* $2mn$ *numbers* $B_{\nu_j,\ \mu_j}(\alpha_i)$ *and* $B'_{\nu_j,\ \mu_j}(\alpha_i)$, $i = 1, \cdots, n$, $j = 1, \cdots, m$, *are algebraically independent.*

From this theorem with $2\mu_j = \nu_j = 2\lambda_j + 1, j = 1, \cdots, m$, *and* $\alpha_i^2 = -4\beta_i^2$, $i = 1, \cdots, m$, follows Theorem 2 of §2, proved by Siegel in 1929.

THEOREM 11. *Suppose* $\nu_j, \mu_j \in \mathbf{Q}$, $\nu_j \neq -k_1, \mu_j, \mu_j - \nu_j \neq k_2$, $j = 1, \cdots, m$; $2m_j - \nu_j - (2\mu_k - \nu_k) \pm (\nu_j \pm \nu_k) \neq 2k_3$, $1 \leqslant j < k \leqslant m$, *where* $k_1, k_2, k_3 \in \mathbf{Z}$, $k_1 \geqslant 0$, *and* $\alpha_1, \cdots, \alpha_n \in \mathbf{A}$ *are linearly independent over* $\mathbf{Q}$. *Then the* $2mn$ *numbers* $A_{\nu_j,\ \mu_j}(\alpha_i)$ *and* $A'_{\nu_j,\ \mu_j}(\alpha_i)$, $i = 1, \cdots, n$, $j = 1, \cdots, m$, *are algebraically independent.*

Theorems 10 and 11 are included in a more general theorem, proved by Belogrivov in the same paper, concerning the values of the set of functions $A_{\nu,\ \mu}(\alpha)$, $A'_{\nu,\ \mu}(\alpha)$, $K_\lambda(\beta)$, $K'_\lambda(\beta)$, e^γ for distinct values of the parameters ν, μ, λ and distinct α, β and γ.

Ju. V. Nesterenko [58] proved some general theorems by means of which the proof of the algebraic independence over $C(z)$ of a set of functions satisfying nonhomogeneous linear differential equations reduces to the analogous problem for the corresponding homogeneous differential equations, moreover with any number of algebraic equations with coefficients in $\mathbf{C}(z)$ relating the functions under consideration. These theorems generalize the results of [41] mentioned at the beginning of this section, where the functions could be related by only one algebraic equation.

With the aid of this method, using the functional part of Siegel's proof of Theorem 2 of §2, it is very easy to prove a theorem on the value of the functions $K_{\lambda,\ \mu}(z)$ (15).

THEOREM 12. *Suppose* $\lambda_i, \mu_i \in \mathbf{Q}$, $\lambda_i - \mu_i \neq (2k_1 + 1)/2$, $k_1 \in \mathbf{Z}$, $i = 1, \cdots, m$, $(\lambda_i - \mu_i) \pm (\lambda_j - \mu_j) \neq k_2$, $k_2 \in \mathbf{Z}$, $1 \leqslant i < j \leqslant m$, *and* $\xi_1, \cdots, \xi_n \in \mathbf{A}$ *are such that* $\xi_1^2, \cdots, \xi_n^2$ *are different from each other and from zero. Then the* $2mn$ *numbers* $K_{\lambda_i,\ \mu_i}(\xi_j)$ *and* $K'_{\lambda_i,\ \mu_i}(\xi_j)$, $i = 1, \cdots, m$, $j = 1, \cdots, n$, *are algebraically independent.*

As we mentioned above, this result, stated in a somewhat lengthier way, was obtained by Šmelev [52] by means of a very complicated argument.

Analogously, using the functional part of Belogrivov's proof of a theorem of this section, it is also easy to prove a theorem on values of the functions $A_{\lambda,\ \mu,\ \nu}(z)$ (37).

THEOREM 13. *Suppose* $\lambda_j, \mu_j, \nu_j \in \mathbf{A}$, $\lambda_j, \mu_j, \nu_j \neq -k_1 - 1$, $\nu_j - \lambda_j \neq k_2$, $\nu_j - \mu_j \neq k_3$, $j = 1, \cdots, m$; $(2\nu_j - \mu_j - \lambda_j) - (2\nu_k - \mu_k - \lambda_k) \pm [(\lambda_j \pm \lambda_k) - (\mu_j \pm \mu_k)] \neq 2k_4$, $1 \leqslant j < k \leqslant m$, *where* $k_1, k_2, k_3, k_4 \in \mathbf{Z}$ *and* $k_1 \geqslant 0$. *Furthermore, suppose* $\xi_1, \cdots, \xi_n \in \mathbf{A}$ *are such that* $\xi_1^2, \cdots, \xi_n^2$ *are different from each other and from zero. Then the* $2mn$ *numbers* $A_{\lambda_j, \mu_j, \nu_j}(\xi_i)$ *and* $A'_{\lambda_j, \mu_j, \nu_j}(\xi_i)$, $i = 1, \cdots, n$, $j = 1, \cdots, m$, *are algebraically independent.*

In 1971 V. H. Salihov proved a general theorem on the algebraic independence of values at algebraic points of a set of E-functions, each of which satisfies a first order linear differential equation with coefficients in $\mathbf{C}(z)$. This theorem easily implies the Lindemann-Weierstrass Theorem and Theorems 10 and 12 of §3.

We state again that for all of the results of this section we can give estimates for the measures of mutual transcendence, using the corresponding general theorems on measure estimates of §6. Such estimates for the functions (39) and (40) are indicated in [56].

§8. Generalizations to other classes of functions and numbers

The general theorems stated in §§3–6 completely solve the arithmetic problem concerning the transcendence and algebraic independence of values at algebraic points of E-functions satisfying linear differential equations of any orders with coefficients in $\mathbf{C}(z)$, and also the problem of estimating from below the measures of the linear independence and mutual transcendence of values of such functions. But, in many cases, for the successful application of these theorems to concrete functions it is necessary to develop the analytic problem; that is, to establish new methods for proving the algebraic independence of functions over $\mathbf{C}(z)$ and determining the basic algebraic equations relating the functions if they are algebraically dependent. There are a number of such methods which are powerful enough to be applicable to many concrete functions. But they are clearly inadequate for application in all cases.

There naturally arises the problem of extending the methods for proving algebraic independence to values of other classes of analytic functions and to values of E-functions or other functions at certain transcendental points as well as algebraic points; in particular, to values at algebraic points of entire hypergeometric functions of the form (5) with irrational algebraic values of the parameters.

In a lecture at the number theory conference at Palanga (1965) this author [59] announced the following result. *If* $\mathbf{K}$ *is an algebraic field over* $\mathbf{Q}$ *and* $\lambda \in \mathbf{K}$, *then all values of the functions* $\varphi_\lambda(z)$ (13), $K_\lambda(z)$ (6) *and* $K'_\lambda(z)$ *for all but finitely many* $z \in \mathbf{K}$ *are irrational.* The proof is obtained by means of the above general method and the application of an additional concept.

V. G. Sprindžuk considered a similar problem in [60] and published the proof in [61]. He investigated entire functions with algebraic power series coefficients satisfying a first order linear differential equation with coefficients in $\mathbf{C}(z)$, from a class wider than the class of E-functions. He called these E^*-functions. They are defined analogously to E-functions, but in conditions 2) and 3) on the quantities $\overline{|c_n|}$ and q_n a significantly greater growth is permitted. The author proved that there can exist only finitely many algebraic numbers of given degree for which such a function assumes algebraic values of degree not exceeding a certain number. He also proved that entire hypergeometric functions of the form (5) with algebraic values of the parameters are E-functions. Hence the stated result is valid for the functions $\varphi_\lambda(z)$ (13) with irrational algebraic values of the parameter λ. In [62] Sprindžuk estimated the least common multiple of the coefficients of hypergeometric functions of the form (5) with irrational algebraic values of the parameters.

From Theorem 8, stated in §6 and proved by Galočkin in [35], there follow easily certain arithmetic results for values of hypergeometric functions with algebraic values of the parameters at algebraic points.

Osgood [37], using an inductive method involving ideas from the Hermite-Lindemann method, proved a number of theorems on Diophantine approximations of values of certain hypergeometric functions of the form (5) with rational and irrational algebraic values of the parameters, namely, the functions

$$f(z)=\sum_{n=1}^{\infty}\frac{z^n}{\prod_{l=1}^{n}P(l)}, \qquad P(z)\in \mathbf{J}[z],$$

and their derivatives at various points of the field $\mathbf{J}$. In [63] he applied this method to more general entire functions.

In his candidate's dissertation Galočkin again considered the functions

$$\varphi(z)=\sum_{n=0}^{\infty}\frac{1}{[\lambda_1+1,\,n]\ldots[\lambda_m+1,\,n]}\left(\frac{z}{m}\right)^{mn},$$

$$\lambda_i\neq -1-2,\ldots, \qquad i=1,\ldots,m,$$

$$\varphi_1(z)=\varphi(z), \qquad \varphi_k(z)=z\frac{d}{dz}\varphi_{k-1}(z), \quad k=1,\ldots,m,$$

and proved a theorem on simultaneous approximations of their parameters by numbers of the field $\mathbf{J}$.

THEOREM 1. *Suppose for some $\xi\in\mathbf{C}$, $\xi\neq 0$, and real $\lambda_1,\cdots,\lambda_m$ that the numbers $\varphi_1(\xi),\cdots,\varphi_m(\xi)$ together with the number 1 are linearly dependent over $\mathbf{J}$. Then for any numbers $b\geqslant 1, b_1\geqslant 2,\cdots,b_m\geqslant 2, a_1,\cdots,a_m$ in $\mathbf{Z}$ and any integer $a\in\mathbf{J}$ we have when $B>B_1(m,\xi,\lambda_1,\cdots,\lambda_m)$ the inequality*

$$\left|\xi^m-\frac{a}{b}\right|+\sum_{k=1}^{m}\left|\lambda_k-\frac{a_k}{b_k}\right|>e^{-c^mB\ln B},$$

where $c = c(\xi, \Lambda)$, $\Lambda_1 = \max(|\lambda_1|, \cdots, |\lambda_m|)$, $B = d^m b b_1 \cdots b_m \ln b_m$, *and* d *is the least common denominator of* $b_1, \cdots, b_m$.

If the numbers $\varphi_1(\xi), \cdots, \varphi_m(\xi)$ are linearly dependent over $\mathbf{J}$ and $\lambda_1\lambda_2 \cdots \lambda_m = 0$, the above estimate can be made somewhat more precise. In particular, we can make the following assertion.

Suppose λ *is a real number such that the logarithmic derivative* $J'_\lambda(z)/J_\lambda(z)$ *of the corresponding Bessel function assumes at the point* $r \in \mathbf{Q}$, $r \neq 0$, *a value in* $\mathbf{Q}$. *Then for any* $p, q \in \mathbf{Z}$, $q \geqslant 2$, *we have the inequality*

$$\left|\lambda - \frac{p}{q}\right| > e^{-\gamma(\lambda, r) q^2 \ln q}.$$

Galočkin [64], [65] established a theorem on the algebraic independence of values of a set of E-functions for certain transcendental values of the argument admitting sufficiently good approximations by numbers of $\mathbf{A}$.

THEOREM 2. *Suppose the set of E-functions* (19) *is algebraically independent over* $\mathbf{C}(z)$ *and comprises a solution of the system of linear differential equations* (12), *and* $\xi \in \mathbf{C}$ *is a root of the equation*

$$P(z, f_1(z), \ldots, f_m(z)) = 0,$$

where $P(x_0, x_1, \cdots, x_m) \not\equiv 0$ *is a polynomial with coefficients in* $\mathbf{A}$. *Then for any fixed positive numbers* ϵ *and* κ *the inequality*

$$|\xi - \theta| < e^{-h^\epsilon} \tag{41}$$

has a finite number of solutions in numbers θ *of* $\mathbf{A}$ *of height* h *and degree at most* κ.

If $P(\xi, x_1, \cdots, x_m) \not\equiv 0$ *and* $\xi T(\xi) \neq 0$, *then for any* $\epsilon > 0$ *and any polynomial* $g(z) \in \mathbf{Z}[z]$, $g(z) \not\equiv 0$, *of height* H *and degree at most* d *we have*

$$|g(\xi)| > ce^{-H^\epsilon}, \tag{42}$$

where $c > 0$ *is a constant depending only on* ϵ, ξ, d *and* $f_1(z), \cdots, f_m(z)$.

COROLLARY. *Suppose the E-functions* (19) *are algebraically independent over* $\mathbf{C}(z)$ *and comprise a solution of the system* (12), *and* $\xi \in \mathbf{C}$ *is such that inequality* (41) *has an infinite number of solutions for certain positive* ϵ *and* κ. *Then the numbers* ξ *and* $f_1(\xi), \cdots, f_m(\xi)$ *are algebraically independent.*

The above theorem is applicable to many E-functions, for example the functions $\psi_k(z)$ (24) and many other functions considered in this paper. The estimate (42), in particular, is valid for the algebraic A-points of the functions $K_\lambda(z)$ (6) and their derivatives when $\lambda \in \mathbf{Q}$, and hence also for the zeros of the Bessel functions $J_\lambda(z)$ and their derivatives.

Siegel [9] indicated that his method could be used to study certain arithmetic properties of the values of another class of analytic functions with a finite radius of convergence of the Taylor series. He called these G-functions.

A function

$$f(z)=\sum_{n=0}^{\infty} c_n z^n \tag{43}$$

is called a G-function if the coefficients c_n satisfy the same three conditions as the numbers c_n in the definition of an E-function. A G-function differs from an E-function in that in the general term of its power series (43) the factor $1/n!$ is absent.

Siegel [9] stated several arithmetic facts concerning the values of various concrete G-functions, but did not give the proofs.

In the papers of M. S. Nurmagomedov [66], [67] Siegel's general method and its generalizations were first applied to the study of arithmetic properties of the values of G-functions.

We call a function (43) a G-function of class $G(K, \lambda, c, \mu, q)$ if it satisfies the following conditions:

1) $c_n \in \mathbf{K}$, $n = 0, 1, 2, \cdots$, where $\mathbf{K}$ is an algebraic field of finite degree over $\mathbf{Q}$.

2) $\overline{|c_n|} \leqslant \lambda c^n$, $n = 0, 1, 2, \cdots$, where c and λ are positive constants.

3) There exists a sequence $\{q_n\}$ of natural numbers such that each $q_n c_k$, $k = 0, 1, 2, \cdots$, is an integer of $\mathbf{K}$, and $q_n \leqslant \mu q^n$, where $\mu \geqslant 1$ and $q \geqslant 1$ are constants.

Note that in view of the slow convergence of the power series of G-functions arithmetic results concerning their values can be obtained only for sufficiently small values of the argument in $\mathbf{A}$.

The following theorems are proved in [66].

THEOREM 3. *Suppose the set of functions*

$$f_k(z)=\sum_{n=0}^{\infty} c_{k,n} z^n, \quad k=1, \ldots, m, \quad m \geqslant 2, \tag{44}$$

of class $G(J, \lambda, c, \mu, q)$ *comprises a solution of the system of homogeneous linear differential equations* (10), *where all* $Q_{k,i} \in \mathbf{C}(z)$, *and is linearly independent over* $\mathbf{C}(z)$; *H and b are natural numbers; $a, u_1, \cdots, u_m$ are any integers of* $\mathbf{J}$; $|u_i| \leqslant H$, $i = 1, \cdots, m$, $|u_1| + \cdots + |u_m| > 0$, *and*

$$\frac{a}{b}T\left(\frac{a}{b}\right)\neq 0,\quad \left|\frac{a}{b}\right|\leqslant\frac{1}{4c}.$$

Then there exists a natural number n_1, *depending only on* m *and the functions* (44), *such that when*

$$b\geqslant\exp\left(\gamma_1\sqrt{\ln n_1}+2m\ln|a|+\frac{1}{n_1}\ln H-\frac{1}{n_1}\ln r\right)$$

the numbers

$$f_1\left(\frac{a}{b}\right),\ldots,f_m\left(\frac{a}{b}\right) \tag{45}$$

are not related by a homogeneous linear equation whose coefficients are integers of J *of modulus at most* H, *and we have the estimate*

$$\left|u_1f_1\left(\frac{a}{b}\right)+\ldots+u_mf_m\left(\frac{a}{b}\right)\right|\geqslant r\exp\left(-\gamma_1n_1\sqrt{\ln n_1}\right)b^{-(2m-1)n_1}, \tag{46}$$

where

$$\gamma_1=m\,(10m-9)\sqrt{\ln(cq_1)},\qquad q_1=q^m,$$
$$r=\min\left(\max_{1\leqslant i\leqslant m}\left|f_i\left(\frac{a}{b}\right)\right|,\ 1\right)\neq 0.$$

Theorem 4 shows that when b varies over a certain interval the estimate (46) is close to the natural limit of the upper bound of the linear form obtained by means of Dirichlet's principle.

THEOREM 4. *Suppose the conditions of Theorem 3 are satisfied and, in addition,* ν *and* θ *are any real numbers such that* $0<\nu\leqslant 2m$ *and* $1\leqslant\theta\leqslant 1+\nu/4m$. *Then there exists a natural number* n_1, *depending only on* m *and the functions* (44), *such that when*

$$b\geqslant\left[\exp\left\{\frac{2m\theta}{4m-\nu}\left(\gamma_1\sqrt{\ln n_1}+2m\ln|a|+\frac{1}{n_1}\ln H-\frac{2m\ln r}{(4m-\nu)\,n_1}\right\}\right]+1$$

we have the estimate

$$\left|u_1f_1\left(\frac{a}{b}\right)+\ldots+u_mf_m\left(\frac{a}{b}\right)\right|\geqslant\gamma_2r^{\frac{4m^2}{4m-\nu}}H^{1-m-\nu},$$

where

$$\gamma_2=\exp\{-(m\,(30m^2-17m-9)\sqrt{\ln(cq_1)}\,n_1\sqrt{\ln n_1}$$
$$+6m^2n_1\ln|a|+(2m-1)\,n_1)\},$$

and γ_1 *and* q_1 *are defined as in Theorem* 3.

THEOREM 5. *Suppose the set of functions* (44) *of class* $G(J, \lambda, c, \mu, q)$ *comprises a solution of the system of linear differential equations* (12), *where all* $Q_{k,i} \in \mathbf{C}(z)$, *and the functions are not related by an algebraic equation of degree* N, $N \geqslant 1$, *over* $\mathbf{C}(z)$. *Suppose also that* H *and* b *are natural numbers and* a *is an integer of* $\mathbf{J}$, *where*

$$\frac{a}{b} T\left(\frac{a}{b}\right) \neq 0, \qquad \left|\frac{a}{b}\right| \leqslant \frac{1}{8c}.$$

Then there exists a natural number n_2, *depending only on the functions* (44), *the numbers* m *and* N *and the field* $\mathbf{J}$, *such that when*

$$b \geqslant \exp\left(\gamma_3 \sqrt{\ln n_2} + 2m_1 \ln |a| + \frac{1}{n_3} \ln H\right)$$

the numbers (45) *are not related by an algebraic equation of degree* N *whose coefficients are integers of* $\mathbf{J}$ *of modulus at most* H, *and we have the estimate*

$$\left|P\left(f_1\left(\frac{a}{b}\right), \ldots, f_m\left(\frac{a}{b}\right)\right)\right| \geqslant \exp\left(-\gamma_3 n_2 \sqrt{\ln n_2}\right) b^{-(2m_1-1)n_2},$$

where $P(z_1, \cdots, z_m) \not\equiv 0$ *is a polynomial of degree* N *whose coefficients are integers of* $\mathbf{J}$ *of modulus at most* H, *and*

$$\gamma_3 = m_1 (10m_1 - 9) \sqrt{\ln (2cq_1)}, \qquad m_1 = \frac{(N+m)!}{N!m!}, \qquad q_2 = q^{m(1+\ln N)}.$$

Next one proves the homogeneous analogue of Theorem 5, in which $P(z_1, \cdots, z_m)$ is a homogeneous polynomial of degree N, and the functions (44) satisfy the homogeneous system (10).

In the same article [66] there are established lower bounds of linear forms and polynomials in the set of G-functions (44), valid on a wider interval of b-values than in the above theorems but only for sufficiently large H.

In the statements of Theorems 3, 4 and 5 only the existence of the natural numbers n_1 and n_2 is asserted. However, it is shown in this article that in certain cases n_1 and n_2 can be effectively calculated. In these cases the corresponding estimates for concrete functions are given with precisely calculated constants.

In [67] there are proved theorems analogous to Theorem 5 and its homogeneous analogue for the case where $\mathbf{J}$ is replaced by an arbitrary field $\mathbf{K}$ of finite degree over $\mathbf{Q}$. However, in this case the constants appearing in the estimates are not calculated, although they could be by careful attention to details.

As simple examples of the application of the general theorems of [66] and [67] consider the functions

$$\varphi_{\lambda, k}(z) = \sum_{n=0}^{\infty} \frac{z^n}{(n+\lambda)^k}, \qquad \lambda \neq -1, -2, \ldots, \quad k = 1, \ldots, m,$$

and

$$\ln(1 - \beta_k z) = -\sum_{n=1}^{\infty} \frac{\beta_k^n}{n} z^n, \qquad k = 1, \ldots, m. \tag{47}$$

We cite only one result.

THEOREM 6. *Suppose H, b and N are natural numbers, $a \neq 0$, $u_0, u_1, \cdots, u_m$ are integers of* $\mathbf{J}$, $|u_i| \leqslant H$, $i = 0, 1, \cdots, m$, $|u_0| + |u_1| + \cdots + |u_m| > 0$, *and the numbers σ_1, σ_2 and n_1 are defined as follows*:

$$\sigma_1 = (m+1)(10m+1)\sqrt{\ln(\beta e^2)}, \qquad \beta = \max_{1 \leqslant i \leqslant m} |\beta_i|,$$

$$\sigma_2 = m_1(10m_1 - 9)\sqrt{\ln(2\beta e^{2m(1+\ln N)})},$$

$$n_1 = \left[\max\left\{\exp((m+1)^2(2m^2+2m+3)\ln(\beta e^2)), \quad \frac{4\ln 4\,(m+1)\,|\,d\,|^{3/2}}{(m+1)\ln(\beta e^2)}\right\}\right] + 1,$$

where d is the discriminant of $\mathbf{J}$. *Then the following assertions are true*:

1) *When*

$$b \geqslant \exp\left(\sigma_1\sqrt{\ln n_1} + 2(m+1)\ln|a| + \frac{1}{n_1}\ln H\right),$$

we have the inequality

$$\left|u_0 + u_1 \ln\left(1 - \frac{\beta_1 a}{b}\right) + \ldots + u_m \ln\left(1 - \frac{\beta_m a}{b}\right)\right| \geqslant \exp(-\sigma_1 n_1 \sqrt{\ln n_1})\, b^{-(2m+1)n_1}.$$

2) *There exists a natural number n_2, depending only on the functions* (47), *the numbers m and N, and the field* $\mathbf{J}$, *such that when*

$$b \geqslant \exp\left(\sigma_2\sqrt{\ln n_2} + 2m_1\ln|a| + \frac{\ln H}{n_2}\right)$$

we have

$$\left|P\left(\ln\left(1 - \beta_1\frac{a}{b}\right), \ldots, \ln\left(1 - \beta_m\frac{a}{b}\right)\right)\right| \geqslant \exp(-\sigma_2 n_2 \sqrt{\ln n_2})\, b^{-(2m_1-1)n_2},$$

where $P(z_1, \cdots, z_m) \not\equiv 0$ is a polynomial of degree N whose coefficients are integers of $\mathbf{J}$ *of modulus at most H.*

The second part of this theorem contains for the first time a lower estimate for polynomials in the values of a set of logarithms at algebraic points.

In joint papers Nurmagomedov and V. G. Čirskiĭ considered the set of functions

$$\psi_k(z)=\sum_{n=0}^{\infty}\frac{\mu(\mu+1)\dots(\mu+n-1)}{\lambda(\lambda+1)\dots(\mu+n-1)}z^n$$

and the general hypergeometric series

$$F(z)=\sum_{n=0}^{\infty}\frac{\mu(\mu+1)\dots(\mu+n-1)\,\nu(\nu+1)\dots(\nu+n-1)}{n!\,\lambda(\lambda+1)\dots(\lambda+n-1)}z^n,\quad \lambda\neq 0,-1,-2,\dots,$$

with distinct values of the parameters $\lambda, \mu, \nu \in \mathbf{Q}$ and at algebraic points, to which the general theorems are applicable.

We remark that the results on the values of G-functions stated without proofs in Siegel's paper [9] have yet to be proved.

Conclusion

Expositions of Siegel's method and his results can be found in the monographs of Schneider [34] and Gel′fond [68]. The generalization of the method discussed in §3 is contained in the notes of Mahler [69], Lang's monograph [70] and the lectures of Mahler [71], in which are also considered the generalizations discussed in §4.

In the paper of Oleĭnikov [72] there are proved auxiliary propositions which can be used to prove the Second Fundamental Theorem of §4 and theorems of a similar type.

Mahler's note [73] contains a variant of the proof of the fundamental lemma used to prove the theorems of §4.

The generalizations of the method considered in §§3–6 were reported on at the Third All-Union Mathematical Congress [74], the Fourth All-Union Mathematical Congress [29] and the International Congress of Mathematicians in Moscow [75].

The papers [4] and [29] contain rather extensive surveys of many questions considered in this present report. In the papers [4], [26], [29] a number of unsolved problems are mentioned.

The printed version of this report is a substantial expansion of the version read at the conference. The basic content has been preserved. But many results have been stated in more detail, and the report now represents a rather complete survey of the literature of the question under discussion.

Because of the lack of space there is no consideration whatever in this report of other methods of the theory of transcendental numbers.

BIBLIOGRAPHY

1. Ch. Hermite, *Sur la fonction exponentielle,* C. R. Acad. Sci. Paris **77** (1873), 18–24, 74–79, 226–233, 285–293.

2. F. Lindemann, *Über die Zahl* π, Math. Ann. **20** (1882), 213–225.

3. K. Weierstrass, *Zu Lindemann's Abhandlung "Über die Ludolph'sche Zahl",* S. -B. Preuss. Akad. Wiss. 1885, 1067–1085.

4. N. I. Fel'dman and A. B. Šidlovskiĭ, *The development and present state of the theory of transcendental numbers,* Uspehi Mat. Nauk **22** (1967), no. 3 (135), 3–81 = Russian Math. Surveys **22** (1967), no. 3, 1–79. MR **35** #5400.

5. E. Borel, *Sur la nature arithmétique du nombre e,* C. R. Acad. Sci. Paris **128** (1899), 596–599.

6. J. Popken, *Sur la nature arithmétique du nombre e,* C. R. Acad. Sci. Paris **186** (1928), 1505–1507.

7. ———, *Zur Transzendens von e,* Math. Z. **29** (1929), 525–541.

8. K. Mahler, *Zur Approximation der Exponentialfunktion und des Logarithmus.* I, J. Reine Angew. Math. **166** (1932), 118–136.

9. C. L. Siegel, *Über einige Anwendungen Diophantischer Approximationen,* Abh. Preuss. Akad. Wiss. Phys. -Math. Kl. **1929**, no. 1, 1–70.

10. W. Maier, *Potenzreihen irrationalen Grenzwertes,* J. Reine Angew. Math. **156** (1927), 93–148.

11. A. Thue, *Über Annäherungswerte algebraischer Zahlen,* J. Reine Angew. Math. **135** (1909), 284–305.

12. C. L. Siegel, *Transcendental numbers,* Ann. of Math. Studies, no. 16, Princeton Univ. Press, Princeton, N. J., 1949. MR **11**, 330.

13. A. B. Šidlovskiĭ, *On transcendence and algebraic independence of the values of entire functions of certain classes,* Dokl. Akad. Nauk SSSR **96** (1954), 697–700. (Russian) MR **16**, 117.

14. ———, *Transcendence and algebraic independence of the values of entire functions of certain classes,* Moskov. Gos. Univ. Uč. Zap. No. 186 (1959), 11–70. (Russian) MR **25** #49.

15. ———, *On a criterion of algebraic independence of the values of a class of entire functions,* Dokl. Akad. Nauk SSSR **100** (1955), 221–224. (Russian) MR **16**, 907.

16. ———, *On criteria for algebraic independence of values of a class of integral functions,* Izv. Akad. Nauk SSSR Ser. Mat. **23** (1959), 35–66; English transl., Amer. Math. Soc. Transl. (2) **22** (1962), 339–370. MR **21** #1295.

17. ———, *On transcendence and algebraic independence of values of some functions,* Trudy Moskov. Mat. Obšč. **8** (1959), 283–320; English transl., Amer. Math. Soc. Transl. (2) **27** (1963), 191–230. MR **21** #6357; **27** #1418.

18. ———, *On a generalization of Lindemann's theorem,* Dokl. Akad. Nauk SSSR **138** (1961), 1301–1304 = Soviet Math. Dokl. **2** (1961), 841–844. MR **24** #A1885.

19. ———, *On algebraic independence of transcendental numbers of a certain class,* Dokl. Akad. Nauk SSSR **108** (1956), 400–403. (Russian) MR **18**, 382.

20. ———, *On the transcendence and algebraic independence of the values of certain E-functions,* Vestnik Moskov. Univ. Ser. I Mat. Meh. **1961**, no. 5, 44–59. (Russian) MR **25** #3010.

21. ———, *Algebraic independence of the values of certain hypergeometric E-functions,* Trudy Moskov. Mat. Obšč. **18** (1968), 55–64 = Trans. Moscow Math. Soc. **18** (1968), 59–70. MR **38** #4419.

22. A. B. Šidlovskiĭ, *On transcendental numbers of certain classes,* Dokl. Akad. Nauk SSSR **103** (1955), 977–980. (Russian) MR **17**, 466.

23. ———, *On the transcendence and algebraic independence of values of the E-functions related to an algebraic equation over the rational function field,* Vestnik Moskov. Univ. Ser. I Mat. Meh. 1960, no. 5, 19–28. (Russian) MR **24** #A89.

24. ———, *On transcendence of the values of a class of entire functions satisfying linear differential equations,* Dokl. Akad. Nauk SSSR **105** (1955), 35–37. (Russian) MR **17**, 947.

25. ———, *On a new criterion of transcendence and algebraic independence of values assumed by a class of entire functions,* Dokl. Akad. Nauk SSSR **106** (1956), 399–400. (Russian) MR **17**, 1057.

26. ———, *Transcendence and algebraic independence of values of E-functions related by an arbitrary number of algebraic equations over the field of rational functions,* Izv. Akad. Nauk SSSR Ser. Mat. **26** (1962), 877–910; English transl., Amer. Math. Soc. Transl. (2) **50** (1966), 141–177. MR **26** #2395.

27. ———, *On a theorem of C. L. Siegel,* Vestnik Moskov. Univ. Ser. I Mat. Meh. **24** (1969), no. 6, 39–42 = Moscow Univ. Math. Bull. **24** (1969), no. 5/6, 86–88. MR **43** #160.

28. ———, *On a general theorem about the algebraic independence of the values of E-functions,* Dokl. Akad. Nauk SSSR **171** (1966), 810–813 = Soviet Math. Dokl. **7** (1966), 1569–1572. MR **34** #4209.

29. ———, *Transcendence of the values of E-functions,* Proc. Fourth All-Union Math. Congr. (Leningrad, 1961), vol. 2, "Nauka", Leningrad, 1964, pp. 147–158. (Russian) MR **36** #2562.

30. S. Lang, *A transcendence measure for E-functions,* Mathematika **9** (1962), 157–161. MR **26** #6124.

31. A. B. Šidlovskiĭ, *Transcendence measure estimates for the values of E-functions,* Mat. Zametki **2** (1967), 33–44 = Math. Notes **2** (1967), 508–513. MR **35** #5399b.

32. ———, *On estimates of the measure of transcendence of the values of E-functions,* Uspehi Mat. Nauk **22** (1967), no. 3 (135), 245–246. (Russian) MR **35** #5399a.

33. A. I. Galočkin, *Estimate of the mutual transcendence measure of the values of E-functions,* Mat. Zametki **3** (1968), 377–386 = Math. Notes **3** (1968), 240–246. MR **37** #5156.

34. Th. Schneider, *Einführung in die transzendenten Zahlen,* Springer-Verlag, Berlin, 1957. MR **19**, 252.

35. A. I. Galočkin, *A lower bound for linear forms in values of certain hypergeometric functions,* Mat. Zametki **8** (1970), 19–28 = Math. Notes **8** (1970), 478–484. MR **42** #3027.

36. N. I. Fel'dman, *Estimates from below for certain linear forms,* Vestnik Moskov. Univ. Ser. I Mat. Meh. **22** (1967), no. 2, 63–72. (Russian) MR **35** #132.

37. C. F. Osgood, *Some theorems on diophantine approximation,* Trans. Amer. Math. Soc. **123** (1966), 64–87. MR **33** #1287.

38. A. I. Galočkin, *The Diophantine approximations of values of an exponential function and the solutions of certain transcendental equations,* Vestnik Moskov. Univ. Ser. I Mat. Meh. **27** (1972), no. 3, 16–23 = Moscow Univ. Math. Bull. **27** (1972), no. 3/4, 12–18. MR **45** #8608.

39. I. Yakabe, *Note on arithmetic properties of Kummer's function,* Mem. Fac. Sci. Kyushu Univ. Ser. A **13** (1959), 49–52. MR **21** #2641.

40. V. A. Oleĭnikov, *The transcendence and algebraic independence of the values of certain E-functions,* Vestnik Moskov. Univ. Ser. I Mat. Meh. **17** (1962), no. 6, 34–38. (Russian) MR **26** #1285.

41. A. B. Šidlovskiĭ, *Transcendence and algebraic independence of values of E-functions satisfying linear nonhomogeneous differential equations of second order,* Dokl. Akad. Nauk SSSR **169** (1966), 42–45 = Soviet Math. Dokl. **7** (1966), 879–883. MR **33** #7302.

42. V. A. Oleĭnikov, *Transcendence and algebraic independence of values of E-functions representing a solution of a third-order linear differential equation,* Dokl. Akad. Nauk SSSR **166** (1966), 540–543 = Soviet Math. Dokl. **7** (1966), 118–121.

43. ———, *Algebraic independence of values of E-functions satisfying linear nonhomogeneous differential equations of third order,* Dokl. Akad. Nauk SSSR **169** (1966), 32–34 = Soviet Math. Dokl. **7** (1966), 869–871. MR **33** #7303.

44. ———, *On the transcendence and algebraic independence of the values of certain entire functions,* Izv. Akad. Nauk SSSR Ser. Mat. **32** (1968), 63–92 = Math. USSR Izv. **2** (1968), 61–88. MR **36** #5085.

45. ———, *The algebraic independence of the values of E-functions,* Mat. Sb. **78 (120)** (1969), 301–306 = Math. USSR Sb. **7** (1969), 293–298. MR **38** #4418.

46. ———, *Certain properties of differential polynomials,* Vestnik Moskov. Univ. Ser. I Mat. Meh. **24** (1969), no. 3, 50–58 = Moscow Univ. Math. Bull. **24** (1969), no. 3/4, 41–47. MR **40** #7233.

47. ———, *On differential irreducibility of a linear nonhomogeneous equation,* Dokl. Akad. Nauk SSSR **194** (1970), 1017–1020 = Soviet Math. Dokl. **11** (1970), 1337–1341. MR **43** #194.

48. ———, *The differential irreducibility of linear equations,* Vestnik Moskov. Univ. Ser. I Mat. Meh. **26** (1971), no. 1, 44–51 = Moscow Univ. Math. Bull. **26** (1971), no. 1/2, 35–41. MR **43** #7419.

49. ———, *A sufficient criterion for differential irreducibility of a linear equation,* Vestnik Moskov. Univ. Ser. I Mat. Meh. **26** (1971), no. 3, 45–52 = Moscow Univ. Math. Bull **26** (1971), no. 3/4, 35–40. MR **43** #7420.

50. T. V. Peršikova, *On the transcendence of values of certain E-functions,* Vestnik Moskov. Univ. Ser. I Mat. Meh. **21** (1966), no. 2, 55–61. (Russian) MR **33** #4012.

51. R. M. Redheffer, *Power series and algebraic numbers,* Amer. Math. Monthly **60** (1953), 25–27. MR **14**, 541.

52. A. A. Šmelev, *Algebraic independence of values of certain E-functions,* Izv. Vysš. Učebn. Zaved. Matematika **1969**, no. 4 (83), 103–112. (Russian) MR **40** #96.

53. I. I. Belogrivov, *Transcendence and algebraic independence of values of certain E-functions,* Vestnik Moskov. Univ. Ser. I Mat. Meh. **22** (1967), no. 2, 55–62. (Russian) MR **35** #1555.

54. K. Mahler, *Applications of a theorem by A. B. Shidlovski,* Proc. Roy. Soc. Ser. A **305** (1968), 149–173. MR **37** #1322.

55. I. I. Belogrivov, *Transcendence and algebraic independence of values of hypergeometric E-functions,* Dokl. Akad. Nauk SSSR **174** (1967), 267–270 = Soviet Math. Dokl. **8** (1967), 610–613. MR **35** #5398.

56. ———, *Transcendence and algebraic independence of the values of some hypergeometric E-functions,* Mat. Sb. **82 (124)** (1970), 387–408 = Math. USSR Sb. **11** (1970), 355–376. MR **43** #3216.

57. ———, *The transcendence and algebraic independence of the values of Kummer functions,* Sibirsk. Mat. Ž. **12** (1971), 961–982 = Siberian Math. J. **12** (1971), 690–705. MR **44** #5279.

58. Ju. V. Nesterenko, *On the algebraic independence of the values of E-functions satisfying nonhomogeneous linear differential equations,* Mat. Zametki **5** (1969), 587–598 = Math. Notes **5** (1969), 352–358. MR **40** #93.

59. A. B. Šidlovskiĭ, *Transcendence of the values of some classes of functions,* Litovsk. Mat. Sb. **6** (1966), 129–130. (Russian)

60. V. G. Sprindžuk, *Finiteness of the number of rational and algebraic points on certain transcendental curves,* Dokl. Akad. Nauk SSSR **177** (1967), 524–527 = Soviet Math. Dokl. **8** (1967), 1452–1455. MR **37** #1323.

61. ———, *The irrationality of the values of certain transcendental functions,* Izv. Akad. Nauk SSSR Ser. Mat. **32** (1968), 93–107 = Math. USSR Izv. **2** (1968), 89–104. MR **36** #5087.

62. ———, *On the theory of Siegel's hypergeometric functions,* Dokl. Akad. Nauk BSSR **13** (1969), 389–391. (Russian) MR **43** #579.

63. C. F. Osgood, *Theorems about the derivatives of certain entire functions at algebraic points.* I, II, Nederl. Akad. Wetensch. Proc. Ser. A **70** = Indag. Math. **29** (1967), 431–443, 547–555. MR **36** #5346.

64. A. I. Galočkin, *The algebraic independence of the values of E-functions at certain transcendental points,* Vestnik Moskov. Univ. Ser. I Mat. Meh. **25** (1970), no. 5, 58–63 = Moscow Univ. Math. Bull. **25** (1970), no. 5/6, 41–45. MR **44** #6614.

65. ———, *On the algebraic independence of the values of E-functions at certain transcendental points,* Uspehi Mat. Nauk **25** (1970), no. 4 (154), 168. (Russian)

66. M. S. Nurmagomedov, *Arithmetic properties of the values of a class of analytic functions,* Mat. Sb. **85 (127)** (1971), 339–365 = Math. USSR Sb. **14** (1971), 329–357. MR **44** #162.

67. ———, *Arithmetical properties of the values of G-functions,* Vestnik Moskov. Univ. Ser. I Mat. Meh. **26** (1971), no. 6, 79–86 = Moscow Univ. Math. Bull. **26** (1971), no. 5/6, 111–116. MR **45** #3332.

68. A. O. Gel'fond, *Transcendental and algebraic numbers,* GITTL, Moscow, 1952; English transl., Dover, New York, 1960. MR **15**, 292; **22** #2598.

69. K. Mahler, *On a theorem of Shidlovsky,* Mimeographed notes, Amsterdam, 1959.

70. S. Lang, *Introduction to transcendental numbers,* Addison-Wesley, Reading, Mass., 1966. MR **35** #5397.

71. K. Mahler, *Lectures on transcendental numbers,* Proc. Sympos. Pure Math., vol. 20, Amer. Math. Soc., Providence, R. I., 1969, pp. 248–274.

72. V. A. Oleĭnkov, *Some properties of algebraically dependent quantities,* Vestnik Moskov. Univ. Ser. I Mat. Meh. **17** (1962), no. 5, 11–17. (Russian) MR **26** #1334.

73. K. Mahler, *Concerning a lemma of A. B. Šidlovskiĭ,* Mat. Zametki **2** (1967), 25–32 = Math. Notes **2** (1967), 504–507. MR **36** #2561.

74. A. B. Šidlovskiĭ, *On a class of transcendental numbers,* Proc. Third All-Union Math. Congress (Moscow, 1956), vol. I, Izdat. Akad. Nauk SSSR, Moscow, 1956, pp. 15–16. (Russian)

75. ———, *Transcendence and algebraic independence of values of E-functions,* Proc. Internat. Congress Math. (Moscow, 1966), "Mir", Moscow, 1968, pp. 299–307; English transl., Amer. Math. Soc. Transl. (2) **70** (1968), 209–215. MR **39** #5481.

Translated by G. A. KANDALL

Trudy Mat. Inst. Steklov.
132 (1973)

Proc. Steklov Inst. Math.
132 (1973)

DIRICHLET CHARACTERS AND POLYNOMIALS *

D. A. BURGESS

The topic of this paper is a survey of the progress to date in the solution of the following problem:

Problem. *To investigate the distribution of the values of*

$$\chi(f(x))$$

where $f(x)$ denotes a fixed polynomial with integer coefficients and χ is any non-principal Dirichlet character of fixed order k to the prime modulus p.

In the special case $f(x)=x$ I have obtained [1] the estimate

$$\sum_{x=N+1}^{N+H} \chi(x) \ll H^{1-1/r} p^{(r+1)/4r^2} \log p, \qquad r=1, 2, \ldots, \tag{1}$$

which implies that the values of $\chi(f(x))$ are uniformly distributed in all intervals $[N+1, N+H]$ with $H \gg p^{(1/4)+\varepsilon}$. The case $r=1$ of (1) is the inequality obtained in about 1918 by Pólya [8] and Vinogradov [9]. For general polynomials we have only the analogue of the inequality of Polya and Vinogradov, that

$$\sum_{x=N+1}^{N+H} \chi(f(x)) \ll p^{1/2} \log p \tag{2}$$

(where $f(x)$ is not a perfect k-th power). This is a consequence of Weil's theorem that the Riemann Hypothesis holds for the ζ-function of the appropriate function field. (2) is sufficient to prove only that the values of $\chi(f(x))$ are uniformly distributed in intervals of length $\gg p^{(1/2)+\varepsilon}$.

Vinogradov gave a simple argument that, combined with his character sum estimate, enabled him to show that if

$$H \gg p^{(1/2\sqrt{e})+\varepsilon} \tag{3}$$

then there exists an x satisfying

$$0 < x < H, \qquad \chi(x) \neq 1.$$

Combining his argument with (1) enables (3) to be strengthened to

$$H \gg p^{(1/4\sqrt{e})+\varepsilon}.$$

This led me to consider the following problem:

Problem. *To find a bound for those H for which*

AMS (MOS) subject classifications (1970). Primary 10G05; Secondary 10A15.

**Editor's note.* The body of this article is an unaltered reprinting of the original paper. Details of the bibliography have been checked and put in standard form, and MR citations have been added.

$$\chi(f(1)) = \chi(f(2)) = \ldots = \chi(f(H)).$$

In 1963 I proved [2]

Theorem 1. *If*

$$f(x) = x^{\alpha_1}(x + a_2)^{\alpha_2} \ldots (x + a_\nu)^{\alpha_\nu}$$

where $a_2, \ldots, a_\nu$ form an increasing sequence of fixed positive integers and $f(x)$ is not a pefect k-th power then (4) *implies that*

$$H \ll p^{(1/4\sqrt{e})+\varepsilon}.$$

In 1967 I discussed irreducible quadratic polynomials $f(x)$ and proved [3]:

Theorem 2. *If $f(x) = x^2 + a$ is an irreducible quadratic then there exists an x satisfying*

$$0 < x \ll p^{(1/2\sqrt{e})+\varepsilon}, \quad \left(\frac{x^2 + a}{p}\right) \neq 1.$$

This theorem depends on the following lemma.

Lemma. *If $(x_1, x_2) = 1$ then there exist integers $r, y_1, \ldots, y_r, \alpha_1, \ldots, \alpha_r$ satisfying*

$$x_1^2 + kx_2^2 = \prod_{i=1}^{r} (y_i^2 + k)^{\alpha_i}$$

and

$$y_i \ll x_1^2 + kx_2^2.$$

The theorem is deduced by the use of this lemma and estimates, similar to (1), for character sums in the quadratic field $\mathbf{Q}(\sqrt{k})$ [4]. Gillett [5] has obtained some generalisation of this to other characters.

To extend the work to irreducible cubics one needs a corresponding lemma for cubics, for suitable character sum estimates have been given by Hudson [6] and by Karacuba [7].

I have obtained such lemmas only under the condition that $f(x)$ has a rational linear factor, and even then only in special cases. As an example of the results I have obtained I present the following:

Theorem 3. *If $f(x) = x(x^2 - c(c+1))$ where c is a fixed integer, then* (4) *implies that*

$$H \ll p^{(1/2\sqrt{e})+\varepsilon}.$$

Proof. The proof depends on the polynomial identity

$$\frac{f(x^2 + x - c(c+1))}{f(x)f(x+1)} = \frac{(x-c)(x+(c+1))}{x(x+1)}.$$

Thus if $x \ll H^{1/2}$ we have

$$\chi\left\{\frac{(x-n)(x+n+1)}{x(x+1)}\right\}$$

is constant for $0 < x < H^{1/2}$. But this implies that

$$\chi\{x(x+n)^{k-1}(x+n+1)^{k-1}(x+2n+1)\}$$

is constant for $0<x<H^{1/2}-n$, and thus Theorem 1 implies Theorem 3.

Theorem 3 can be extended to many other reducible cubics. The simplest example for which I have not yet obtained such a result is

$$f(x)=x(x^2+10).$$

If one considers just those integers x for which

$$x\equiv 1 \pmod{(a+1)}$$

then for the polynomial

$$f(x)=x(x^2+a)$$

one obtains the identity

$$\frac{f\left(\frac{(x^2+a)}{1+a}+x\right)f(1)^3}{f(x)f(x+a+1)}=\frac{(x+a)(x+1)}{x(x+a+1)}.$$

Thus one obtains that if (4) holds then for x satisfying

$$x\ll H^{1/2}, \quad x\equiv 1 \pmod{(a+1)}$$

we have

$$\chi\left(\frac{(x+a)(x+1)}{x(x+a+1)}\right)$$

is constant. It seems likely that this may be sufficient to imply a suitable bound on H.

BIBLIOGRAPHY

1. D. A. Burgess, *On character sums and L-series*. II, Proc. London Math. Soc. (3) 13 (1963), 524–536. MR 26 #6133.

2. ———, *On Dirichlet characters of polynomials*, Proc. London Math. Soc. (3) 13 (1963), 537–548. MR 26 #6134.

3. ———, *On the quadratic character of a polynomial*, J. London Math. Soc. 42 (1967), 73–80. MR 34 #5765.

4. ———, *Character sums and primitive roots in finite fields*, Proc. London Math. Soc. (3) 17 (1967), 11–25. MR 36 #2597.

5. J. R. Gillett, *Character sums in algebraic number theory*, Ph. D. Thesis, Nottingham, 1968.

6. Margaret C. Hudson, *Some arithmetical problems in Galois fields*, Ph. D. Thesis, Nottingham, 1968.

7. A. A. Karacuba, *Character sums and primitive roots in finite fields*, Dokl. Akad. Nauk SSSR 180 (1968), 1287–1289 = Soviet Math. Dokl. 9 (1968), 755–757. MR 37 #5188.

8. G. Pólya, *Über die Verteilung der quadratischen Reste und Nichtreste*, Nachr. Ges. Wiss. Göttingen Math.-Phys. Kl. 1918, 21–29.

9. I. M. Vinogradov, *Sur la distribution des résidus et des non-résidus des puissances*, J. Phys.-Math. Soc. Perm. 1 (1918), 94–98.

Trudy Mat. Inst. Steklov.
132 (1973)

Proc. Steklov Inst. Math.
132 (1973)

ON THE EXTENDED HECKE THETA-FORMULA*

T. TATUZAWA

ABSTRACT. Hecke's theta-formula is deduced in generalized form. This permits one to apply the formula to prove functional equations for Hecke's L-functions with complex characters.

In this opportunity, I should like to look at again the precious result of the theta-formula due to Hecke. Though Landau's exposition was precise, he did not use the fractional ideals. Later on Artin constructed the ideal theory including fractional ideals, and Suetuna reintroduced Landau's exposition based on this new ideal theory (printed in Japanese). By using the Poisson summation formula (see [1—5]), we shall deduce the formula in a slightly extended form which can be used for the proof of the functional equations of Hecke's L-functions with complex Grössen-characters.

Let k be an algebraic number field of degree $n=r_1+2r_2$, $k^{(p)}$ $(1 \leqslant p \leqslant r_1)$ be r_1 real conjugate fields and $k^{(p)}$, $k^{(p+r_2)}$, $(r_1+1 \leqslant p \leqslant r_1+r_2)$ be r_2 pairs of complex conjugate fields. Let $\mathfrak{O}$ be the integral domain of all integers in k, d be the discriminant of k and $\mathfrak{D}$ be the different of k, which can be characterised by

$$\mathfrak{D}^{-1}=\{\gamma;\ \gamma \in k,\ \operatorname{trace}(\gamma\alpha)=S(\gamma\alpha)=\text{rational integer, for } \forall\alpha \in \mathfrak{O}\},$$

so that $N\mathfrak{D}=|d|$. Let $\mathfrak{M}$ be an integral ideal in k and $\mathfrak{M}_\infty$ be some finite formal product of infinite places $\mathfrak{x}_\infty^{(p)}$ $(1 \leqslant p \leqslant r_1)$. A formal product

$$\tilde{\mathfrak{M}}=\mathfrak{M}\cdot\mathfrak{M}_\infty=\mathfrak{M}\prod \mathfrak{x}_\infty^{(p)}$$

is called a generalized module or an integral divisor. For $\alpha \in k$, the congruence

$$\alpha \equiv 1 \bmod \tilde{\mathfrak{M}}$$

means that $\alpha \equiv 1 \bmod \mathfrak{M}$ in multiplicative congruence and $\alpha^{(p)}>0$ for p appearing in $\mathfrak{M}_\infty=\prod \mathfrak{x}_\infty^{(p)}$. Let $S=S_{\tilde{\mathfrak{M}}}$ be the group of all principal ideals (α), generated by α satisfying $\alpha \equiv 1 \bmod \tilde{\mathfrak{M}}$. Let $A=A_{\tilde{\mathfrak{M}}}$ be the group of all ideals $\mathfrak{A}$ prime to $\mathfrak{M}$ in k. A character

$$\chi \bmod \tilde{\mathfrak{M}}$$

can be defined in natural way from an abelian character of A/S. We write, for brevity,

$$\tilde{\mathfrak{f}}/\tilde{\mathfrak{M}}$$

AMS (MOS) subject classifications (1970). Primary 10G05; Secondary 10H10.

**Editor's note.* The body of this article is an unaltered reprinting of the original paper. Details of the bibliography have been checked and put in standard form, and MR citations have been added.

if $\tilde{\mathfrak{M}}$ is divisible by $\tilde{\mathfrak{f}}$. The character χ mod $\tilde{\mathfrak{M}}$ is called to be definable by $\tilde{\mathfrak{f}}$, whenever the assumption

$$\alpha \equiv 1 \bmod \tilde{\mathfrak{f}}, \qquad (\alpha, \mathfrak{M}) = \mathfrak{O}$$

entails the conclusion

$$\chi(\alpha) = 1.$$

There exists the minimal generalized module which defines χ. This is called the conductor of χ. If the conductor of χ mod $\tilde{\mathfrak{M}}$ is $\tilde{\mathfrak{M}}$ itself, then χ is called a primitive character mod $\tilde{\mathfrak{M}}$.

Let $\tilde{\mathfrak{f}}$ be a generalized module in k. In $(\mathfrak{f}\mathfrak{D})^{-1}$ there exist

$$\rho_j \equiv 1 \bmod \mathfrak{f}_\infty \qquad (1 \leqslant j \leqslant s = N(\mathfrak{f})),$$

satisfying further conditions: all the pairs of ρ_j are not congruent mod $\mathfrak{D}^{-1}$ and if $\rho \in (\mathfrak{f}\mathfrak{D})^{-1}$ then $\rho \equiv \rho_j \bmod \mathfrak{D}^{-1}$ for some ρ_j. The set of these ρ_j is called a complete residue system mod $\mathfrak{D}^{-1}$ in $(\mathfrak{f}\mathfrak{D})^{-1}$. We now define

$$G(\chi) = \sum_{j=1}^{s} \chi(\rho_j \mathfrak{D}\mathfrak{f})\, e^{2\pi i \operatorname{trace}(\rho_j)}.$$

Let $\mu \neq 0$ and $\mu \in (\mathfrak{f}\mathfrak{D})^{-1}$. We further define

$$G(\mu, \chi) = \sum_{\beta} \chi(\beta)\, e^{2\pi i \operatorname{trace}(\beta\mu)},$$

where β runs over a complete residue system mod $\mathfrak{f}$ satisfying $\beta \equiv 1 \bmod \mathfrak{f}_\infty$. If we take $\eta = \eta(\mu)$ such that

$$\eta \equiv 1 \bmod \mathfrak{f}, \qquad \eta \equiv \mu \bmod \mathfrak{f}_\infty,$$

then we can prove

$$G(\mu, \chi) = \bar{\chi}(\eta(\mu)\mu\mathfrak{D}\mathfrak{f})\, G(\chi)$$

and

$$\chi(\eta(\mu)) = \prod_p (\operatorname{sgn} \mu^{(p)}),$$

where p runs overi ndexes appearing in $\mathfrak{f}_\infty = \prod \mathfrak{x}_\infty^{(p)}$. Further we can infer that

$$\sum_{j=1}^{s} | G(\rho_j, \chi) |^2 = \varphi(\mathfrak{f}) | G(\chi) |^2 = \varphi(\mathfrak{f}) N(\mathfrak{f}).$$

Consequently, if we define

$$I(\chi) = (-i)^q G(\chi) / \sqrt{N(\mathfrak{f})},$$

q being the number of infinite places in $\tilde{\mathfrak{f}}$, then we obtain (see [4])

$$| I(\chi) | = 1, \qquad \overline{I(\chi)} = I(\bar{\chi}).$$

Let $\mathfrak{A}$ be an ideal in k and $\mu \in \mathfrak{A}$. Define

$$\psi(\mathfrak{A}, \mu) = \begin{cases} \chi\left(\frac{\mu}{\mathfrak{A}}\eta(\mu)\right) & \text{if } \mu \neq 0, \\ 0 & \text{if } \mu = 0,\ \mathfrak{f} \neq \mathfrak{O}, \\ \bar{\chi}(\mathfrak{A}) & \text{if } \mu = 0,\ \mathfrak{f} = \mathfrak{O}, \end{cases}$$

and put $\psi(\mu) = \psi(\mathfrak{O}, \mu)$. If we write

$$c = c(\mathfrak{A}) = \{|d|\, N(\mathfrak{A})^2 N(\mathfrak{f})\}^{-\frac{1}{n}},$$

then

$$c' = c((\mathfrak{A}\mathfrak{f})\mathfrak{D}^{-1}) = c^{-1}.$$

Now the Hecke theta-series is defined by

$$\theta(t;\ \mathfrak{A},\ \chi) = \sum_{\mu \in \mathfrak{A}} \psi(\mathfrak{A}, \mu) P(\mu) \exp\left\{-\pi c \sum_{p=1}^{n} t_p |\mu^{(p)}|^2\right\},$$

where t_p are real variables satisfying

$$t_p = t_{p+r_2}\ (r_1 + 1 \leqslant p \leqslant r_1 + r_2),$$

and

$$P(\mu) = \begin{cases} 1 & \text{if } \mathfrak{f}_\infty = \varnothing \text{ (vacuous)}, \\ \prod \mu^{(p)} & \text{if } \mathfrak{f}_\infty = \prod \mathfrak{x}_\infty^{(p)}. \end{cases}$$

Then the Hecke theta-formula is stated as follows:

$$\theta(t;\ \mathfrak{A},\ \chi) = c^{-q} P(t)^{-1} (t_1 \ldots t_n)^{-1/2} I(\chi)\, \theta\left(\frac{1}{t};\ \frac{1}{\mathfrak{A}\mathfrak{f}\mathfrak{D}},\ \bar{\chi}\right).$$

If we replace $P(\mu)$ by the new

$$P(\mu) = \begin{cases} 1 & \text{if } \mathfrak{f}_\infty = \varnothing, \\ \left(\prod \mu^{(p)}\right)\left(\mu^{(r_1+1)}\right)^{b_{r_1+1}} \ldots \left(\mu^{(r_1+r_2)}\right)^{b_{r_1+r_2}} & \text{if } \mathfrak{f}_\infty = \prod \mathfrak{x}_\infty^{(p)}, \end{cases}$$

b_{r_1+j} $(1 \leqslant j \leqslant r_2)$ being some non-negative integers, then the new theta-formula becomes

$$\theta(t;\ \mathfrak{A},\ \chi) = c^{-q} (ic)^{-b_{r_1+1} - \cdots - b_{r_1+r_2}} P(t)^{-1} (t_1 \ldots t_n)^{-\frac{1}{2}} I(\chi)\, \theta\left(\frac{1}{t};\ \frac{1}{\mathfrak{A}\mathfrak{f}\mathfrak{D}},\ \bar{\chi}\right),$$

which we wish to prove here.

To deduce this formula, we need the following lemmas.

L e m m a 1. *If we write* $c_0 = c_0(\mathfrak{A}) = (|d|\, N(\mathfrak{A})^2)^{-\frac{1}{n}}$ *and* $c_0' = c_0((\mathfrak{A}\mathfrak{D})^{-1})$, *then* $c_0 c_0' = 1$ *and*

$$\sum_{\mu \in \mathfrak{A}} P(\mu + \beta) \exp\left\{-\pi c_0 \sum_{p=1}^{n} |\mu^{(p)} + \beta^{(p)}|^2 t_p\right\}$$

$$= (ic_0)^{-a} P(t)^{-1} (t_1 \ldots t_n)^{-\frac{1}{2}} \sum_{M \in (\mathfrak{A}\mathfrak{D})^{-1}} \bar{P}(M) \exp\left\{-\pi c_0' \sum_{p=1}^{n} \frac{|M^{(p)}|^2}{t_p} + 2\pi i S(\beta M)\right\}$$

$$(a = q + b_{r_1+1} + \ldots + b_{r_1+r_2}).$$

Lemma 2.

$$\sum_{\mu\in\mathfrak{A}} \psi(\mathfrak{A}, \mu) P(\mu) \exp\left\{-\pi c \sum_{p=1}^{n} t_p |\mu^{(p)}|^2\right\}$$

$$= c^{-q}(ic)^{-b_{r_1+1}-\cdots-b_{r_1+r_2}} P(t)^{-1}(t_1 \ldots t_n)^{-\frac{1}{2}} I(\chi)$$

$$\times \sum_{M\in(\mathfrak{A}\mathfrak{f}\mathfrak{D})^{-1}} \bar{\psi}\left(\frac{1}{\mathfrak{A}\mathfrak{f}\mathfrak{D}}, M\right) \bar{P}(M) \exp\left\{-\pi c' \sum_{p=1}^{n} \frac{|M^{(p)}|^2}{t_p}\right\}.$$

Here we shall sketch the proof of Lemma 1. Following in Landau's wake (see [3]), we can deduce Lemma 2 and also the new theta-formula.

Let $\alpha_1, \ldots, \alpha_n$ be a basis of $\mathfrak{A}$. If we determine $A_1, \ldots, A_n$ by the simultaneous equations

$$\alpha_p = \sum_{q=1}^{n} S(\alpha_p \alpha_q) A_q \qquad (1 \leqslant p \leqslant n),$$

then these become a basis of $(\mathfrak{A}\mathfrak{D})^{-1}$ and satisfy the relations

$$S(\alpha_p A_q) = \delta_{pq} \qquad (1 \leqslant p,\ q \leqslant n).$$

If $l_j\ (1 \leqslant j \leqslant n)$ run over all rational integers independently, then

$$\mu = l_1\alpha_1 + \ldots + l_n\alpha_n, \qquad M = l_1 A_1 + \ldots + l_n A_n$$

move over all numbers of $\mathfrak{A}$ and $(\mathfrak{A}\mathfrak{D})^{-1}$ respectively. We write

$$\beta = \alpha_1 u_1 + \ldots + \alpha_n u_n,$$

then

$$l_1 u_1 + \ldots + l_n u_n = M^{(1)}\beta^{(1)} + \ldots + M^{(n)}\beta^{(n)}.$$

We write further

$$f(u_1, \ldots, u_n) = \exp\left\{-\pi c_0 \sum_{p=1}^{n} |\alpha_1^{(p)} u_1 + \ldots + \alpha_n^{(p)} u_n|^2 t_p \right.$$

$$\left. + 2\pi i \sum_{p=1}^{r+1} (\alpha_1^{(p)} u_1 + \ldots + \alpha_n^{(p)} u_n) s_p\right\},$$

where s_p and $t_p > 0$ are real variables satisfying $t_p = t_{p+r_2}$ $(r_1 + 1 \leqslant p \leqslant r_1 + r_2)$. By using Poisson's summation formula, we have

$$\sum_{l_j=-\infty}^{\infty} \ldots \sum^{\infty} f(u_1 + l_1, \ldots, u_n + l_n) = \sum_{l_j=-\infty}^{\infty} \ldots \sum^{\infty} \exp\{-2\pi i (l_1 u_1 + \ldots$$

$$\ldots + l_n u_n\} \int_{-\infty}^{\infty} \ldots \int_{-\infty}^{\infty} f(u_1, \ldots, u_n) \exp\{2\pi i (l_1 u_1 + \ldots + l_n u_n)\} du_1 \ldots du_n.$$

We make change of variables in the following way,

$$\alpha_1^{(j)}u_1+\ldots+\alpha_n^{(j)}u_n=v_j \qquad (1\leqslant j\leqslant r_1),$$

$$\left.\begin{aligned}(\Re\alpha_1^{(r_1+j)})\,u_1+\ldots+(\Re\alpha_m^{(r_1+j)})\,u_m&=\frac{1}{\sqrt{2}}\,v_{r_1+j},\\ (\Im\alpha_1^{(r_1+j)})u_1+\ldots+(\Im\alpha_m^{(r_1+j)})\,u_m&=\frac{1}{\sqrt{2}}\,v_{r_1+r_2+j}.\end{aligned}\right\}\quad(1\leqslant j\leqslant r_2)$$

Then we have

$$l_1u_1+\ldots+l_nu_n=M^{(1)}v_1+\ldots+M^{(r_1)}v_{r_1}+\sqrt{2}\,(\Re M^{(r_1+1)}v_{r_1+1}$$
$$-\Im M^{(r_1+1)}v_{r_1+r_2+1})+\ldots+\sqrt{2}\left(\Re M^{(r_1+r_2)}v_{r_1+r_2}-\Im M^{(r_1+r_2)}v_{r_1+2r_2}\right)$$

and

$$\left|\frac{\partial\,(u_1,\ldots,u_n)}{\partial\,(v_1,\ldots,v_n)}\right|=\frac{1}{N\mathfrak{A}\sqrt{|d|}}.$$

Hence

$$f(u_1,\ldots,u_n)\exp\{2\pi i\,(l_1u_1+\ldots+l_nu_n)\}$$
$$=\prod_{p=1}^{r_1}\exp\{-\pi ct_pv_p^2+2\pi i\,(s_p+M^{(p)})\,v_p\}$$
$$\times\prod_{p=r_1+1}^{r_1+r_2}\exp\left\{-\pi ct_pv_p^2+2\pi i\left(\frac{1}{\sqrt{2}}\,s_p+\sqrt{2}\,\Re M^{(p)}\right)v_p\right\}$$
$$\times\prod_{p=r_1+r_2+1}^{r_1+2r_2}\exp\{-\pi ct_pv_p^2-\sqrt{2}\,\pi s_{p-r_2}v_p-2\pi i\,(\sqrt{2}\,\Im M^{(p)})\,v_p\}$$

and consequently

$$\int_{-\infty}^{\infty}\cdots\int_{-\infty}^{\infty}f(u_1,\ldots,u_n)\exp\{2\pi i\,(l_1u_1+\ldots+l_nu_n)\}\,du_1\ldots du_n$$
$$=\frac{1}{N\mathfrak{A}\sqrt{|d|}}\prod_{p=1}^{r_1}\frac{1}{\sqrt{c_0t_p}}\exp\left\{-\frac{\pi\,(s_p+M^{(p)})^2}{c_0t_p}\right\}\prod_{p=r_1+1}^{r_1+r_2}\frac{1}{\sqrt{c_0t_p}}$$
$$\times\exp\left\{\frac{-\pi\left(\frac{1}{\sqrt{2}}\,s_p+\sqrt{2}\,\Re M^{(p)}\right)^2}{c_0t_p}\right\}\prod_{p=r_1+r_2+1}^{r_1+2r_2}\frac{1}{\sqrt{c_0t_p}}$$
$$\times\exp\left\{-\frac{2\pi\,(\Im M^{(p)})^2}{c_0t_p}+\frac{\pi s_{p-r_2}^2}{2c_0t_p}+\pi i\,\frac{2s_{p-r_2}\Im M^{(p)}}{c_0t_p}\right\}.$$

It follows from Poisson's summation formula that

$$\sum_{\mu\in\mathfrak{A}}\exp\left\{-\pi c_0\sum_{p=1}^{n}t_p\,|\mu^{(p)}+\beta^{(p)}|^2+2\pi i\sum_{p=1}^{r+1}s_p\,(\mu^{(p)}+\beta^{(p)})\right\}$$
$$=\frac{1}{\sqrt{t_1\ldots t_n}}\sum_{M\in(\mathfrak{A}\mathfrak{D})^{-1}}\exp\left\{-\pi c_0'\sum_{p=1}^{r_1}\frac{(s_p+M^{(p)})^2}{t_p}\right.$$
$$\left.-2\pi c_0'\sum_{p=r_1+1}^{r_1+r_2}\frac{M^{(p)}s_p+|M^{(p)}|^2}{t_p}-2\pi iS\,(\beta M)\right\}.$$

By differentiating once by $s_1, \ldots, s_q$ and b_j times by s_j $(r_1+1 \leqslant j \leqslant r_1+r_2)$ and then putting $s_j=0$, we get Lemma 1.

BIBLIOGRAPHY

1. S. Bochner, *Vorlesungen über Fouriersche Integral,* Akad. Verlag., Leipzig, 1932; English transl., Princeton Univ. Press, Princeton, N. J., 1959. MR 21 #5851.

2. E. Hecke, *Über die L-Funktionen und den Dirichletschen Primzahlsatz für einen beliebigen Zahlkörper,* Nachr. Ges. Wiss. Göttingen Math.-Phys. Kl. 1911, 299–318.

3. E. Landau, *Über Ideale und Primideale in Idealklasse,* Math. Z. 2 (1918), 52–154.

4. T. Tatuzawa, *On the Hecke-Landau L-series,* Nagoya Math. J. 16 (1960), 11–20. MR 22 #1552.

5. ———, *Some results in the Fourier analysis,* Nagoya Math. J. 27 (1966), 55–59. MR 33 #1652.

Trudy Mat. Inst. Steklov.
132 (1973)

Proc. Steklov Inst. Math.
132 (1973)

MODULAR CORRESPONDENCES, HEIGHTS AND ISOGENIES OF ABELIAN VARIETIES

A. N. PARŠIN

ABSTRACT. A connection is established between Tate's conjectures in the theory of abelian varieties and the behavior of the height function on moduli schemes, and possible ways of proving these conjectures are discussed.

Bibliography: 20 items.

Introduction

In this paper we study different approaches to finiteness conjectures for isogenies of abelian varieties. These conjectures arose in Tate's investigations of algebraic cycles on schemes defined over fields of finite type over a prime subfield [1], [2]. In particular, if X and Y are abelian varieties over such a field K, Tate has conjectured that the canonical mapping

$$\mathrm{Hom}(X,\ Y)\otimes \mathbf{Z}_l \to \mathrm{Hom}(T_l(X),\ T_l(Y)) \tag{1}$$

is bijective (here $T_l(X)$, $T_l(Y)$ are the respective Tate modules). This conjecture was proved by Tate for finite fields [1]. The essential part of his proof was the following fact about isogenies over the field K.

T. There exist, up to K-isomorphism, only a finite number of abelian varieties Y that are isogenous to a given variety X and have a polarization of given degree d^2.

For a global field K, the role of this assertion in Diophantine geometry is discussed in [3]. For finite fields K, the statement T is true for trivial reasons. In other cases, for example, if K is a number field, T has been proved only for varieties of dimension 1 (I. R. Šafarevič; see [2], [3]). But if K is a function field of finite characteristic p, it is necessary to restrict attention to isogenies whose degree is prime to p (see the example at the end of §3 of [2]). The conjecture T in this form will hold if $\dim X \leqslant 2$ and $(d, p) = 1$ [2].

The meaning of these results can be explained by Serre's remark that conjecture T for varieties of dimension 2 implies the bijectivity of (1) for varieties of dimension 1. For a proof of this see §4 of [2] (as far as I know, Serre's proof is unpublished). We will now give some variants of conjecture T, each of which is sufficient to prove the bijectivity of (1) in the one-dimensional case (see Remark 4 in §4 of [2]). The definition of a Tate isogeny is given below (Definition 4 of §1).

AMS (MOS) subject classifications (1970). Primary 14K15, 14K10, 14G25, 14K99; Secondary 14D20, 14C20, 14K30, 10B99.

Let X be an abelian variety over the field K; $l, l_1, \cdots, l_s$ prime numbers; $d \geqslant 1$ (if $p = \text{char } K > 0$, it is assumed that all these numbers are prime to p); λ a polarization on X of degree d^2.

There exist only a finite number of pairs (Y, μ), up to K-isomorphism, where Y is an abelian variety over K, and μ is a polarization on Y of degree d^2, such that there exists a Tate isogeny $X \to Y$ of degree that

$T(X, K, \lambda, d, l)$: *is l;*

$T(X, K, \lambda, d, l_1, \cdots, l_s)$: *is the product of the degrees l_i, $i = 1, 2, \cdots, s$;*

$T(X, K, \lambda, d)$: *is prime to p.*

We assume that all these assertions are true if K is an algebraic number field or an algebraic function field with a finite field of constants. In order to establish a connection between them, we need

LEMMA. *Let λ be a polarization of degree d^2 on X, $(d, p) = 1$ and $f: X \to Y$ an isogeny. Then the following conditions are equivalent.*

1. *On Y there exists a polarization μ such that*

$$\hat{f} \circ \mu \circ f = m\lambda, \qquad \deg \mu = \left(\frac{m^{\dim X}}{\deg f}\right)^2 \deg \lambda.$$

2. $\operatorname{Ker} f \subset \operatorname{Ker}(m\lambda)$ *and is the isotropy subgroup of the canonical bilinear form $e_{m\lambda}$ on* $\operatorname{Ker}(m\lambda)$.

For the definition of the form e_λ, its connection with the Weil pairing and a proof of the lemma, see [4], and also the end of §1. We now denote by $\Rightarrow$, & and $\forall$ the standard logical symbols.

PROPOSITION. *The following implications hold*:

1. $T(X, K, \lambda, d) \Rightarrow T(X, K, \lambda, d, l_1, \cdots, l_s) \Rightarrow T(X, K, \lambda, d, l)$.

2. $\forall X, \lambda, T(X, K, \lambda, d, l_1)$ & $\cdots$ &$\forall X, \lambda, T(X, K, \lambda, d, l_s) \Rightarrow \forall X, \lambda,$ $T(X, K, \lambda, d, l_1, \cdots, l_s)$.

3. *If* $(l, d) = 1$, *then* $\forall X, \lambda, T(X, K, \lambda, 1, l) \Rightarrow \forall X, \lambda, T(X, K, \lambda, d, l)$.

PROOF. Assertion 1 is obvious. In order to prove 2, we consider an isogeny $X \to Y$ of the condition of the conjecture $T(X, K, \lambda, d, l_1, \cdots, l_s)$. It can be represented in the form

$$X \xrightarrow{f_1} X_1 \to \cdots \to X_{s-1} \xrightarrow{f_s} Y,$$

where $\deg f_i = l_i$, $i = 1, \cdots, s$. Using the lemma, it is easy to construct polarizations of degree d^2 on the varieties X_i such that the f_i become Tate isogenies. By hypothesis each variety X_i defines the succeeding X_{i+1} up to a finite number of possibilities. This obviously gives the requirement.

We now turn to the isogeny $f\colon X \to Y$ of the conditions of the conjecture $T(X, K, \lambda, d, l)$. Using the lemma, we can construct varieties X' and Y', a diagram

$$\begin{array}{ccc} X & \xrightarrow{f} & Y \\ {\scriptstyle\varphi}\downarrow & & {\scriptstyle\psi}\downarrow \\ X' & & Y' \end{array} \tag{2}$$

and polarizations λ' and μ' of degree 1 on X' and Y' such that

$$\hat{\varphi} \circ \lambda' \circ \varphi = \lambda, \qquad \deg \varphi = d,$$
$$\hat{\psi} \circ \mu' \circ \psi = \mu, \qquad \deg \psi = d.$$

To do this, it suffices to take as $\operatorname{Ker} \varphi$ the isotropy subgroup of e_λ of order d in $\operatorname{Ker} \lambda$ and to put $X' = X/\operatorname{Ker} \varphi$. We treat the morphism ψ analogously.

Since $(l, d) = 1$, from the relations

$$\hat{f} \circ \mu \circ f = m\lambda, \qquad m^{\dim X} = \deg f$$

we will get that $f(\operatorname{Ker} \lambda) \subset \operatorname{Ker} \mu$. Moreover, the bilinear forms e_λ and e_μ are connected by the relation

$$e_\mu(fu, fv) = m e_\lambda(u, v)$$

(see [4]). We can therefore fix the morphism φ on X, and on all varieties Y of the condition of the conjecture we can set

$$\operatorname{Ker} \psi = f(\operatorname{Ker} \varphi).$$

This is an isotropy subgroup of order d. From this it follows that for each isogeny f of the condition of the hypothesis $T(X, K, \lambda, d, l)$ diagram (2) can be completed to a commutative diagram

$$\begin{array}{ccc} X & \xrightarrow{f} & Y \\ {\scriptstyle\varphi}\downarrow & & {\scriptstyle\psi}\downarrow \\ X' & \xrightarrow{f'} & Y', \end{array}$$

in which f' is a Tate isogeny (relative to the polarizations λ' and μ') of degree $\deg f$. Applying $T(X', K, \lambda', 1, l)$ to Y', we get the desired result.

Beginning in §2 of this paper, we will consider abelian varieties with polarizations of degree 1. This greatly simplifies the presentation, and passing to the general case presents no difficulties (see §2). Aside from this, in the case of dimension 1 we can limit ourselves to such polarizations (see §1). Finally, in view of what we just proved, results about polarizations of degree 1 automatically extend to a wide class of polarizations of degree > 1. However, we note that if we are interested in applying the finiteness conjecture to a proof of the bijectivity of (1), it is necessary to take polarizations of arbitrary degree into account (see Theorem 1 and the proof of Proposition 4 in §4 of [2]).

We proceed to a description of the contents of the various sections of this paper. §1 contains a brief survey of the theory of moduli of abelian varieties and isogenies of them over **Z**. The results given here are basically due to Mumford [5] or are obtained by combining the constructions in his book with the general theory of Hilbert schemes.

In §2 the analytic approach to moduli is presented in a parallel way. The basic source here is the Cartan seminar of 1957/58 [6]. For the case of elliptic curves the material of §§1 and 2 are also presented in Deligne's lecture [7].

In §3 we introduce the important sheaf L on moduli schemes, and with its help we construct compactifications of the moduli, both analytic and algebraic. Another theme of this section is the explanation of the behavior of the sheaf L relative to modular correspondences connected with isogenies. The invariance properties obtained here are the main results of §4.

This section begins with a reformulation of the finiteness conjecture given above in terms of assertions about rational points of moduli schemes. We discuss two approaches to a proof of the finiteness of the sets of rational points obtained in this way. The first is based on an application of the usual theory of height [8] to modular correspondences and to the sheaf L. This gives an estimate from above for the growth of the height on a set of rational points representing isogenous abelian varieties. In case of dimension 1, combining this with Mumford's estimate [9] of the growth of the height on the set of all rational points, we will obtain conjecture $T(X, K, \lambda, 1, l)$ for number fields K. As we have already observed, in this case the stronger assertion $T(X, K, \lambda, 1)$ (and even T) holds. The proof of this, due to Šafarevič, is considerably more difficult than ours, since it uses Siegel's theorem on integer points [8].

The second approach studied in §4 consists of constructing a new height, which behaves invariantly relative to modular correspondences, resembling the Tate height in the category of abelian varieties. At first it appeared that such a height could be constructed by using the Tate height [10]. My paper [11] was based on this. However, Manin observed [12] that this is false. In §4 we propose a definition for the local components of an invariant height, corresponding to archimedean valuations: an arithmetic index of intersection at the infinite places.(1) The question of the existence of nonarchimedean components remains open. [11] serves as a model for the arguments carried out in §§2–4 (some inaccuracies occurring there are corrected in this paper).

In order to carry the first approach over to abelian varieties of dimension > 1, we need a height estimate à la Mumford on the moduli of these varieties. This reduces to a number of interesting problems discussed in §5. There are several difficulties which do not allow us to generalize Mumford's estimate to higher-dimensional varieties,

(1) *Added in proof.* Using this definition, S. Ju. Arakelov has constructed an elegant theory of intersections on arithmetic surfaces. In particular, he proved an analogue of the adjunction formula in this situation.

in particular to moduli varieties. The main reason is that the irregularity of these last is equal to zero. This does not allow us to use the theory of heights on abelian varieties.

In conclusion we give another approach to Tate's conjecture, which, although it does not yield the desired result, nevertheless has appeared rather interesting and has been studied in recent years. The question is that of proving the conjecture of the finiteness of (the number of isomorphism classes of) abelian varieties with given places of bad reduction (the formulation, discussion and connection of this conjecture with conjecture T is contained in [3]). At first glance, this conjecture seems much more complicated than T. However, in the case of dimension 2, which is all that is needed for application to the conjecture about the homomorphisms (1), there is the following argument. If λ is a polarization of degree 1 on such an abelian variety, then it is either a product of elliptic curves, or the Jacobian of a curve of genus 2, and λ is its canonical polarization ([2], §3). Šafarevič proved that there exist only a finite number, up to isomorphism, of curves of genus 1 and 2 which are defined over a number field and have given places of bad reduction [3] (if the genus is 1, it is assumed that the curve has a rational point). This gives the finiteness conjecture for products of elliptic curves, and in the second case there is the difficulty that a curve can have bad reduction at some place, while its Jacobian does not. See [2], §1, for a description of such distressing examples. In order to obviate this difficulty, we must know how to estimate such degenerations by means of the remaining ones. In the function field case this is done in [13]. In the number field case there is an appropriate generalization of Šafarevič's theorem for curves of genus 2 ([2], §1), which is apparently not sufficient to apply to Tate's conjecture.

As a rule, the notations used in this paper are the conventional ones. Thus, O_X is the structure sheaf of the scheme X, $p^{-1}(X)$ is the inverse image of the scheme X under the morphism p, $\hat{f}$ is the dual morphism of abelian schemes, 1 is the identity morphism, deg is the degree of an isogeny, an invertible sheaf on a curve, or a polynomial. Because of the lack of a conventional term, we use the expression "abelian scheme" also for demarcation of the corresponding analytic object.

§1. Moduli. Algebraic theory

If X is an abelian variety defined over an algebraically closed field k, then for any n prime to char k the finite abelian group X_n of points of order n can be equipped with a nondegenerate skew-symmetric bilinear form e_n, which takes values in $\mathbf{Z}/n\mathbf{Z}$. See [4] for the definition and properties of e_n.

DEFINITION 1. Let $p: X \to S$ be an abelian scheme, g its relative dimension and n an integer prime to the characteristics of the residue fields of all the points of the scheme S. A *Jacobian rigidity of level* n is a set of $2g$ sections σ_i, $i = 1, \cdots, 2g$, of the morphism p which satisfy the following conditions:

1. $n\sigma_i = \epsilon$, where $n: X \to X$ is multiplication by n and ϵ is the identity section.
2. For all geometric points $s \in S$ the points $\sigma_i(s)$ form a basis of the group

$(X_s)_n$, and the matrix $e_n(\sigma_i(s), \sigma_j(s))$ has the form

$$\left(\begin{array}{c|c} 0 & \begin{matrix} 0 & & 1 \\ & \cdot^{\cdot^{\cdot}} & \\ 1 & & 0 \end{matrix} \\ \hline \begin{matrix} 0 & & -1 \\ & \cdot^{\cdot^{\cdot}} & \\ -1 & & 0 \end{matrix} & 0 \end{array}\right).$$

In general, on a given abelian scheme there cannot exist a Jacobian rigidity; however, it is always possible to introduce one by making an étale base change. We also recall the definition of a polarization on an abelian scheme $p: X \to S$. If L is an invertible sheaf on X, then it determines the sheaf $s^*(L) \otimes p_1^*(L)^{-1} \otimes p_2^*(L)^{-1}$, where $s: X \times_S X \to X$ is the multiplication, and $p_{1,2}: X \times_S X \to X$ are the projections, on the scheme $X \times_S X$. By the definition of the Picard scheme we can associate this sheaf to a morphism $X \to \operatorname{Pic} X/S$, and it is easily verified that in fact this morphism goes from X into the connected component of the identity of the group scheme $\operatorname{Pic} X/S$. This last is an abelian scheme, denoted by X and is called the dual scheme to X. The morphism thus constructed from X into X is denoted by $\Lambda(L)$ (for details see [5], Chapter 6.2).

DEFINITION 2. A *polarization* of an abelian scheme $p: X \to S$ is a homomorphsim $\lambda: X \to \hat{X}$ such that the induced morphism $\bar{\lambda}$ on each geometric fiber $\overline{X}_s$ has the form $\Lambda(L)$ for some ample sheaf on $\overline{X}_s$. The degree of the morphism λ is the degree of the polarization.

If the scheme X is one-dimensional over S, there exists a canonical polarization on X. It is equal to $\Lambda(O_X(\epsilon(S)))$, where $\epsilon: S \to X$ is the identity section of the scheme X. This polarization has degree 1.

The basic object studied in the theory of moduli of abelian varieties is the following functor.

DEFINITION 3. Let S be a scheme, and g, d and n integers. Put $\mathfrak{M}_{g,d,n}(S) =$ {the set of abelian schemes over S of relative dimension g, provided with a Jacobian rigidity of level n and a polarization of degree d^2 and considered up to S-isomorphism}.

In view of the remarks made above, in case $g = 1$ there is a canonical polarization, and therefore it is more natural to consider $\mathfrak{M}_n(S) =$ { the set of abelian schemes over S of relative dimension 1 possessing a Jacobian rigidity of level n, considered up to S-isomorphism}.

If we take into account how the structures involved in these definitions vary under base change, then we see that $\mathfrak{M}_{g,d,n}$ and $\mathfrak{M}_n$ are functors from the category of schemes into the category of sets. It is not hard to show that there exists a canonical isomorphism of the functors $\mathfrak{M}_{1,1,n}$ and $\mathfrak{M}_n$.

PROPOSITION 1. *The functors* $\mathfrak{M}_{g,d,n}$ *(and hence* $\mathfrak{M}_n$*) are representable if* $n \geqslant 3$.

The schemes $M_{g,d,n}$ and M_n representing these functors are called moduli schemes of abelian varieties, or, in the latter case, of elliptic curves. Over them there are universal abelian schemes which are respectively denoted by $A_{g,d,n}$ and A_n (or simply A).

DEFINITION 4. Let $f\colon X \to Y$ be an isogeny of abelian schemes over some scheme S; λ and μ are polarizations of the schemes X and Y. Then f is called a *Tate isogeny* if $\deg f = m^{\dim X}$ and $\hat{f} \circ \mu \circ f = m\lambda$. When $\dim X = \dim Y = 1$ and the polarizations are taken to be the canonical polarizations occurring in this case, then every isogeny that has a cyclic group as kernel is a Tate isogeny and vice versa. Such isogenies are sometimes called *cyclic*.

DEFINITION 5. Let S be a scheme, g, d, n and m integers. Put $\mathcal{H}_{g,d,n,m}(S) =$ {the set of sequences $(X, \lambda, \sigma_1, \cdots, \sigma_{2g}, Y, \mu, \tau_1, \cdots, \tau_{2g}, f)$, where X and Y are abelian schemes of dimension g over S, λ and μ are polarizations of them of degree d^2, $\sigma_1, \cdots, \sigma_{2g}, \tau_1, \cdots, \tau_{2g}$ are Jacobian rigidities of level n and $f\colon X \to Y$ is a Tate isogeny of degree m^g such that $f \circ \sigma_i = \tau_i, i = 1, \cdots, 2g$}.

In a natural way one defines two morphisms of functors $p_{1,2}\colon \mathcal{H}_{g,d,n,m} \to \mathfrak{M}_{g,d,n}$. If $g = 1$, then as above, it is possible to consider only the canonical polarizations. The corresponding functor classifying cyclic isogenies is denoted by $\mathcal{H}_{n,m}$ $(\cong \mathcal{H}_{1,1,n,m})$. The natural morphisms of it into $\mathfrak{M}_n$ are also denoted by p_1 and p_2.

PROPOSITION 2. *The functors* $\mathcal{H}_{g,d,n,m}$ *(and* $\mathcal{H}_{n,m}$*) are representable if* $n \geqslant 3$.

The schemes representing these functors will be denoted by $H_{g,d,n,m}$ and $H_{n,m}$ respectively. The morphisms $p_{1,2}$ define morphisms of the corresponding schemes, which will also be denoted by $p_{1,2}$. By Definition 4, over the scheme $H_{g,d,n,m}$ (and analogously $H_{n,m}$) there is a universal isogeny

$$\begin{array}{ccc} p_1^{-1}(A_{q,d,n}) & \xrightarrow{\ \Phi\ } & p_2^{-1}(A_{q,d,n}) \\ & \searrow \quad \swarrow & \\ & H_{q,d,n,m} & \end{array} \tag{3}$$

of degree m^g.

To shorten the statements in this paper we will always omit the indices g, d, n, m introduced in this section, and write simply M, A, H assuming thereby that the assertion under consideration holds for any possible set of indices.

PROPOSITION 3. *The schemes* $M \otimes \mathbf{Q}$, $A \otimes \mathbf{Q}$ *and* $H \otimes \mathbf{Q}$ *are quasi-projective.*

Using Grothendieck's local deformation theory, we can study the properties of smoothness of moduli schemes. Namely they are smooth over Spec $\mathbf{Z}$ at points x for which char $k(x)$ is prime to d, n and m. In particular, the schemes of Proposition 3 are always smooth (see [14]).

The group $\mathrm{Sp}(2g, \mathbf{Z}/n\mathbf{Z})$ acts on the schemes $M_{g,d,n}$ and M_n. In order to define this action, it suffices to describe it on the sets $\mathfrak{M}_{g,d,n}(S)$. Let X be an abelian scheme over S, (σ_i) a Jacobian rigidity of it, and λ a polarization, i.e. a point of $\mathfrak{M}_{g,d,n}(S)$. If $\alpha \in \mathrm{Sp}(2g, \mathbf{Z}/n\mathbf{Z})$, then α does not change the scheme X and the polarization λ, but acts only on the rigidity. Namely, if

$$\alpha \equiv \begin{pmatrix} \alpha_{11} \cdots \alpha_{1,\,2g} \\ \cdot\ \cdot\ \cdot\ \cdot\ \cdot\ \cdot \\ \alpha_{2g,\,1} \cdots\ \cdots \end{pmatrix} \bmod n, \quad \alpha_{ij} \in \mathbf{Z}, \quad i,\ j = 1, \ldots, 2g,$$

then

$$(\alpha\,(\sigma_i))_j = \sum_{i=1}^{2g} \alpha_{ji} \sigma_i, \qquad j = 1,\ 2,\ \ldots\ 2g.$$

PROPOSITION 4. *For each* $\alpha \in \mathrm{Sp}(2g, \mathbf{Z}/n\mathbf{Z})$ *there exists an automorphism* α' *making the diagram*

$$\begin{array}{ccc} A_{g,\,d,\,n} & \xrightarrow{\alpha'} & A_{g,\,d,\,n} \\ \downarrow & & \downarrow \\ M_{g,\,d,\,n} & \xrightarrow{\alpha} & M_{g,\,d,\,n} \end{array}$$

commutative, where the vertical arrows are the natural projections.

α' can be informally defined as follows. If $x \in M_{g,d,n}$, then the fibers A_x and $A_{\alpha(x)}$ are identical abelian varieties, "differing" only by their Jacobian rigidities. Thus $\alpha'|_{A_X} = 1$.

The last remark which we make is relevant to the structural morphism $M \to \mathrm{Spec}\,\mathbf{Z}$ of moduli schemes. Put

$$(n) = \mathrm{Spec}\,\mathbf{Z}\left[e^{\frac{2\pi i}{n}},\ \frac{1}{n}\right]. \tag{4}$$

If $M = M_{g,d,n}$ or M_n, the structural morphism can be represented in the form $M \to (n) \to \mathrm{Spec}\,\mathbf{Z}$.

In fact, if $p\colon X \to S$ is an abelian scheme, the Weil pairing consists of giving a bilinear morphism of finite group schemes $X_n \times \hat{X}_n \to \mu_n$, where X_n is the scheme of points of order n, and μ_n are the nth roots of unity. The polarization $\lambda\colon X \to \hat{X}$ defines a morphism $X_n \to X_n$ and, hence, a bilinear morphism $X_n \times X_n \to \mu_n$ (the corresponding bilinear form on the group of points is denoted by e_λ). The fixing of a Jacobian rigidity reduces to fixing an isomorphism $X_n \approx (\mathbf{Z}/n\mathbf{Z})_S^{2g}$. From this we will obtain an isomorphsim between $\mathbf{Z}/n\mathbf{Z}$ and μ_n The scheme of isomorphisms between these groups is (n), so that a Jacobian rigidity defines a mapping $S \to (n)$, as was required. An analogous remark is true for A and H.

§2. Moduli. Analytic theory

The following notation will be constantly used in what follows: $g(x)$ is the action of an element g of a group on a variety, U/G is the quotient space modulo the action of the group G, ${}^tA$ is the transposed matrix of A, and E is the identity matrix.

Let $\mathfrak{S}_g$ be the set of matrices Z of order g with complex coefficients and satisfying the conditions ${}^tZ = Z$, with the imaginary part $\operatorname{Im} Z$ positive definite.

$\mathfrak{S}_g$ is an open subset of $\mathbf{C}^{g(g+1)/2}$ and is therefore provided with a complex-analytic structure. The symplectic group $\mathrm{Sp}(2g, \mathbf{R})$ acts on $\mathfrak{S}_g$ in an analytic way according to the formula

$$g(Z) = (AZ + B)(CZ + D)^{-1}, \quad g = \begin{pmatrix} A & B \\ C & D \end{pmatrix}. \tag{5}$$

The complex manifold $\mathfrak{S}_g$ is called the Siegel upper half-plane. It can be realized as a bounded domain in affine space.

DEFINITION 1. Let n be an integer. The *modular group of level* n is the group $\Gamma(n) \subset \mathrm{Sp}(2g, \mathbf{R})$, consisting of matrices with integer coefficients and congruent to E modulo n.

$\Gamma(n)$ is a discrete subgroup of the group $\mathrm{Sp}(2g, \mathbf{R})$ and, if $n \geqslant 3$, acts on $\mathfrak{S}_g$ without fixed points.

DEFINITION 2. $\mathbf{M}_{g,n} = \mathfrak{S}_g/\Gamma(n)$. This is a complex space and nonsingular for $n \geqslant 3$. There exists a canonical family of abelian varieties of dimension g over it. In general, it has degenerate fibers, but if $n \geqslant 3$, this family is an abelian scheme. To construct it we consider the group $\mathfrak{D}_n$ consisting of pairs (γ, m), $\gamma \in \Gamma(n)$, $m \in \mathbf{Z}^{2g}$, with the composition law

$$(\gamma, m) \cdot (\gamma', m') = (\gamma\gamma', m \cdot {}^{\iota}\gamma' + m'),$$

where

$${}^{\iota}\begin{pmatrix} A & B \\ C & D \end{pmatrix} = \begin{pmatrix} D & C \\ B & A \end{pmatrix}.$$

We have an exact sequence

$$1 \to \mathbf{Z}^{2g} \xrightarrow{i} \mathscr{D}_n \xrightarrow{u} \Gamma(n) \to 1, \tag{6}$$

in which the morphism u possesses a section

$$v : \Gamma(n) \to \mathscr{D}_n, \qquad v(\gamma) = (\gamma, 0).$$

The group $\mathfrak{D}_n$ is a semidirect product of the subgroups $i(\mathbf{Z})^{2g}$ and $v(\Gamma(n))$.

We now define an action of the group $\mathfrak{D}_n$ on the manifold $\mathfrak{S}_g \times \mathbf{C}^g$. If

$$(\gamma, m) \in \mathscr{D}_n, \quad \gamma = \begin{pmatrix} A & B \\ C & D \end{pmatrix}, \qquad m = (M, N), \ M, \ N \in \mathbf{Z}^g$$

and $(\mathbf{Z}, \omega) \in \mathfrak{S}_g \times \mathbf{C}^g$, then

$$(\gamma, m)(Z, \omega) = (\gamma(Z), (\omega + M + NZ)(CZ + D)^{-1}). \tag{7}$$

DEFINITION 3. $\mathbf{A}_{g,n} = \mathfrak{S}_g \times \mathbf{C}^g / \mathfrak{D}_n$.

From the sequence (6) and formulas (5), (7) we find that there exists a projection q: $\mathbf{A}_{g,n} \to \mathbf{M}_{g,n}$. This is the desired family. Since for $n \geqslant 3$ the group $\Gamma(n)$ and hence $\mathfrak{D}_n$ acts without fixed points, we will obtain that in this case q is an abelian scheme of dimension g. Its identity section is given by the trivial imbedding of the variety $\mathfrak{S}_g = \mathfrak{S}_g \times (0)$ into $\mathfrak{S}_g \times \mathbf{C}^g$.

PROPOSITION 1. *The variety $\mathbf{M}_{g,n}$, $n \geqslant 3$, represents a functor from the category of analytic spaces to the category of sets, which associates to a space S the set of abelian schemes over S of dimension g, equipped with polarizations of degree 1 and a Jacobian rigidity of level n. $\mathbf{A}_{g,n}$ is the universal abelian scheme under this functor.*

We denote by X^{an} the analytic space associated by GAGA to a scheme X of finite type over $\mathbf{C}$. From Proposition 1 it is easy to derive the following corollary, establishing a connection between the analytic theory of moduli and the algebraic theory.

COROLLARY. *If $n \geqslant 3$, (n) is the scheme (4) and $\mathbf{C}$ is considered as a scheme over (n) relative to the canonical embedding $(n) \hookrightarrow \mathbf{C}$, then the natural morphism*

$$\varphi : (M_{g,1,n} \otimes_{(n)} \mathbf{C})^{an} \to \mathbf{M}_{g,n}$$

is an isomorphism.

We note that $\varphi^{-1}(\mathbf{A}_{g,n}) = (A_{g,1,n} \otimes_{(n)} \mathbf{C})^{an}$.

Here and in what follows we restrict ourselves to the case $d = 1$ (principal polarizations). This simplifies the study, but generalizing to arbitrary d does not present any particular difficulties (instead of the modular groups $\Gamma(n)$ we must use the so-called paramodular groups and the associated analogues of the Siegel upper half-plane; see [15]).

Before we proceed to a consideration of isogenies, we state an analytic analogue of Proposition 4 of §1. Since

$$\mathrm{Sp}(2g, \mathbf{Z})/\Gamma(n) \xrightarrow{\sim} \mathrm{Sp}(2g, \mathbf{Z}/n\mathbf{Z}),$$

the group $\mathrm{Sp}(2g, \mathbf{Z}/n\mathbf{Z})$ acts on $\mathbf{M}_{g,n}$. An analogous isomorphism for the groups $\mathfrak{D}_1$ and $\mathfrak{D}_n$ allows us to define an action of $\mathrm{Sp}(2g, \mathbf{Z}/n\mathbf{Z})$ on $\mathbf{A}_{g,n}$. It is easy to verify that these actions are compatible with each other (also as in Proposition 4 of §1; cf. the proof of Proposition 2 of §1 in [11]). One can also show that the isomorphism φ of the corollary transposes the action of $\mathrm{Sp}(2g, \mathbf{Z}/n\mathbf{Z})$ defined purely algebraically in Proposition 4 of §1, into the action which we have defined analytically.

DEFINITION 4. Let m be an integer; $\Gamma_0(m)$ is the subgroup of $\mathrm{Sp}(2g, \mathbf{R})$ consisting of matrices of integral coefficients of the form $\begin{pmatrix} A & B \\ C & D \end{pmatrix}$, where A, B, C, D are of order g and $C \equiv 0 \bmod m$. It is a discrete subgroup having finite index in $\mathrm{Sp}(2g, \mathbf{Z})$.

DEFINITION 5. $\mathbf{H}_{g,n,m} = \mathfrak{S}_g/\Gamma_0(m) \cap \Gamma(n)$.

Thus we obtain a complex space that is nonsingular for $n \geqslant 3$. The inclusion of the group $\Gamma_0(m) \cap \Gamma(n)$ in $\Gamma(n)$ induces a projection $p_1 : \mathbf{H}_{g,n,m} \to \mathbf{M}_{g,n}$. In order to define another projection p_2, we introduce an automorphism τ of the upper half-plane $\mathfrak{S}_g$:

$$\tau(Z) = -m^{-1}Z^{-1}. \tag{8}$$

The following relations hold in the group of automorphisms of the upper half-plane:

$$\tau^2 = 1,$$
$$\tau^{-1}\Gamma_0(m)\tau = \Gamma_0(m),$$
$$\tau^{-1}(\Gamma_0(m) \cap \Gamma(n))\tau = \Gamma_0(m) \cap \Gamma(n), \quad \text{if} \quad (m, n) = 1.$$

From this we see that τ defines an involution of the variety $\mathbf{H}_{g,n,m}$ if m and n are relatively prime. In what follows we shall always assume that this condition holds. The automorphism obtained in this way will also be denoted by τ. Put $p_2 = p_1 \circ \tau$.

It remains for us to consider the commutative diagram

$$\begin{array}{ccc} p_1^{-1}(\mathbf{A}_{g,n}) & \xrightarrow{\Psi} & p_2^{-1}(\mathbf{A}_{g,n}) \\ & \searrow \quad \swarrow & \\ & \mathbf{H}_{g,n,m} & \end{array} \tag{9}$$

in which Ψ is a Tate isogeny of degree m^g preserving the Jacobian rigidity. To construct this diagram we introduce the group $\mathfrak{D}_{n,m}$ consisting of pairs (γ, r), $\gamma \in \Gamma_0(m) \cap \Gamma(n)$, $r \in \mathbf{Z}^{2g}$, with composition law and properties the same as those of the group $\mathfrak{D}_n$ constructed above. The group $\mathfrak{D}_{n,m}$ acts on $\mathfrak{S}_g \times \mathbf{C}^g$ according to formula (7), and

$$p_1^{-1}(\mathbf{A}_{g,n}) = (\mathfrak{S}_g \times \mathbf{C}^g)/\mathfrak{D}_{n,m}.$$

We consider the mapping δ of the variety $\mathfrak{S}_g \times \mathbf{C}^g$ into itself defined by the relation

$$\delta(Z, \omega) = (\tau(Z), \omega Z^{-1}).$$

It is easy to verify that in the group of automorphisms of $\mathfrak{S}_g \times \mathbf{C}^g$ the mapping δ commutes with the group $\mathfrak{D}_{n,m}$, namely,

$$\delta \circ (\gamma, M, N) = (\gamma', M', N') \circ \delta,$$

where

$$\gamma = \begin{pmatrix} A & B \\ C & D \end{pmatrix}, \quad \gamma' = \begin{pmatrix} D & -m^{-1}C \\ -mB & A \end{pmatrix}, \quad M, N \in \mathbf{Z}^g, \quad M' = N, \quad N' = -mM.$$

From this it follows that δ defines a morphism of the quotient spaces, which we denote by Ψ. It is the required isogeny.

PROPOSITION 2. *The variety* $\mathbf{H}_{g,n,m}$, $n \geqslant 3$, $(n, m) = 1$, *represents a functor from the category of analytic spaces to the category of sets, which associates to a space* S *the set of pairs of abelian schemes over* S, *equipped with a polarization of degree* 1 *and a Jacobian rigidity, and a Tate isogeny preserving the rigidity* (*cf. Definition* 5 *of* §1).

Here diagram (9) is the universal isogeny.

As above, we will obtain a

COROLLARY. *If* $n \geqslant 3$, $(m, n) = 1$, (n) *is the scheme* (4) *and* $\mathbf{C}$ *is considered as a scheme over* (n) *relative to the canonical inclusion* $(n) \hookrightarrow \mathbf{C}$, *then the natural morphism*

$$(H_{g,1,n,m} \otimes_{(n)} \mathbf{C})^{an} \to \mathbf{H}_{g,n,m}$$

is an isomorphism.

This morphism takes diagram (9) into diagram (3) and, in particular, the projections p_1 and p_2 of this section into the projections p_1 and p_2 of §1.

Diagram (9) defines a correspondence on the variety $\mathbf{H}_{g,n,m}$. It is the simplest, and at the same time a rather interesting, example of a modular correspondence. Similar correspondences can be constructed for an arbitrary arithmetic group acting in a bounded symmetric domain.

§3. The sheaf L. Construction of a compactification and invariance properties

In this section we introduce on the moduli schemes a certain invertible sheaf L, which is the "principal hero" in the studies that follow. This sheaf can be constructed on the basis of any abelian scheme. We let $\Lambda^n F$ denote the nth exterior power of a locally free sheaf F.

DEFINITION 1. Let $p: X \to S$ be an abelian scheme of relative dimension g, ϵ: $S \to X$ its identity section. Then

$$L = \bigwedge^g \varepsilon^* \Omega^1_{X/S}.$$

This is the conormal sheaf to the image $\epsilon(S)$ of the identity section. The following properties of the sheaf L are trivial, but very useful.

If $a: S' \to S$ is a base change and L' is the sheaf corresponding to the scheme $X \times_S S'$, then

$$a^*L \cong L'. \tag{10}$$

If $f: X \to Y$ is an étale isogeny of schemes over S, and L_1 and L_2 are the corresponding sheaves, then the canonical morphism

$$L_2 \rightrightarrows L_1 \tag{11}$$

is an isomorphism. This last property holds for arbitrary isogenies if the residue fields of all points of the scheme S have characteristic zero.

Applying Definition 1 to the schemes and varieties of the preceding sections, we will obtain canonical invertible sheaves on them, denoted by $\mathbf{L}$ on $\mathbf{M}_{g,n}, \cdots$ and by L on $M_{g,1,n}, H_{g,1,n,m}, \cdots$. The "analytic" sheaves $\mathbf{L}$ are connected with the "algebraic" ones L just as the moduli schemes are so connected in the corollaries of §2.

In order not to cloud the essence of the matter with indices, we consider an arbitrary subgroup $\Gamma \subset \mathrm{Sp}(2g, \mathbf{Z})$ of finite index and without fixed points on $\mathfrak{S}_g$. With respect to it we can construct a group $\mathfrak{D}$ just as we did in §2. Hence we have two quotient spaces $\mathbf{M}(\Gamma) = \mathfrak{S}_g/\Gamma$ and $\mathbf{A} = (\mathfrak{S}_g \times \mathbf{C}^g)/\mathfrak{D}$ and a projection $g\colon \mathbf{A} \to \mathbf{M}(\Gamma)$ turning $\mathbf{A}$ into an abelian scheme over $\mathbf{M}(\Gamma)$. Definition 1 gives a sheaf $\mathbf{L}$ on $\mathbf{M}(\Gamma)$.

LEMMA. *There exists an isomorphism*

$$\mathbf{L}^{g+1} \cong \bigwedge^{\frac{g(g+1)}{2}} \Omega^1_{\mathbf{M}(\Gamma)}.$$

It commutes with the action of $\mathrm{Sp}(2g, \mathbf{Z})/\Gamma$ *on* $\mathbf{M}(\Gamma)$.

IDEA OF THE PROOF. Let T denote the tangent sheaf and consider the Kodaira-Spencer map

$$\theta : T_{\mathbf{M}(\Gamma)} \to R^1 q_* T_{\mathbf{A}/\mathbf{M}(\Gamma)}.$$

Since $T_{\mathbf{A}/\mathbf{M}(\Gamma)} \cong q^*\epsilon^* T_{\mathbf{A}/\mathbf{M}(\Gamma)}$ (a general property of abelian schemes), we obtain that the morphism θ is a mapping into $R^1q_*O_{\mathbf{A}} \otimes R^1q_*O_{\mathbf{A}}$, and it is well known that it is injective, and its image coincides with the symmetric second tensor power of $R^1q_*O_{\mathbf{A}}$. (This can be proved by analytic means, using the Riemann-Frobenius conditions, or by purely algebraic ones, by studying the obstruction to lifting the polarizations of abelian varieties; see [14].) The requirement follows from this if we take into account that the sheaves Ω^1 and T are dual to each other.

PROPOSITION 1. *The following assertions hold.*

1. *The sheaf* $\mathbf{L}$ *is ample.*
2. *If* $g > 1$, *then for all* $n \geq 0$ *we have* $\dim H^0(\mathbf{M}(\Gamma), \mathbf{L}^{\otimes n}) < \infty$, *and the ring* $\bigotimes_{n \geq 0} H^0(\mathbf{M}(\Gamma), \mathbf{L}^{\otimes(g+1)n})$ *is an integrally closed integral domain of finite type over* $\mathbf{C}$.

PROOF. By the lemma it suffices to verify the analogous assertions for the sheaf $\Lambda^{g(g+1)/2}\Omega^1_{\mathbf{M}(\Gamma)}$. This is the sheaf of differential forms of maximal degree, and its sections over $\mathbf{M}(\Gamma)$ are interpreted as modular forms, i.e. as holomorphic solutions $f(Z)$ of the equation

$$f(\gamma(Z)) = \det(CZ + D)^2 f(Z), \qquad \gamma = \begin{pmatrix} A & B \\ C & D \end{pmatrix} \in \Gamma.$$

Our assertions then become the fundamental theorems of the theory of modular forms [6].

We can now construct a compactification of $\mathbf{M}(\Gamma)$.

DEFINITION 2. Let $g > 1$. Then

$$\overline{M}(\Gamma) = \operatorname{Proj}\Big(\bigoplus_{n \geqslant 0} H^0(\mathbf{M}(\Gamma), \mathbf{L}^{\otimes(g+1)n})\Big).$$

The sheaf $\mathbf{L}$ defines the canonical sheaf L on $\overline{M}(\Gamma)$.

$\overline{M}(\Gamma)$ is a normal projective variety containing an open nonsingular subvariety $M(\Gamma)$ such that the natural imbedding $\mathbf{M}(\Gamma) \hookrightarrow \overline{M}(\Gamma)^{an}$ defines an isomorphism

$$M(\Gamma)^{an} \cong \mathbf{M}(\Gamma). \tag{12}$$

The boundary $\overline{M}(\Gamma) - M(\Gamma)$ has codimension g. If $g = 1$, then the second half of Proposition 1 will cease to hold, so that the procedure given above is inapplicable. The situation in this case is essentially simpler, since for each one-dimensional complex manifold $\mathbf{X}$ there exists at most one algebraic curve X with $X^{an} \cong \mathbf{X}$. In our case such a curve actually exists, which gives the possibility of introducing

DEFINITION 2$'$. Let $g = 1$. Then $M(\Gamma)$ is a curve with $M(\Gamma)^{an} \cong \mathbf{M}(\Gamma)$, and $\overline{M}(\Gamma)$ is the projective nonsingular closure of $M(\Gamma)$.

Considering the abelian scheme $q\colon A \to M(\Gamma)$, $A^{an} = \mathbf{A}$, we can consider its minimal model over the curve $\overline{M}(\Gamma)$. It represents a nonsingular projective ($g = 1$!) surface equipped with an epimorphism $\overline{q}\colon \overline{A} \to \overline{M}(\Gamma)$, coinciding with q over $M(\Gamma)$ and not having exceptional curves of the first kind in the degenerate fibers.

DEFINITION 2$''$. Let $g = 1$. Then

$$L = \varepsilon^* \Omega^1_{\overline{A}/\overline{M}(\Gamma)},$$

where ϵ is the identity section of the scheme $\overline{A}$.

Now consider two subgroups $\Gamma' \supset \Gamma$ of finite index in $\mathrm{Sp}(2g, \mathbf{Z})$, acting on $\mathfrak{S}_g$ without fixed points. We have an étale morphism $t_0\colon \mathbf{M}(\Gamma) \to \mathbf{M}(\Gamma')$, and if Γ is a normal subgroup of Γ', an automorphism $\sigma_0\colon \mathbf{M}(\Gamma) \to \mathbf{M}(\Gamma)$. By (10) and (11) these mappings preserve the sheaves $\mathbf{L}$ and hence define morphisms

$$t : \overline{M}(\Gamma) \to \overline{M}(\Gamma') \quad \text{and} \quad \sigma : \overline{M}(\Gamma) \to \overline{M}(\Gamma).$$

In case $\Gamma = \Gamma_0(m) \cap \Gamma(n)$, the automorphism τ of (8) of the variety $\mathbf{M}(\Gamma) = \mathbf{H}_{g,n,m}$ also preserves the sheaf $\mathbf{L}$ (this follows from diagram (9)) and gives an automorphism of the scheme $\overline{M}(\Gamma)$, also denoted by τ.

PROPOSITION 2. *The sheaf L has the following properties*:

1. $t^*L' \cong L$.
2. $\sigma^*L \cong L$.
3. $\tau^*L_{n,m} \cong L_{n,m}$.

Here L, L' and $L_{n,m}$ are the canonical invertible sheaves on the schemes $\overline{M}(\Gamma)$, $\overline{M}(\Gamma')$ and $\overline{M}(\Gamma_0(m) \cap \Gamma(n))$ respectively.

For the case $g = 1$ this proposition is analyzed in [11]. If $g > 1$, then it follows from the functorial properties of the operation Proj.

REMARK 1. Above we introduced three sorts of objects: the algebraic moduli $M_{g,d,n}$, M_n, $H_{g,d,n,m}, \cdots$ — these are schemes over $\mathbf{Z}$ (§1); the analytic moduli

$\mathbf{M}_{g,n}, \cdots$ and more generally $\mathbf{M}(\Gamma)$ – these are analytic varieties (§2); and the varieties $M(\Gamma) \subset \overline{M}(\Gamma)$ – these are schemes over $\mathbf{C}$. We summarize the connections between them (some were already considered in (7) and the corollaries of §2):

$$M \text{ (or } H) \otimes_{(n)} \mathbf{C} \cong M(\text{the corresponding group } \Gamma), \tag{13}$$

and the composition of this isomorphism with (12) coincides with the morphisms of the corollaries of §2.

Since the morphism (13) takes the algebraic sheaves L on the schemes M and H of §1 into the sheaves L on the schemes $M(\Gamma)$, we can define a compactification of the moduli schemes $M \otimes \mathbf{Q}$ and $H \otimes \mathbf{Q}$ in the following way. In contrast to the definitions given above these will be schemes over $\mathbf{Q}$.

DEFINITION 3. If $g > 1$, then

$$\overline{M}_{g,1,n} = \operatorname{Proj}\Big(\bigoplus_{k\geqslant 0} H^0(M_{g,1,n} \otimes \mathbf{Q}, L^{\otimes (g+1)k})\Big),$$

$$\overline{H}_{g,1,n,m} = \operatorname{Proj}\Big(\bigoplus_{k\geqslant 0} H^0(H_{g,1,n,m} \otimes \mathbf{Q}, L^{\otimes (g+1)k})\Big).$$

If $g = 1$, then $\overline{M}_n$ and $\overline{H}_{n,m}$ are the projective nonsingular closures of the curves $M_n \otimes \mathbf{Q}$ and $H_{n,m} \otimes \mathbf{Q}$.

From Proposition 2 we obtain a corollary about the behavior of the canonical sheaves $L_{g,1,n}$ on these schemes (constructed just as in Definitions 2, 2′ and 2″ above) relative to the morphisms p_1 and p_2 (obtained by extending the morphisms p_1 and p_2 of §1 to the schemes $\overline{M}$ and $\overline{H}$).

COROLLARY.

$$p_1^* L_{g,1,n} \cong p_2^* L_{g,1,n},$$

$$\sigma^* L_{g,1,n} \cong L_{g,1,n} \quad \textit{for} \quad \sigma \in \operatorname{Sp}(2g, \mathbf{Z}/n\mathbf{Z}).$$

REMARK 2. If we turn to analogues of Proposition 2 and its corollary for the sheaf L on the moduli schemes M of §1, which are schemes over Spec $\mathbf{Z}$, then properties 1 and 2 are preserved (the latter follows from Proposition 4 of §1). As for property 3 (or the corollary), the isogeny Φ in diagram (3) is not étale, so that it ceases to be true. When $g = 1$, this is true from the description of the degenerate fibers of the scheme $H_{1,1,n,p}$ given in [7] (p a prime, $(p, n) = 1$). The fiber $H_{1,1,n,p} \otimes \mathbf{F}_p$ over p consists of two components, one of which classifies the separable isogenies, and the other the purely inseparable ones. Therefore property 3 fails at the points of the second component (and only at them).

§4. Moduli. Arithmetic

1. *Finiteness conjectures and the moduli variety.* Let M be the moduli scheme constructed in §1, and let K be a field. We consider the conjectures stated in the Introduction, and we translate them into the language of rational points of the

scheme M. Let X be an abelian variety over K, $\dim X = g$, λ a polarization of X of degree d^2 and $f: X \to Y$ the Tate isogeny of the conditions of conjecture $T(X, K, \lambda, d, l)$.

We choose $n \geqslant 3$ such that $(l, n) = 1$ and $(\text{char } K, n) = 1$.

LEMMA 1. *If $E \supset K$ is a finite separable extension, then, up to K-isomorphism, there exist only a finite number of pairs (Y, μ), where Y is an abelian variety and μ is a polarization of Y such that $(X, \lambda) \otimes E$ and $(Y, \mu) \otimes E$ are isomorphic.*

For a proof see [2], §3. Now suppose $E \supset K$ is such that all the points of order n of the variety X are rational over E. Since f induces an isomorphism $X_n \xrightarrow{\sim} Y_n$, we will obtain that this is also true for all varieties Y from the hypothesis of the conjecture. We now choose a Jacobian rigidity of level n on the variety $X \otimes E$ and translate it to the variety Y by means of the isogeny f.

From Proposition 1 of §1 it follows that a point y of $M_{g,d,n}(E)$ corresponds to each Y. Thus the set of varieties from the hypothesis of the conjecture $T(X, K, \lambda, d, l)$ is connected with a certain set $\{y\} = \mathfrak{S} \subset M_{g,d,n}(E)$. From Proposition 1 of §1 and Lemma 1 it follows that the finiteness of the set $\mathfrak{S}$ implies $T(X, K, \lambda, d, l)$.

We now turn to diagram (3). The projections p_1 and p_2 define the following correspondence:

$$\begin{array}{c} H_{g,d,n,l} \\ p_1\downarrow \;\; \downarrow p_2 \\ M_{g,d,n}. \end{array}$$

By means of the lemma of the Introduction it is not hard to show that every Tate isogeny of degree l^{mg} splits into a product of m Tate isogenies of degree l^g. This is true over the original field E. In fact, if f is a Tate isogeny of degree l^{mg}, its kernel $\operatorname{Ker} f$ is a direct sum of g groups $\mathbf{Z}/l^m\mathbf{Z}$ (over an algebraic closure of E). The group $\operatorname{Ker} f$ has a filtration

$$\operatorname{Ker} f \supset l \operatorname{Ker} f \supset \ldots \supset l^{m-1} \operatorname{Ker} f \supset (0),$$

whose successive quotients are defined over E and have order l^g. This filtration gives a splitting of f into isogenies of degree l^g. That they are Tate isogenies follows from the lemma of the Introduction. Combining this fact with Proposition 2 of §1, we find that for all $y \in \mathfrak{S}$ there exist sequences of points $x_0, x_1, \cdots, x_m = y \in M(E)$ and $z_1, \cdots, z_m \in H(E)$ such that

$$p_1(z_i) = x_{i-1}, \quad p_2(z_i) = x_i, \quad i = 1, 2, \ldots, m, \tag{14}$$

and x_0 is the point corresponding to the variety X.

Analogously we can restate the other finiteness conjectures mentioned in the Introduction. In what follows it will be more suitable to use the compactifications of the schemes M and H over $\mathbf{Q}$ instead of M and H themselves; these compactifications were defined in §3 (Definition 3). Hence we restrict ourselves to the case $d = 1$. We also assume that K is an algebraic number field.

The varieties $\overline{M}_{g,1,n}$ and $\overline{H}_{g,1,n,l}$ are projective and normal. There is a correspondence (as above)

$$\begin{array}{c} \overline{H}_{g,1,n,l} \\ p_1\downarrow \ \downarrow p_2 \\ \overline{M}_{g,1,n}. \end{array} \tag{15}$$

The set $\mathfrak{S}$ defines an analogous set in $\overline{M}(E)$, also denoted by $\mathfrak{S}$, and property (14) holds for (15).

2. *Height.* A *category of schemes* over K will be a pair consisting of a category C and a functor $i\colon C \to \mathrm{Sch}/K$ into the category of all proper schemes over K such that $\mathrm{Hom}(X, Y)$ is contained in $\mathrm{Hom}(i(X), i(Y))$ for all $X, Y \in \mathrm{Ob}(C)$. In other words, a category of schemes is a category whose objects are schemes over K with additional structures, and whose morphisms are usual morphisms of schemes (but not necessarily all of them).

In such a category there are three functors with abelian groups for values: Pic, associating to a scheme X its Picard group; F, associating to a scheme X the group (under composition) of $\mathbf{R}$-valued functions on the set $X(K)$; and a subfunctor B of F, "consisting" of the bounded functions on $X(K)$. Denote by F/B the appropriate quotient functor and assume that K is a number field.

DEFINITION 1. A *height* h in a category of schemes C is a morphism of functors $h\colon \mathrm{Pic} \to F/B$ such that the set of points $x \in X(K)$ satisfying $h(L)(x) \leqslant c$ is finite for all $X \in \mathrm{Ob}(C)$, for any ample sheaf $L \in \mathrm{Pic}\, X$ and for all $c \in \mathbf{R}$.

Such a height is called a *Weil height.* Denote by $\mathfrak{V}_K$ the category of projective schemes over K with arbitrary morphisms.

PROPOSITION 1. *A Weil height exists in the category* $\mathfrak{V}_K$.

The explicit construction of such a height is well known; see, for example, Lang's book [8]. We shall call it ordinary height or simply height, and denote it by h.

It would be interesting to describe the set of Weil heights in the category $\mathfrak{V}_K$. It is perhaps more natural to consider, instead of arbitrary heights, only those which have a local decomposition (see §4.5 below), and to explain uniqueness for local quasifunctions satisfying an axiomatic definition of the type of Definition 1.(2)

3. *Correspondence.* As usual, a correspondence is a diagram of the form

(2) Interesting examples, different from the usual height, are constructed in [12]. They represent functions with values in $\mathbf{C}$.

$$\begin{array}{c} Z \\ p_1\downarrow\ \downarrow p_2 \\ X. \end{array} \qquad (16)$$

Assume that everything belongs to the category $\mathcal{O}_K$, and denote by $\approx$ algebraic equivalence of sheaves.

PROPOSITION 2. *Let $L \in \operatorname{Pic} X$ and $x_0, x_1, \cdots, x_m \in X(K)$, and assume that for each i there exists a $z_i \in Z(K)$ such that $p_1(z_i) = x_{i-1}$ and $p_2(z_i) = x_i$, $i = 1, \cdots, m$. Then there exist a constant $c_1 > 0$ and a function $c_2(\epsilon) > 0$, $\epsilon > 1$, both depending only on X and L, satisfying the following conditions:*

1) If $p_1^*L \cong p_2^*L$, then $h(L)(x_i) \leqslant h(L)(x_0) + c_1 i$, $i = 1, \cdots, m$.

2) If $p_1^*L \approx p_2^*L$, then $h(L)(x_i) \leqslant h(L)(x_0) + c_2(\epsilon)\epsilon^i$, $i = 1, \cdots, m$, for all $\epsilon > 1$.

The proof trivially follows from Definition 1 in the first case. In the second case we must use the following lemma, which is a formal consequence of Definition 1 ([8], p. 68).

LEMMA 2. *If $X \in \operatorname{Ob} \mathcal{O}_K$, $L_1, L_2 \in \operatorname{Pic} X$ and $L_1 \approx L_2$, then for all $\epsilon > 0$ there exist constants c_1 and c_2 such that*

$$(1-\varepsilon)\, h(L_2) + c_1 \leqslant h(L_1) \leqslant (1+\varepsilon)\, h(L_2) + c_2.$$

If X is a curve of genus > 1, $L \in \operatorname{Pic} X$ an ample sheaf and the points $(x_i) \in X(K)$, $i = 1, 2, \cdots$, are numbered in order of increasing height $h(L)$, Mumford showed [9] that there exist $a > 1$ and $b > 0$ such that $h(L)(x_i) > ba^i$, $i = 1, 2, \cdots$. Comparing this estimate with the assertions of Proposition 2 (in case 2 it is necessary to put $\epsilon < a$), we obtain

PROPOSITION 3. *Let $X \in \operatorname{Ob} \mathcal{O}_K$ be a curve of genus > 1, L an ample sheaf on X such that $p_1^*L \approx p_2^*L$, $x_0 \in X(K)$ and $\mathfrak{E} \subset X(K)$ a set such that for all $y \in \mathfrak{E}$ there exist $x_1, \cdots, x_{m-1} \in X(K)$ and $z_1, \cdots, z_m \in Z(K)$ such that $p_1(z_i) = x_{i-1}$, $p_2(z_i) = x_i$ and $x_m = y$, $i = 1, \cdots, m$. Then the set $\mathfrak{E}$ is finite.*

REMARK. If the correspondence (16) has the form

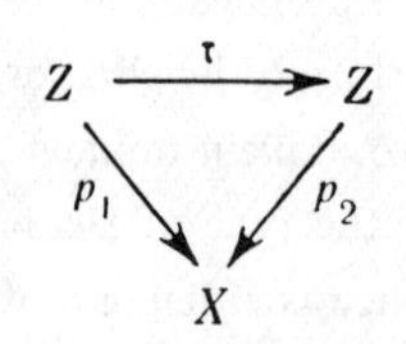

$$p_2 = p_1 \circ \tau, \quad \deg \tau = 1,$$

then the condition $p_1^*L \approx p_2^*L$ holds for any sheaf L, since $\deg p_1^*L = \deg p_2^*L$, and on a curve the degree of a sheaf determines its algebraic equivalence class.

4. *Application to isogenies.* If $\overline{M}_n$ is the projective closure of the variety of moduli of elliptic curves (over $\mathbf{Q}$), then for large n the genus of the curve $\overline{M}_n$ is greater than 1 [11].

Therefore from §§4.1 and 4.3 we get that the conjectures $T(X, K, \lambda, 1, l)$ hold for number fields K and $\dim X = 1$. More precisely, we have

THEOREM 1. *Let X be an elliptic curve over a number field K, and l a prime number. There exist, up to K-isomorphism, only a finite number of elliptic curves Y over K for which there exists an isogeny $X \to Y$ of degree l^m, $m \in \mathbf{Z}$.*

We observe that to obtain this theorem we do not need to use the canonical sheaf L on the moduli scheme and its invariance properties (§3) as was done in [11]. Since the proof is based on Proposition 2.2, here we can use any invertible sheaf L of positive degree. If $g > 1$, this is not so (the Néron-Severi group of the variety $\overline{M}_{g,1,n}$ is essentially greater than $\mathbf{Z}$). In this case, from (10), (11), the corollary to Proposition 2 of §3, and Proposition 2.1 in an application to the set $\mathfrak{S} \subset \overline{M}_{g,1,n}$ corresponding to the varieties of the conditions of the conjecture $T(X, K, \lambda, 1, l)$, $\dim X = g$ (§4.1), we will obtain the following result:

THEOREM 2. *Let $g > 1$ and $n \geqslant 3$; let $L \in \operatorname{Pic} \overline{M}_{g,1,n}$ be the canonical invertible sheaf (Definition 3 of §3). If the conjecture $T(X, K, \lambda, 1, l)$ does not hold, then there exist a constant $c > 0$ and arbitrarily long sequences of points $x_1, \cdots, x_m$ of $\overline{M}_{g,1,n}(K)$ for which $h(L)(x_i) \leqslant ci$.*

In order to obtain conjecture $T(X, K, \lambda, 1, l)$ from this, we must have an estimate of Mumford's type for the rational points of $\overline{M}$ (see §5).

5. *Tate height.* The worst inadequacy of the above considerations is the presence of the uncontrollable constants c_1 and c_2. In general they depend on l, which obstructs the proving of the stronger assertion $T(X, K, \lambda, 1)$ for $\dim X = 1$ by such methods. A possible escape from this is suggested by recent work of Tate [4] and Néron [10].

DEFINITION 2. Let C be a category of schemes. A *Tate height* in C is a morphism $\hat{h}$ of functors, giving a commutative diagram

$$\begin{array}{ccc} F & \longrightarrow & F/B \\ {}_{\hat{h}}\nwarrow & & \nearrow_{h} \\ & \mathrm{Pic} & \end{array}$$

We let $\mathfrak{A}$ denote the category of abelian varieties and homomorphisms of them.

PROPOSITION 4 (*Tate; see* [10]). *A Tate height exists in* $\mathfrak{A}$.

This is unique in a sense (see Theorem 1 in §7 of [10]). In the case $C = \mathcal{O}_K$ there does not exist a Tate height. For example, if we consider the category of punctual schemes containing $\mathbf{P}^1$, then by the properties of the action of the group PGL(1) on $\mathbf{P}^1$ we will obtain that $\hat{h}(L) = \text{const}$, for all $L \in \operatorname{Pic}(\mathbf{P}^1)$, which contradicts Definition 1. The motivation for introducing the Tate height is the following fact:

PROPOSITION 5. *Suppose there is a Tate height* $\hat{h}$ *in the category* C. *Suppose*

$$p_1(i) \downdownarrows p_2(i), \qquad i \in I, \qquad Z_i \rightrightarrows X$$

is a family of correspondences in C, $L \in \operatorname{Pic} X$ *an ample sheaf such that* $p_1^*(i)L \cong p_2^*(i)L$ *for all* $i \in I$, *and* $\mathfrak{S} \subset X(K)$ *a set such that for all* $x, y \in \mathfrak{S}$ *there exist* $i \in I$ *and* $z \in Z_i$ *such that* $p_1(i)(z) = x$ *and* $p_2(i)(z) = y$. *Then* $\mathfrak{S}$ *is finite.*

PROOF. From Definition 2 we get that $\hat{h}(L) = \text{const}$ on $\mathfrak{S}$. Hence $h(L)$ is bounded on $\mathfrak{S}$, and Definition 1 gives the result.

The question of the existence of a Tate height in one or another category of schemes naturally leads to a local problem: the consideration of canonical quasifunctions in the sense of Néron. The definition of quasifunctions can be found at the beginning of [10].

Let K be a number field, and let v run through all normalizations of K. Denote by K_v the corresponding completions, by $|\ |_v$ the absolute value on K_v defined by v, and by C a category of schemes over K.

To each $X \in \operatorname{Ob} C$ we associate a set $X(K_v)$. The quasifunctions on this set form a group $Q(X)$ under composition. We denote the subgroup of bounded quasifunctions by $BQ(X)$ and the group of Cartier divisors by $\operatorname{Div}(X)$.

We now introduce a category C_v of schemes over K_v, obtained from C by "augmentation" to K_v and the canonical inclusion $i_v\colon C \to C_v$. The category C contains the functors Pic, F and B, and the category C_v contains Div, Q and BQ. Distinguishing the principal divisors by Div, we obtain a functor Div_l. Each quasifunction has a divisor (see [10]). This gives a morphism $\delta\colon Q \to \operatorname{Div}$, whose kernel coincides with BQ.

If $d \in \operatorname{Div}_l(X)$ is the divisor of a function f, it corresponds to a quasifunction $|f(x)|_v$, $x \in X(K_v)$. In this way we obtain a morphism of functors: $\operatorname{Div}_l \to Q$.

DEFINITION 3. A *local Tate height* in the category C_v (or any category of schemes over K_v) is a morphism of functors h_v giving a commutative diagram

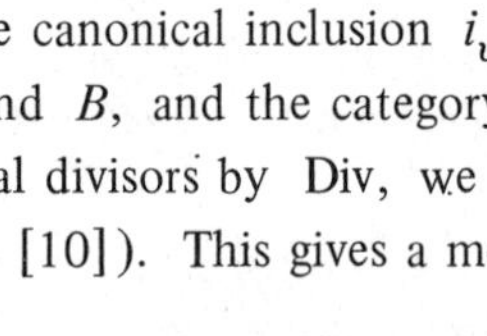

and such that for an effective divisor $d \in \mathrm{Div}(X)$ the function $\hat{h}_v(d)$ is bounded below on $X(K_v)$.

If (h_v) is a set of local heights, then it forms a *local decomposition* of a Tate height $\hat{h}$ if $\hat{h}_v(d)(x) = 0$ for almost all v, for all $d \in \mathrm{Div}(X) \subset \mathrm{Div}(i_v(X))$, and for all $x \in X(K) \subset i_v(X)(K_v)$, and if

$$\hat{h}(L)(x) = \sum_v \hat{h}_v(d)(x),$$

where $L = O_X(d)$.

From the point of view developed in this report, the question of the existence of a Tate height (local and global) in a category containing the correspondences (15) is of fundamental significance. This would give a proof of the finiteness conjectures mentioned in the Introduction (via Proposition 5). We restrict consideration here to archimedean valuations v. In the nonarchimedean case the question remains open.

6. *Local Tate height for* $K_v = \mathbf{C}$. We denote by $\mathfrak{F}_\mathbf{C}$ the category whose objects are pairs consisting of a punctual nonsingular scheme X over $\mathbf{C}$ plus a Kähler metric on X^{an}. The morphisms are surjective étale maps which preserve selected points.

PROPOSITION 6. *A local Tate height exists in* $\mathfrak{F}_\mathbf{C}$.

IDEA OF THE PROOF. Let $X \in \mathrm{Ob}\ \mathfrak{F}_\mathbf{C}$ and $d \in \mathrm{Div}\ X$, and let f be a quasifunction on $X(\mathbf{C})$ with divisor d. There are many such quasifunctions (namely, $f + BQ(X)$). Since X^{an} has a Kähler metric, we can apply the Laplace operator Δ to f. It runs out that in the set $f + BQ(X)$ (of quasifunctions with divisor d) there is, up to an additive constant, only one quasifunction f such that

$$\Delta f = \mathrm{const.} \tag{17}$$

Normalizing its value at a selected point on X, we obtain $\hat{h}_v(d)$.

The category $\mathfrak{F}_\mathbf{C}$ contains compact quotient spaces of bounded domains modulo discrete groups acting without fixed points (there is a canonical Kähler metric on them), and correspondences between them are of the type considered in §2. This is inapplicable to the correspondences (15), since the varieties included in them have singularities, and, more importantly, have the canonical Kähler metric defined only on open subsets (M and H). Since our interest is only in the values of the quasifunctions at points of $M_{g,1,n} \otimes \mathbf{C} \subset \overline{M}_{g,1,n}$ (and analogously for H), we could attempt to carry the construction of Proposition 6 over to this case. This reduces to fine questions about the uniqueness of solutions of the equation (17) on noncompact manifolds (the operator Δ is elliptic inside the manifolds and degenerate on their boundaries). There are apparently no general results in this direction.

In the special case of the correspondences (15) there is a procedure which circumvents this difficulty. We can limit ourselves to the canonical sheaf L on these varieties and interpret its sections (or those of the divisor d) as differential forms ω of maximal degree (the lemma of §3). If θ is the volume form in the Kähler metric on $\mathfrak{S}_g$, it then descends to all the quotient spaces $\mathbf{M}(\Gamma)$, and we can put

$$\hat{h}_v(d) = \log \frac{\theta}{\omega \wedge \bar{\omega}}$$

with a suitable normalization.

This is a solution of equation (17) in the case under consideration, and it satisfies Definition 3 if ω is a cusp form. There are sufficiently many such forms, so that we get a local height that is completely suitable for our applications (see Theorem 2 of [3]).

We also observe that the quasifunction $\hat{h}_v$ is invariant under the action of the groups $\mathrm{Sp}(2g, \mathbf{Z}/n\mathbf{Z})$ on the moduli variety (see §1.2) and represents an intrinsic invariant of abelian varieties.

§5. Estimates on the height

In this section we shall study the following property of algebraic varieties defined over a number or a function field K. We denote by h a height in the category $\mathfrak{V}_K$.

DEFINITION 1. Let X be a projective variety over K; X satisfies *property* (M) or *is of type* (M) if for any finite extension $E \supset K$ one of the following equivalent conditions holds.

1. There exists an ample sheaf $L \in \mathrm{Pic}\, X$ such that for the sequence (x_i) of all points of $X(K)$, ordered according to increasing height $h(L)$, we have for some $b > 0$, $a > 1$ and $\alpha > 0$

$$h(L)(x_i) > ba^{i^\alpha}. \tag{18}$$

2. For any ample sheaf $L \in \mathrm{Pic}\, X$ the sequence of all points of $X(K)$ ordered according to increasing height $h(L)$ satisfies (18).

PROPOSITION 1. *Conditions* 1 *and* 2 *of Definition* 1 *are equivalent.*

The first such estimates appeared in Mumford's paper [9]. He proved at one stroke

PROPOSITION 2. *If X is a curve, then X is of type (M) if and only if the genus of X is > 1.*

The fact that curves of genus 0 and 1 are not of type (M) follows from an estimate on the height obtained by Schanuel and Néron (see [8], [16], and formulas (21) and (22) below).

The proof of Proposition 1 and Proposition 3 (stated below) is based on the following lemma.

LEMMA 1. *Let* $X \in \mathrm{Ob}\,\mathcal{V}_K$, *and let* $L, M \in \mathrm{Pic}\, X$ *with* L *ample. Then there exist* $a > 0$ *and* b *such that* $h(M) \leqslant ah(L) + b$.

PROPOSITION 3. *Let* $X, Y \in \mathrm{Ob}\,\mathcal{V}_K$, *and let* $f\colon X \to Y$ *be a surjective morphism. If the variety* Y *and all the fibers* X_y, $y \in Y$, *are of type* (M), *then* X *is also of type* (M).

In spite of its triviality, this proposition allows us to construct interesting examples of varieties which satisfy condition (M); see below.

COROLLARY 1. *A product* $X \times Y$ *of varieties of type* (M) *is also of type* (M).

COROLLARY 2. *If* $f\colon X \to Y$ *is a finite covering, then* Y *is of type* $(M) \Rightarrow X$ *is of type* (M).

We see from Proposition 2 that the converse is false; however, we do have

PROPOSITION 4. *If* $X, Y \in \mathrm{Ob}\,\mathcal{V}_K$ *and if* $f\colon X \to Y$ *is an étale covering, then* X *is of type* $(M) \Rightarrow Y$ *is of type* (M).

The proof is based on the following result.

LEMMA 2. *Under the hypotheses of Proposition* 4 *there exist a natural number* n *and a finite set* S *of places of the field* K *such that for all* $y \in Y(K)$ *and for all* $x \in f^{-1}(y)$ *the residue field* $K(x)$ *is only ramified over* K *at the places of* S *and* $[K(x)\colon K] \leqslant n$.

Here is another trivial fact about the property (M).

PROPOSITION 5. *If* $X \in \mathrm{Ob}\,\mathcal{V}_K$ *is of type* (M), *so is any subvariety of* X.

COROLLARY. *A variety* X *of type* (M) *does not contain curves of genus* 0 *and* 1.

Finally, from Proposition 2 of §4 we get

PROPOSITION 6. *Let* $X \in \mathrm{Ob}\,\mathcal{V}_K$ *be of type* (M). *Then a connected algebraic group of automorphisms cannot act on* X.

If there were such a group, its elements would act trivially on the Néron-Severi group of X. From this we would obtain that if $L \in \mathrm{Pic}\, X$ is an ample sheaf and g is an element of the group, then $g^*L \approx L$. Since there are always elements of infinite order in an algebraic group over K, we obtain the required contradiction by Proposition 2 of §4.

Proposition 2 shows that in the case of curves the validity of property (M) depends on purely geometric properties of a curve. If X is a surface of type (M), the necessary geometric conditions on X are easily indicated by means of Proposition 6 and the corollary to Proposition 5. Namely, either X is a K3 surface, or the canonical class of X is ample. In both cases it is necessary to exclude surfaces of genus 0 or 1.

Still another condition, but valid for varieties of arbitrary dimension, is suggested by Grauert's proof of Mordell's conjecture for function fields [3]. This condition consists of requiring the ampleness of the sheaf Ω^1. In the function field case it plays an important role in translating Grauert's proof to varieties of dimension > 1. If a surface satisfies this condition, then its canonical class is ample and there are no curves of genus 0 or 1 on it. On the other hand, there exist varieties of type (M) which do not satisfy this condition (for example, products of curves of genus > 1).

We give yet another useful reformulation of property (M). For this we introduce for $h \in \mathbf{R}$ the function $\varphi(h) = \{$the number of points $x \in X(K)$ with $h(L)(x) \leqslant h\}$. It depends on an invertible ample sheaf L. Suppose all the points $(x_i) \in X(K)$ are ordered according to increasing height $h(L)$. We consider the estimates

$$h(L)(x_i) > f(i) \tag{19}$$

and

$$\varphi(h) < g(h), \tag{20}$$

where f and g are strictly monotone functions. Then (20) follows from (19) with $g = f^{-1}$, and (19) follows from (20) with $f = g^{-1}$.

In particular, property (M), i.e. (18), in terms of the function $\varphi(h)$ appears as $\varphi(h) < c_1 \log h + c_2$, $c_1 > 0$. Here and below c_1, c_2 and c are constants greater than 1.

If $X = \mathbf{P}^n$, then

$$c_1 \cdot c^{(n+1)h} < \varphi(h) < c_2 \cdot c^{(n+1)h}, \tag{21}$$

and in case X is an abelian variety

$$c_1 h^{r/2} < \varphi(h) < c_2 h^{r/2} \tag{22}$$

(r is the rank of the group $X(K)$) (see [8], [10], [16]). Finally, for any variety X of $\mathfrak{V}_K$ of dimension n there is an estimate from above

$$\varphi(h) < c_1 \cdot c^{(n+1)h}.$$

This is obtained if X is represented as a finite covering of $\mathbf{P}^n$.

All the examples of varieties of type (M) known to the author are obtained from curves of genus > 1 by the constructions described above. The irregularity of such varieties is always > 0. At the same time we know that the moduli varieties $\overline{M}_{g,1,n}$, $g > 1$, have irregularity zero. Therefore, if estimate (18) also holds for them (say for large n), completely new arguments are necessary to obtain it. The following example exhibits the difficulties arising here.

EXAMPLE. Let X be a smooth projective curve of genus 2,

$$\pi : X' \to X \tag{23}$$

a two-sheeted étale covering of X, and let $J = \mathrm{Pic}^0 X$ and $J' = \mathrm{Pic}^0 X'$ be the Jacobian varieties of X and X'. The genus of X' equals 3, and the abelian variety $P = J'/\mathrm{Im}\ \pi^*$ is an elliptic curve ($\pi^*: J \to J'$ is the canonical morphism). A Jacobian rigidity of level n, $(2, n) = 1$, on J' defines Jacobian rigidities on J and P. There is a canonical bijection between the set of coverings (23) and the set of nonzero points on J_2. Each point $a \in J_2$, $a \neq 0$, therefore defines the varieties J' and P. We consider the moduli scheme $M^c_{2,n}$ (over $\mathbf{Q}$) of curves of genus 2, classifying the following objects: a curve X of genus 2 plus a Jacobian rigidity of level 2 on J plus a Jacobian rigidity of level n on J' (as the point $a \in J_2$ determining J' we choose the first point of the base giving the rigidity on J).

This scheme exists, and we have a morphism

$$t : M^c_{2,n} \to M_n \tag{24}$$

into the moduli scheme of elliptic curves, associating a curve X to the corresponding curve P. The group $\mathrm{Sp}(4, \mathbf{Z}/2\mathbf{Z})$ acts on the set of Jacobian rigidities of level 2 on J. This action never carries over to the scheme $M^c_{2,n}$; however, a finite group G equipped with an epimorphism $\varphi: G \to \mathrm{Sp}(4, \mathbf{Z}/2\mathbf{Z})$ acts on $M^c_{2,n}$. Under this the Jacobian rigidities of level 2 on the curves x and $\sigma(x)$, $\sigma \in G$, differ by the automorphism $\varphi(\sigma)$. The mapping

$$\prod_{\sigma \in G} t \circ \sigma : M^c_{2,n} \to \prod_{\sigma \in G} M_n \tag{25}$$

is an inclusion.

If n is sufficiently large, then the genus of the curve $\overline{M}_n$ is greater than 1 [11] and we obtain an inclusion of the variety $M^c_{2,n}$ into a product of such curves. The closure of $M^c_{2,n}$ in this product is a projective variety satisfying property (M) (Proposition 5).

Now let $M_{2,n}$ be the moduli scheme of abelian varieties of dimension 2 over $\mathbf{Q}$ (see §3). The natural morphism $b: M^c_{2,n} \to M_{2,n}$ is étale, and its image "consists" of the Jacobian varieties. But if we extend b to a covering of all of $M_{2,n}$ or $\overline{M}_{2,n}$, branching occurs, so that Proposition 4 cannot be used in this situation to obtain property (M) for $M_{2,n}$.

We also note a purely geometric property connected with the construction given above.(3) Let U be a Teichmüller domain in $\mathbf{C}^{3g-3}$, and let Γ be the Teichmüller group. There is a subgroup Γ' of finite index such that

$$(M^c_{2,n} \otimes \mathbf{C})^{an} \cong U/\Gamma'.$$

From (25) it follows that the group $H^1(\Gamma', \mathbf{Z})$ is infinite for n such that the genus of $\overline{M}_n > 0$. This shows an essential distinction between Γ and discrete groups in homogeneous symmetric domains such as the groups of §2 (for them the group $H^1(\Gamma)$ is always finite for $g > 1$).

(3) A. N. Tjurin mentioned to me that this is a special case of the so-called Prym mapping.

In conclusion we stop to compare the height estimates of integer points on algebraic varieties.

Let $\Lambda \subset K$ be the ring of integers of the number field K, $X \in \mathrm{Ob}\,\mathcal{C}_K$, and L a very ample sheaf on X.

DEFINITION 2. Let $(s_0, s_1, \cdots, s_n)$ be elements of the group $H^0(X, L)$, and d the divisor of zeros of the section s_0. A point $x \in (X - d)(K)$ is said to be *integral* if $(s_i/s_0)(x) \in \Lambda$ for all $i > 0$.

Height estimates, and in particular the calculation of the analogue φ' of the function $\varphi(h)$, were treated in this situation by Skolem [17] and Dörge [18]–[20]. Their results relate to the case $\Lambda = \mathbf{Z}$. In [19] Dörge proved that for the irreducible curves $F(x, y) = g(x, y)$, where $F(x, y)$ is a form with nonzero discriminant, $g(x, y)$ is a polynomial, and the degree of F in x and in y is not less than $\deg(g)$, the estimates

$$\varphi'(h) < c \cdot h, \qquad \deg(F) - \deg(g) = 1,\ 2,$$
$$\varphi'(h) < c \cdot \log h, \qquad \deg(F) - \deg(g) \geqslant 3,$$

hold. His method, based on an expansion of y in a power series in x, is very simple, but it does not work in the case of dimension > 1. Skolem [17] obtained some results in the higher-dimensional case. Namely, he proved that if there are sufficiently "many" integral points, then the variety is rational. On the contrary, his condition is expressed in terms which are not invariant, and are not reformulated in the form of height estimates for $\varphi'(h)$.

All these results arise in studying the Hilbert irreducibility theorem [8]. They are also connected with problems of the lifting of rational or integral points (see [16], [20] and the literature mentioned there). This last consists of a quantitative description of the set of the rational (integral) points $X(K)$ which lift to rational (integral) points in $Y(K)$ if we are given a surjective morphism $X \to Y$.

In the one-dimensional case, using Siegel's theorem, we can obtain a conclusive answer about the behavior of $\varphi'(h)$. Denoting by g the genus of the curve X, and by δ the number of points of the divisor d, we obtain a table

g	δ	$\varphi'(h)$
>0	>0	$<$ const
0	>2	$<$ const
0	2	$<$ const $\cdot\, h$
0	1	$< e^{\text{const}\,\cdot\, h}$

If the sets of integral points are not empty, then each estimate from above in this table can be supplemented by an analogous estimate from below. In place of

Siegel's theorem we can use Mumford's estimate (Proposition 2). This replaces the first and second rows by

$$\varphi'(h) < \text{const} \cdot \log h.$$

It would be interesting to find a direct proof of this fact.

We observe also that, although the integral points make up a part of the rational points, the problem of estimating $\varphi(h)$ can be reduced by a formal method to the analogous problem for integral points, i.e. for $\varphi'(h)$. Namely, if $L \to X$ is a very ample bundle over the variety X, then $\varphi(h)$ on X is computed by means of $\varphi'(h)$ on L.

BIBLIOGRAPHY

1. J. Tate, *Endomorphisms of abelian varieties over finite fields,* Invent. Math. **2** (1966), 134–144. MR **34** #5829.

2. A. N. Paršin, *Minimal models of curves of genus* 2 *and homomorphisms of abelian varieties defined over a field of finite characteristic,* Izv. Akad. Nauk SSSR Ser. Mat. **36** (1972), 67–109 = Math. USSR Izv. **6** (1972), 65–108.

3. ———, *Quelques conjectures de finitude en géométrie diophantienne,* Proc. Internat. Congress Math. (Nice, 1970), vol. 1, Gauthier-Villars, Paris, 1971, pp. 467–472.

4. D. Mumford, *Abelian varieties,* Tata Inst. of Fund. Res. Studies in Math., no. 5, Oxford Univ. Press, London, 1970; Russian transl., "Mir", Moscow, 1971. MR **44** #219.

5. ———, *Geometric invariant theory,* Ergebnisse der Mathematik und ihrer Grenzgebiete, N. F., Band 34, Academic Press, New York; Springer-Verlag, Berlin, 1965. MR **35** #5451.

6. *Séminaire Henri Cartan* 1957/58. *Fonctions automorphes,* Secrétariat mathématique, Paris, 1958. MR **21** #2750.

7. P. Deligne, *Formes modulaires et représentations l-adiques,* Séminaire Bourbaki 1968/69, Exposé 355, Lecture Notes in Math., vol. 179, Springer-Verlag, Berlin, 1971. MR **42** #7460.

8. S. Lang, *Diophantine geometry,* Interscience Tracts in Pure and Appl. Math., no. 11, Interscience, New York, 1962. MR **26** #119.

9. D. Mumford, *A remark on Mordell's conjecture,* Amer. J. Math. **87** (1965), 1007–1016. MR **32** #4083.

10. A. Néron, *Quasi-fonctions et hauteurs sur les variétés abéliennes,* Ann. of Math. (2) **82** (1965), 249–331. MR **31** #3424.

11. A. N. Paršin, *Isogenies and torsion of elliptic curves,* Izv. Akad. Nauk SSSR Ser. Mat. **34** (1970), 409–424 = Math. USSR Izv. **4** (1970), 415–430. MR **43** #218.

12. Ju. I. Manin, *The refined structure of the Néron-Tate height,* Mat. Sb. **83** (125) (1970), 331–348 = Math. USSR Sb. **12** (1970), 325–342. MR **43** #1992.

13. S. Ju. Arakelov, *Families of algebraic curves with fixed degeneracies,* Izv. Akad. Nauk SSSR Ser. Mat. **35** (1971), 1269–1293 = Math. USSR Izv. **5** (1971), 1277–1302.

14. F. Oort, *Finite group schemes, local moduli for abelian varieties, and lifting problems,* Compositio Math. **23** (1971), 265–296. MR **46** #186.

15. U. Christina, *Einführung in die Theorie der paramodularen Gruppen,* Math. Ann. **168** (1967), 59–104. MR **34** #4215.

16. Ju. I. Manin, *Two theorems on rational surfaces,* Rend. Mat. e Appl. (5) **25** (1966), 198–207.

17. Th. Skolem, *Untersuchungen über die möglichen Verteilungen ganzzahliger Lösungen gewisser Gleichungen,* Norske Vid. Selsk. Skr. **1921**, no. 17.

18. K. Dörge, *Ein Beitrag zur Theorie der diophantischen Gleichungen mit zwei Unbekannten,* Math. Z. **24** (1926), 193–198.

19. ————, *Über die Seltenheit der reduziblen Polynome und der Normalgleichungen,* Math. Ann. **95** (1926), 247–256.

20. A. Schinzel, *Reducibility of polynomials,* Proc. Internat. Congress Math. (Nice, 1970), vol. 1, Gauthier-Villars, Paris, 1971, pp. 491–496.

Translated by J. S. JOEL

Trudy Mat. Inst. Steklov.
132 (1973)

Proc. Steklov Inst. Math.
132 (1973)

CONSTRUCTIVE METHOD IN THE THEORY OF EQUATIONS OVER FINITE FIELDS

S. A. STEPANOV

ABSTRACT. In this paper further development is presented of the method by which the author has obtained an elementary proof of A. Weil's theorem on estimating trigonometric sums with a rational function.

Bibliography: 18 items.

1. In the theory of numbers, the solutions of arithmetic problems are often found by the application of mathematical methods which are connected only indirectly with the formulation of the problems. Thus interest naturally arises in the discovery of straightforward methods which are based directly on the arithmetical nature of the question under investigation.

In problems of the theory of congruences modulo a prime number, and more generally in problems of the theory of equations over an arbitrary finite field, the methods which have been most effective up to now are those of algebraic geometry. In 1968 I developed a constructive method in the theory of equations over a finite field, by means of which it was possible to give elementary proofs of theorems which had previously been proved only by the use of the contemporary machinery of algebraic geometry.

2. Historically, the congruences with a prime modulus

$$F(x,\ y)\equiv 0 \pmod p, \tag{1}$$

were the objects of the first investigations in the theory of algebraic equations over a finite field, and the basic difficulty arises precisely in this particular case. The passage to an arbitrary finite field does not give rise to noticeable further difficulties.

Isolated facts related to special cases of the congruence (1) have been known for a long time. In order to prove the theorem that every natural number can be represented as the sum of four squares, Lagrange made use of the assertion that the congruence

$$x^2+y^2+1\equiv 0 \pmod p$$

is solvable. We turn our attention to the proof of this assertion, given by Lagrange. Assume that the given congruence is not solvable. Then, by Euler's criterion, the congruence

AMS (MOS) subject classifications (1970). Primary 12C05; Secondary 02E99.

$$1+(-1-x^2)^{\frac{p-1}{2}}\equiv 0 \pmod p$$

must have p solutions. But the degree of this latter congruence is $p-1$, and by a theorem of Lagrange it cannot have more than $p-1$ solutions. This contradiction gives the desired result.

N. S. Aladov [2], in connection with the study of the distribution of pairs of quadratic residues and nonresidues, extended the method of Lagrange, and for the number I_p of solutions of the congruence

$$y^2\equiv x(x+1) \pmod p$$

found the expression

$$I_p=p-(-1)^{\frac{p-1}{2}}.$$

Some special cases of the congruence

$$y^2\equiv f(x) \pmod p \tag{2}$$

where the polynomial $f(x)$ is of the third degree, were studied by Jacobsthal (see [3], §10).

Gauss [4] stated that the number I_p of solutions of the biquadratic congruence

$$x^2y^2+x^2+y^2\equiv 1 \pmod p)$$

for p of the form $4k+1$ and $p=a^2+4b^2$ is given by

$$I_p=(a-1)^2+4b^2-4.$$

The proof of this formula was given by Herglotz [5] and required the use of the elliptic functions of Weierstrass. However, it is not hard to derive this result in an elementary manner, as was done by Hasse [3].

The problem of the number of solutions of the congruence (2) may be easily reduced to the problem of estimating sums of Lagrange symbols

$$\sum_{x=0}^{p-1}\left(\frac{f(x)}{p}\right). \tag{3}$$

If $f(x)$ is a polynomial of the second degree, then an exact formula is known for the sum (3). If $f(x)$ has degree greater than two, the situation is nontrivial. Artin [6] conjectured that if $f(x)$ is a polynomial of degree n which is not the square of another polynomial mod p, then the number I_p of solutions of the congruence (2) satisfies the estimate

$$|I_p-p|\leqslant 2\left[\frac{n-1}{2}\right]\sqrt{p}.$$

This conjecture of Artin can be expressed as the inequality

$$\left|\sum_{x=0}^{p-1}\left(\frac{f(x)}{p}\right)\right| \leqslant 2\left[\frac{n-1}{2}\right]\sqrt{p}.$$

Hopf [7] in estimating sums of the form (3) used the method of multiple sums. In this method, the sum being estimated is broken up into several similar sums, but with other parameters, and then an average is taken over all polynomials of a given degree. Later this method was extended by Davenport [8] and Mordell [9].

However, the method of multiple sums not only did not yield a proof of the hypothesis of Artin, but it did not even give the correct order of p. As a matter of fact, in problems of congruences for a prime modulus, the method of multiple sums amounts to the use of a general method of analytic number theory, namely the availability of the Frobenius endomorphism, without taking the specific finite field into account.

This explains the fact that the method of multiple sums does not give the best possible results.

In 1933 Hasse [10], in developing the theory of algebraic functions, showed that the number I_p of solutions of the elliptic congruence

$$y^2 \equiv x^3 + ax + b \pmod p$$

satisfies the estimate

$$|I_p - p| < 2\sqrt{p}.$$

An elementary proof of this theorem of Hasse was given by Manin [11].

A. Weil [12] gave an extensive generalization of Hasse's theorem, and in the case of absolutely irreducible polynomials $F(x, y)$ he found that the number I_q of solutions of the equation

$$F(x, y) = 0 \tag{4}$$

in the Galois field $GF(q)$ consisting of $q = p^n$ elements, satisfies the estimate

$$|I_q - q| \leqslant 2g\sqrt{q}, \tag{5}$$

where g is the genus of the curve (4). This enabled him to prove the Riemann hypothesis for Z-functions of algebraic curves defined over finite fields. The estimate given by Weil [13] for rational exponential sums with denominator of the first degree, and in particular Kloosterman sums, followed as a corollary.

The proof given by Weil of the estimate (5) is quite complicated and requires the use of the contemporary machinery of algebraic geometry.

In 1968, in [14], I was able to show in an elementary fashion that the number I_p of solutions of the hyperelliptic congruence

$$y^2 \equiv f(x) \pmod p,$$

in the case when the polynomial $f(x)$ has odd degree, satisfies the estimate

$$|I_p - p| \leqslant \sqrt{3n}\,n\sqrt{p}.$$

Later I extended this method to the congruence

$$y^n \equiv f(x) \pmod p$$

and the equation

$$y^p - y = g(x)$$

in the Galois field $GF(p^n)$ (see [15], [16]).

The resulting asymptotic formula for the number of solutions of the latter equation gave rise to an elementary proof of Weil's theorem on the estimates for rational exponential sums. In addition, making use of the estimates mentioned above for the number of solutions of a hyperelliptic congruence, it was possible to give a completely elementary proof of the theorem of Burgess on the least quadratic nonresidue.

In 1971 N. M. Korobov, using methods based on my work in [14], found that for a polynomial $f(x)$ of odd degree n the number I_p of solutions of the congruence (2) satisfies the estimate

$$|I_p - p| \leqslant (n-1)\sqrt{p - \frac{(n-3)(n-4)}{4}}.$$

3. We consider the general case. Let $m, n \geqslant 2$ be relatively prime natural numbers; $p > 196\, m^3 n(n-1)^2$ a prime number, $F(x, y) = y^n + A_1(x)y^{n-1} + \cdots + A_{n-1}(x)y + A_n(x)$ a polynomial in x and y with integral rational coefficients, and $f(x, y) \doteq y^n + a_1(x)y^{n-1} + \cdots + a_{n-1}(x)y + a_n(x)$ the polynomial arising from $F(x, y)$ by reduction $\bmod p$. Assume that $F(x, y)$ is irreducible $\bmod p$, and that the degrees of the polynomials $a_i(x)$, $i = 1, \cdots, n$, satisfy the following conditions:

$$\begin{aligned} &\deg a_n(x) = m, \\ &n \deg a_i(x) < im, \qquad i = 1, 2, \ldots, n-1. \end{aligned} \tag{6}$$

Let I_p denote the number of solutions of the congruence

$$F(x, y) \equiv 0 \pmod p. \tag{7}$$

We shall prove the following theorem.

THEOREM. *Let $F(x, y)$ satisfy the conditions* (6). *Then the quantity I_p satisfies the inequality*

$$|I_p - p| \leqslant 14\sqrt{mn}\, m(n-1)\sqrt{p}.$$

Note that in this inequality the constant is somewhat worse than the constant in Weil's inequality. To get the same constant as Weil, it is sufficient to transfer the proof of the theorem to an arbitrary Galois field $GF(p^r)$ and then employ the theory of Z-functions of algebraic curves. We restrict ourselves to the proof of the stated inequality with somewhat poorer constants, and for an illustration of the technique of transferring the proof to an arbitrary finite field we refer to my papers [16], [18].

The congruence (7) may be interpreted as the equation

$$f(x, y) = 0 \tag{8}$$

over the finite prime field $\mathbf{Z}_p$ with p elements, and the quantity I_p as the number of rational points on the curve (8).

We introduce the following notation: $\mathbf{Z}_p[x]$ is the ring of polynomials in x with coefficients in $\mathbf{Z}_p$; $\mathbf{Z}_p(x)$ is the quotient field of $\mathbf{Z}_p[x]$. Let $y_s(x)$, $s = 1, \cdots, n$, be roots of equation (8). We designate by $\mathbf{Z}_p(x, y_s)$ the field of algebraic functions which arises from $\mathbf{Z}_p(x)$ by adjoining the elements $y_s(x)$, and by $\mathbf{Z}_p[x, y_s]$ the ring of integral elements of $\mathbf{Z}_p(x, y_s)$. Let $D(x)$ denote the discriminant of the polynomial $f(x, y)$. We separate the elements of $\mathbf{Z}_p$ into classes A_τ, $\tau = 0, 1, \cdots, n$, and B as follows: the class A_τ consists of those elements $x \in \mathbf{Z}_p$ such that $a_n(x) \neq 0$, $D(x) \neq 0$, and equation (8) has exactly τ solutions in y; and the class B consists of those elements $x \in \mathbf{Z}_p$ such that either $a_n(x) = 0$ or $D(x) = 0$.

Let $\mathfrak{I}_\tau$ $(\tau = 0, 1, \cdots, n)$ be the number of elements of the class A_τ, and $\mathfrak{I}$ the number of elements of the class B. It is clear that

$$\sum_{\tau=0}^{n} I_\tau + I = p.$$

In addition,

$$I_p = \sum_{\tau=1}^{n} \tau I_\tau + r,$$

where $0 \leqslant r \leqslant 2mn^2$.

Let $y_s(x)$ be any root of the polynomial $f(x, y)$ and let $x \in \mathbf{Z}_p$. A necessary and sufficient condition for $y_s(x) \in \mathbf{Z}_p$ is that

$$y_s^p(x) - y_s(x) = 0.$$

We shall find a necessary and sufficient condition for $y_s(x) \notin \mathbf{Z}_p$ when $x \in \mathbf{Z}_p$. We have

$$y_s^n + a_1(x) y_s^{n-1} + \ldots + a_{n-1}(x) y_s + a_n(x) = 0,$$

$$y_s^{np} + a_1(x) y_s^{(n-1)p} + \ldots + a_{n-1}(x) y_s^p + a_n(x) = 0.$$

Thus

$$(y_s^{np} - y_s^n) + a_1(x)(y_s^{(n-1)p} - y_s^{n-1}) + \ldots + a_{n-1}(x)(y_s^p - y_s) = 0.$$

It follows from this that $y_s(x) \notin \mathbf{Z}_p$ if and only if

$$\sum_{i=0}^{n-1} b_{s,i} y_s^{i(p-1)} = 0,$$

where

$$b_{s,i} = \frac{y_s^{n-i-1} + a_1 y_s^{n-i-2} + \ldots + a_{n-i-1}}{y_s^{n-i-1}}.$$

Finally, we introduce a differential operator in $\mathbf{Z}_p(x, y_s)$:

$$D = \frac{d}{dx}.$$

Let s be any of the integers $1, \cdots, n$ and let $N < p/2$ be a natural number. We consider in $\mathbf{Z}_p$ the expression

$$L_0(x, y_s) = \sum_{i=0}^{n-1} b_{s,i} y_s^{i(p-1)} \sum_{j=1}^{N} r_{s,j}^{(0)} (x^p - x)^{j-1} + \sum_{i=0}^{n-2} y_s^{i(p-1)} \sum_{j=1}^{N} t_{s,i,j}^{(0)} (x^p - x)^j$$

with undefined coefficients $r_{s,j}^{(0)}, t_{s,i,j}^{(0)} \in \mathbf{Z}_p[x, y_s]$. It is clear that $L_0(x, y_s) = 0$ for those values $x \in \mathbf{Z}_p$ for which $y_s(x) \notin \mathbf{Z}_p$. Put

$$L_k(x, y_s) = D^k L_0(x, y_s), \qquad k = 1, 2, \ldots,$$

and choose $r_{s,j}^{(0)}$ and $t_{s,i,j}^{(0)}$ in such a way that $L_k(x, y_s)$, $k = 1, 2, \cdots, N + [(N-1)/(n-1)]$, has the same property.

Let $r_{s,j}^{(k)}$ and $t_{s,i,j}^{(k)}$ be defined by the recurrence relations

$$r_{s,j}^{(k)} = Dr_{s,j}^{(k-1)} - jr_{s,j+1}^{(k-1)} - (n-1)\frac{Dy_s}{y_s} r_{s,j}^{(k-1)},$$

$$k = 1, 2, \ldots; \qquad j = 1, 2, \ldots, N,$$

$$t_{s,i,j}^{(k)} = Dt_{s,i,j}^{(k-1)} - (j+1)t_{s,i,j+1}^{(k-1)} - i\frac{Dy_s}{y_s} t_{s,i,j}^{(k-1)}$$

$$+ \left(Db_{s,i} + (n-i-1) b_{s,i} \frac{Dy_s}{y_s}\right) r_{s,j+1}^{(k-1)}, \tag{9}$$

$$k = 1, 2, \ldots; \qquad i = 0, 1, \ldots, n-2; \qquad j = 1, 2, \ldots, N,$$

with initial values $r_{s,j}^{(0)}$ and $t_{s,i,j}^{(0)}$, where we may assume that $r_{s,j}^{(0)} = 0$ and $t_{s,i,j}^{(0)} = 0$ for $j > N$.

We say that the expression $L_k(x, y_s)$ has "necessary" form if it can be represented in the form

$$L_k(x, y_s) = \sum_{i=0}^{n-1} b_{s,i} y_s^{i(p-1)} \sum_{j=1}^{N} r_{s,j}^{(k)} (x^p - x)^{j-1} + \sum_{i=0}^{n-2} y_s^{i(p-1)} \sum_{j=1}^{N} t_{s,i,j}^{(k)} (x^p - x)^j,$$

where $r_{s,j}^{(k)}$ and $t_{s,i,j}^{(k)}$ are defined by the recurrence relations (9).

Assume that $L_{k-1}(x, y_s)$ has "necessary" form. Starting from the definitions of $L_k(x, y_s)$, $r_{s,j}^{(k)}$ and $t_{s,i,j}^{(k)}$ it is easy to prove that a sufficient condition for the expression $L_k(x, y_s)$ to have "necessary" form is that

$$t_{s,i,1}^{(k-1)} = \left(Db_{s,i} + (n-i-1) b_{s,i} \frac{Dy_s}{y_s}\right) r_{s,1}^{(k-1)}, \qquad i = 0, 1, \ldots, n-2. \tag{10}$$

Let the algebraic functions $G_{s,i,k}^{(j)} \in \mathbf{Z}_p(x, y_s)$, $j = 1, 2, \cdots; k = 1, \cdots; i = 0, \cdots, n-2$, be given by the recurrence relations

$$G^{(1)}_{s,i,1}=Db_{s,i}+(n-i-1)\,b_{s,i}\frac{Dy_s}{y_s},\quad i=0,\,1,\,\ldots,\,n-2,$$

$$G^{(j)}_{s,i,k}=DG^{(j-1)}_{s,i,k}+(k-1)\,G^{(j-1)}_{s,i,k-1}+(n-i-1)\frac{Dy_s}{y_s}G^{(j-1)}_{s,i,k},$$

$$j=2,\,3,\,\ldots;\qquad k=1,\,2,\,\ldots,\,j-1;\qquad i=0,\,1,\,\ldots,\,n-2, \tag{11}$$

$$G^{(j)}_{s,i,j}=(j-1)\,G^{(j-1)}_{s,i,j-1}+(j-1)!\,G^{(1)}_{s,i,1},$$

$$j=2,\,3,\,\ldots;\qquad i=0,\,1,\,\ldots,\,n-2.$$

By induction on j, we have from (9), (10) and (11) that a sufficient condition that $L_k(x, y_s)$, for $k = 0, 1, \cdots, N + [(N-1)/(n-1)]$, have "necessary" form is that

$$j!\,t^{(0)}_{s,i,j}=\sum_{k=1}^{j}G^{(j)}_{s,i,k}r^{(0)}_{s,k},\qquad i=0,\,1,\,\ldots,\,n-2;\ \ j=1,\,2,\,\ldots,\,N, \tag{12}$$

$$0=\sum_{k=1}^{N}G^{(j)}_{s,i,k}r^{(0)}_{s,k},\qquad i=0,\,1,\,\ldots,\,n-2;\ \ j=N+1,\,\ldots,\,N+\left[\frac{N-1}{n-1}\right]. \tag{13}$$

We shall find a nontrivial solution of the system (13) in elements $r^{(0)}_{s,k}$ of the ring $\mathbf{Z}_p[x, y_s]$.

It follows from (11) that the algebraic functions $G^{(j)}_{s,i,k}$ may be represented in the form

$$G^{(j)}_{s,i,k}=\frac{g^{(j)}_{1,i,k}y_s^{n-1}+\ldots+g^{(j)}_{n,i,k}}{y_s^{n-i+j-k-1}\left(\dfrac{\partial f}{\partial y_s}\right)^{2(j-k)+1}},$$

where the degrees of the polynomials $g^{(j)}_{\rho,i,k}$ do not exceed, respectively, the quantities $\alpha^{(j)}_{\rho,i,k} = 5m(j-k+1)$. Put

$$r^{(0)}_{s,k}=y_s^{n+N-k-1}\left(\frac{\partial f}{\partial y_s}\right)^{2(N-k)+1}r_{s,k}$$

We want $r_{s,k}$ in the form

$$r_{s,k}=\sum_{\rho=1}^{n}R_{\rho,k}y_s^{n-\rho},$$

where $R_{\rho,k}\in\mathbf{Z}_p[x]$.

Then the system (13) may be written as

$$\sum_{k=1}^{N}\sum_{\sigma=1}^{n}\sum_{\rho=1}^{n}A^{(j,\sigma)}_{\rho,i,k}R_{\rho,k}y_s^{n-\sigma}=0,\qquad i=0,\,1,\,\ldots,\,n-2;$$

$$j=N+1,\,\ldots,\,N+\left[\frac{N-1}{n-1}\right],$$

where the degrees of the polynomials $A^{(j,\sigma)}_{\rho,i,k}$ do not exceed, respectively, the quantities $5m(j-k+1)+2m$. This last system is solved in a nontrivial way by polynomials $R_{\rho,k}$ whose degrees do not exceed $5m(2N^2-N+k)$.

Now let $t^{(0)}_{s,i,j}$, $i = 0, \cdots, n-2$; $j = 1, \cdots, N$, be given by the equations (12). Put

$$b_{s,i}y_s^{i(p-1)}r_{s,k}^{(0)}(x^p-x)^{k-1}=\sum_{\sigma=1}^{n}B_{i,k}^{(\sigma)}y_s^{n-\sigma},\qquad i=0,1,\ldots,n-1;$$
$$k=1,2,\ldots,N,$$

$$y_s^{q(p-1)}t_{s,q,j}^{(0)}(x^p-x)^j=\sum_{\nu=1}^{n}C_{q,j}^{(\nu)}y_s^{n-\nu},\qquad q=0,1,\ldots,n-2;\ j=1,2,\ldots,N.$$

Assume that $N\leqslant\sqrt{p/11mn}$ and let k_0, q_0 and j_0 be values of k, q and j for which $r_{s,k}^{(0)}\neq 0$, $t_{s,q,j}^{(0)}\neq 0$, and the values of the expressions kp and $jp+(qmp/n)$ are maximal. From (6), the conditions $(m,n)=1$ and $p>196m^3n(n-1)^2$, and the choice of $r_{s,k}^{(0)}$ and $t_{s,q,j}^{(0)}$ it follows either that from some σ_0 we have

$$\deg B_{n-1,k_0}^{(\sigma_0)}>\deg B_{i,k}^{(\sigma)},\qquad \deg B_{n-1,k_0}^{(\sigma_0)}>\deg C_{q,j}^{(\nu)},$$

for all $\nu,\sigma=1,\cdots,n$; $k,j=1,\cdots,N$; $i=0,\cdots,n-1$; $q=0,\cdots,n-2$; or that for some ν_0 we have

$$\deg C_{q_0,j_0}^{(\nu_0)}>\deg C_{q,j}^{(\nu)},\qquad \deg C_{q_0,j_0}^{(\nu_0)}>\deg B_{i,k}^{(\sigma)}$$

for all $\nu,\sigma=1,\cdots,n$; $k=1,\cdots,N$; $i=0,\cdots,n-1$; $q=0,\cdots,n-2$. Hence in the expression $L_0(x,y_s)$ the terms $b_{s,i}y_s^{i(p-1)}r_{s,k}^{(0)}(x^p-x)^{k-1}$ and $y_s^{q(p-1)}t_{s,q,j}^{(0)}(x^p-x)^j$ do not cancel each other, and hence $L_0(x,y_s)\neq 0$.

We note further that the $L_0(x,y_s)$, $s=1,\cdots,n$, are conjugate over $\mathbf{Z}_p(x)$, so that $L_0(x)=\Pi_{s=1}^n L_0(x,y_s)$ is a polynomial of $\mathbf{Z}_p[x]$. Estimating the degree of $L_0(x)$, we have

$$\deg L_0(x)\leqslant nNp+mnp+10mnN^2+12mn.$$

From the construction of $L_0(x,y_s)$ it is clear that the elements of the class A_τ, $\tau=0,1,\cdots,n-1$, are roots of the polynomial $L_0(x)$ of multiplicity at least $(n-\tau)(N+[(N-1)/(n-1)])$, and since the number of roots of a polynomial does not exceed its degree, we have the inequality

$$\left(N+\left[\frac{N-1}{n-1}\right]\right)\sum_{\tau=0}^{n-1}(n-\tau)\,I_\tau\leqslant nNp+mnp+10mnN^2+12mn.$$

Take $N=[\sqrt{p/11mn}]$. Then

$$\sum_{\tau=0}^{n-1}(n-\tau)\,I_\tau\leqslant(n-1)\,p+12\sqrt{mn}\,m\,(n-1)\sqrt{p}$$

and therefore

$$I_p\geqslant p-14\sqrt{mn}\,m\,(n-1)\sqrt{p}.$$

In order to derive the desired estimate for the quantity I_p we construct, for each natural number

$$N\leqslant\frac{1}{n-1}\sqrt{\frac{p}{6mn}}$$

a polynomial $U_0(x)$, not equal to zero in $\mathbf{Z}_p[x]$, of degree not exceeding

$$nNp+mnp+5mn(n-1)^2N^2+11mn$$

such that all elements of A_τ, $\tau = 1, \cdots, n$, are roots of this polynomial of multiplicity at least $\tau(nN-1)$.

The polynomial $U_0(x)$ will be represented in the form $U_0(x) = \Pi_{s=1}^n U_0(x, y_s)$, where $U_0(x, y_s)$ in turn is to be found in the form

$$U_0(x, y_s) = \sum_{i=1}^{n-1}(1-y_s^{i(p-1)}) \sum_{j=1}^{N} r_{s,i,j}^{(0)}(x^p-x)^{j-1} + \sum_{j=1}^{N} t_{s,j}^{(0)}(x^p-x)^j$$

with undefined coefficients $r_{s,i,j}^{(0)}$, $t_{s,j}^{(0)} \in \mathbf{Z}_p[x, y_s]$.

Let $r_{s,i,j}^{(k)}$ and $t_{s,j}^{(k)}$ be given by the recurrence relations

$$\begin{gathered} r_{s,i,j}^{(k)} = Dr_{s,i,j}^{(k-1)} - jr_{s,i,j+1}^{(k-1)} - i\frac{Dy_s}{y_s} r_{s,i,j}^{(k-1)}, \\ i=1, 2, \ldots, n; \quad j=1, 2, \ldots, N; \quad k=1, 2, \ldots, \\ t_{s,j}^{(k)} = Dt_{s,j}^{(k-1)} - (j+1)t_{s,j+1}^{(k-1)} + \frac{Dy_s}{y_s}\sum_{i=1}^{n-1} ir_{s,i,j+1}^{(k-1)}, \\ j=1, 2, \ldots, N; \quad k=1, 2, \ldots, \end{gathered} \tag{14}$$

with initial values $r_{s,i,j}^{(0)}$ and $t_{s,j}^{(0)}$, where we take $r_{s,i,j}^{(0)} = 0$ and $t_{s,j}^{(0)} = 0$ for $j > N$. Now let

$$U_k(x, y_s) = D^k U_0(x, y_s), \qquad k = 1, 2, \ldots.$$

We say that the expression $U_k(x, y_s)$ has "necessary" form if it can be represented in the following manner:

$$U_k(x, y_s) = \sum_{i=1}^{n-1}(1-y_s^{i(p-1)}) \sum_{j=1}^{N} r_{s,i,j}^{(k)}(x^p-x)^{j-1} + \sum_{j=1}^{N} t_{s,j}^{(k)}(x^p-x)^j,$$

where $r_{s,i,j}^{(k)}$ and $t_{s,j}^{(k)}$ are defined by the recurrence relations (14).

We now choose the coefficients $r_{s,i,j}^{(0)}$ and $t_{s,j}^{(0)}$ in such a way that all the expressions $U_k(x, y_s)$, $k = 0, \cdots, nN-1$, have "necessary" form. For this it is sufficient to have the relations

$$j!\,t_{s,j}^{(0)} = \sum_{i=1}^{n-1}\sum_{k=1}^{j} H_{s,i,k}^{(j)} r_{s,i,k}^{(0)}, \qquad j = 1, 2, \ldots, N, \tag{15}$$

$$0 = \sum_{i=1}^{n-1}\sum_{k=1}^{N} H_{s,i,k}^{(j)} r_{s,i,k}^{(0)}, \qquad j = N+1, \ldots, nN-1, \tag{16}$$

where $H_{s,i,k}^{(j)}$ are given by the recurrence relations

$$H_{s,i,1}^{(1)} = i\frac{Dy_s}{y_s}, \qquad i = 1, 2, \ldots, n-1,$$

$$H_{s,i,k}^{(j)} = DH_{s,i,k}^{(j-1)} + (k-1)H_{s,i,k-1}^{(j-1)} + i\frac{Dy_s}{y_s}H_{s,i,k}^{(j-1)},$$

$$i=1,2,\ldots,n-1;\quad j=2,3,\ldots;\quad k=1,2,\ldots,j-1,$$
$$H^{(j)}_{s,i,j}=(j-1)H^{(j-1)}_{s,i,j-1}+(j-1)!\,H^{(1)}_{s,i,1},$$
$$i=1,2,\ldots,n-1;\quad j=2,3,\ldots.$$

Put

$$r^{(0)}_{s,i,k}=\sum_{\sigma=1}^{n} R_{\sigma,i,k}y_s^{n-\sigma}.$$

The system (16) has a nontrivial solution in elements $r^{(0)}_{s,i,k}$, where we may choose the polynomials $R_{\sigma,i,k}$ in such a way that their degrees do not exceed the quantity $5m(n-1)^2N^2+10m$. We define $t^{(0)}_{s,j}$ by the relations (15). From the conditions

$$p>196m^3n(n-1)^2,\quad N\leqslant\frac{1}{n-1}\sqrt{\frac{p}{6mn}}\quad\text{and}\quad(m,n)=1$$

we find that in $U_0(x,y_s)$ the terms $(1-y_s^{i(p-1)})r^{(0)}_{s,i,j}(x^p-x)^{j-1}$ and $t^{(0)}_{s,k}(x^p-x)^k$ cannot cancel each other, and therefore $U_0(x,y_s)\neq 0$.

Since the $U_0(x,y_s)$, $s=1,\cdots,n$, are conjugate over $\mathbf{Z}_p(x)$, we have that $U_0(x)=\Pi_{s=1}^{n}U_0(x,y_s)$ is a polynomial. In addition, the degree of the polynomial $U_0(x)$ does not exceed the quantity

$$nNp+mnp+5mn(n-1)^2N^2+11mn,$$

and since all the derivatives of the expressions $U_0(x,y_s)$, $s=1,\cdots,n$, up to order $nN-1$, have "necessary" form, it follows that the elements of A_τ, $\tau=1,\cdots,n$, are roots of the polynomial $U_0(x)$ of multiplicity at least $\tau(nN-1)$. Equating the number of roots of the polynomial $U_0(x)$ to its degree, we have that

$$(nN-1)\sum_{\tau=1}^{n}\tau I_\tau\leqslant nNp+mnp+5mn(n-1)^2N^2+11mn.$$

Now take

$$N=\left[\frac{1}{n-1}\sqrt{\frac{p}{6mn}}\right].$$

Then

$$\sum_{\tau=1}^{n}\tau I_\tau\leqslant p+12\sqrt{mn}\,m(n-1)\sqrt{p}$$

and therefore

$$I_p\leqslant p+14\sqrt{mn}\,m(n-1)\sqrt{p}.$$

The theorem is proved.

BIBLIOGRAPHY

1. J. L. Lagrange, *Démonstration d'un théorème d'arithmétique*, Nouv. Mém. Acad. Roy. Sci. Belles-Lettres Berlin, 1770; reprinted in *Oeuvres de Lagrange*, Vol. 3, Gauthier-Villars, Paris, 1869, pp. 189–201.

2. N. S. Aladov, *On the distribution of quadratic residues and quadratic nonresidues of a prime number* p *in the series* $1,2,\cdots,p-1$, Mat. Sb. **18** (1896), 61–75. (Russian)

3. H. Hasse, *Vorlesungen über Zahlentheorie*, Die Grundlehren der math. Wissenschaften, Band 59, Springer-Verlag, Berlin, 1950; Russian transl., IL, Moscow, 1953. MR **14**, 534; **16**, 569.

4. C. F. Gauss, *Werke,* Band 10.1, Teubner, Leipzig, 1917, p. 571.

5. G. Herglotz, *Zur letzten Eintragung im Gaussschen Tagebuch*, Ber. Math.-Phys. Kl. Sächs. Akad. Wiss. Leipzig **73** (1921), 271–276.

6. E. Artin, *Quadratische Körper im Gebiete der höheren Kongruenzen.* II, Math. Z. **19** (1924), 207–246.

7. H. Hopf, *Über die Verteilung quadratischen Reste*, Math. Z. **32** (1930), 222–231.

8. H. Davenport, *On the distribution of quadratic residues* mod p, J. London Math. Soc. **6** (1931), 49–54.

9. L. J. Mordell, *The number of solutions of some congruences in two variables*, Math. Z. **37** (1933), 193–209.

10. H. Hasse, *Abstrakte Begründung der komplexen Multiplikation und Riemannsche Vermutung in Funktionenkörper*, Abh. Math. Sem. Univ. Hamburg **10** (1934), 325–348.

11. Ju. I. Manin, *On cubic congruences to a prime modulus*, Izv. Akad. Nauk SSSR Ser. Mat. **20** (1956), 673–678; English transl., Amer. Math. Soc. Transl. (2) **13** (1960), 1–7. MR **18**, 380; **22** #3711.

12. A. Weil, *Sur les courbes algébriques et les variétés qui s'en déduisent*, Actualités Sci. Indust., no. 1041, Hermann, Paris, 1948. MR **10**, 262.

13. ———, *On some exponential sums*, Proc. Nat. Acad. Sci. U.S.A. **34** (1948), 204–207. MR **10**, 234.

14. S. A. Stepanov, *On the number of points of a hyperelliptic curve over a finite prime field*, Izv. Akad. Nauk SSSR Ser. Mat. **33** (1969), 1171–1181 = Math. USSR Izv. **3** (1969), 1103–1114. MR **40** #5620.

15. ———, *Elementary method in the theory of congruences for a prime modulus*, Acta Arith. **17** (1970), 231–247. MR **42** #4552.

16. ———, *On estimating rational trigonometric sums with prime denominator*, Trudy Mat. Inst. Steklov. **112** (1971), 346–371 = Proc. Steklov Inst. Math. **112** (1971), 358–385.

17. N. M. Korobov, *Estimate of a sum of Legendre symbols*, Dokl. Akad. Nauk SSSR **196** (1971), 764–767 = Soviet Math. Dokl. **12** (1971), 241–245. MR **43** #220.

18. S. A. Stepanov, *An elementary proof of the Hasse-Weil theorem for hyperelliptic curves*, J. Number Theory **4** (1972), 118–143. MR **45** #3418.

Translated by J. E. and T. P. VAUGHAN

ON SUMS OF REAL CHARACTERS *

M. JUTILA

1. As is well known, the theory of Fourier series has found numerous important applications in number theory. In this note we shall consider a recent new application of a Fourier series lemma of I. M. Vinogradov to the estimation of character sums.

Given any real numbers α, β, Δ such that $0<\Delta<{}^1/_4$, $\Delta\leqslant\beta-\alpha\leqslant 1-\Delta$, the lemma in question expresses a certain function $\psi(x)$ of period 1 with the properties

$$\psi(x)=1 \quad \text{for} \quad \alpha+\frac{1}{2}\Delta\leqslant x\leqslant\beta-\frac{1}{2}\Delta,$$

$$0<\psi(x)<1 \quad \text{for} \quad \alpha-\frac{1}{2}\Delta<x<\alpha+\frac{1}{2}\Delta,$$

and

$$\text{for} \quad \beta-\frac{1}{2}\Delta<x<\beta+\frac{1}{2}\Delta,$$

$$\psi(x)=0 \quad \text{for} \quad \beta+\frac{1}{2}\Delta\leqslant x\leqslant 1+\alpha-\frac{1}{2}\Delta$$

as an absolutely convergent Fourier series with rapidly decreasing coefficients (see [2], Ch. 2, Lemma 2). As noted by I. M. Vinogradov more than fifty years ago, this lemma can be applied to character sums

$$S=\sum_{1\leqslant n\leqslant Y}\chi(n),$$

where χ is a primitive character to a modulus $k>1$. Indeed, we have

$$\psi(x)=\beta-\alpha+\sum_{m=1}^{\infty}\left(a_m e^{2\pi imx}+\bar{a}_m e^{-2\pi imx}\right), \tag{1}$$

and taking

$$\Delta=k^{-1/2},\qquad \alpha=\frac{1}{2}\Delta,\qquad \beta=Yk^{-1}+\frac{1}{2}\Delta$$

(we suppose that $k^{1/2}\leqslant Y\leqslant\frac{1}{2}k$ and that k is sufficiently large), we can write the sum S as follows:

AMS (MOS) subject classifications (1970). Primary 10G05.

**Editor's note.* The body of this article is an unaltered reprinting of the original paper. Details of the bibliography have been checked and put in standard form, and MR citations have been added.

$$S=\sum_{n=1}^{k}\psi\left(\frac{n}{k}\right)\chi(n)+\sum_{1\leqslant n\leqslant k^{1/2}}\left(1-\psi\left(\frac{n}{k}\right)\right)\chi(n)$$

$$-\sum_{Y<n\leqslant Y+k^{1/2}}\psi\left(\frac{n}{k}\right)\chi(n)=\tau_k\sum_{m=1}^{\infty}(a_m\bar{\chi}(m)+\bar{a}_m\bar{\chi}(-m))+S'+S'', \qquad (2)$$

say, where τ_k is a Gaussian sum with $|\tau_k|^2=k$. Taking $r=2$ in Vinogradov's lemma, we have

$$a_m\ll\min(\beta-\alpha,\ m^{-1},\ \Delta^{-2}m^{-3}),$$

whence (2) easily yields the following so-called Polya—Vinogradov estimate

$$S\ll k^{1/2}\log k. \qquad (3)$$

2. Now consider the following problem: let $\mathscr{D}$ be the set of integers defined from the following conditions:

(i) D is no square,

(ii) $D\equiv 1 \pmod 4$, or $D=4N$, $N\equiv 1$, 2, or $3 \pmod 4$, let $\left(\frac{D}{n}\right)$ stand for the Kronecker symbol, let $X\geqslant 3$, $Y\geqslant 1$ be real numbers, and consider the value of the double sum

$$S(X,\ Y)=\sum_{|D|\leqslant X}{}'\left|\sum_{1\leqslant n\leqslant Y}\left(\frac{D}{n}\right)\right|^2, \qquad (4)$$

where $\sum'$ denotes a summation restricted to the set $\mathscr{D}$.

By (3) we see that

$$S(X,\ Y)\ll X^2\log^2 X. \qquad (5)$$

Another estimate is obtained if the square in (4) is written out as a double sum, and the summation over D is carried out first. Using the reciprocity law for real characters and the estimate (3), we obtain the result

$$S(X,\ Y)\ll XY\log^3 Y+Y^3\log Y. \qquad (6)$$

In a forthcoming paper [1] we prove by a suitable combination of Vinogradov's lemma and (3) the following result:

$$S(X,\ Y)\ll XY\log^8 X. \qquad (7)$$

This is contained in (5) and (6) in the cases $Y\ll X^{1/2}$ and $Y\gg X$, and so the novelty lies in the values $X^{1/2}\ll Y\ll X$.

It may be noted that if we restrict in (4) the sum over n to fundamental discriminants or to the negatives of fundamental discriminants, it effects only a minor loss in the logarithmic factor in (7).

3. Before sketching the proof of (7), we note on some interesting applications.

For a fundamental discriminant d, write $\chi^{(d)}(n) = \left(\frac{d}{n}\right)$. For the values of $L(1, \chi^{(d)})$ we have the following result: if $k \geqslant 1$ is an integer, $e = \pm 1$, and $X \geqslant 3$, then [1]

$$\sum_{d \in [e, eX]}^{*} L^k(1, \chi^{(d)}) = b(k) X + O(X^{1/2} (\log X)^{c(k)}), \tag{8}$$

where Σ^* denotes a summation over the fundamental discriminants, and $b(k)$, $c(k)$, and the O-constant are functions of k.

In view of the well-known connection between the class number $h(d)$ of the imaginary quadratic field with the discriminant $d < 0$ and $L(1, \chi^{(d)})$, we easily obtain from (8) that

$$\sum_{-X \leqslant d < 0}^{*} h^k(d) = a(k) X^{\frac{k+2}{2}} (1 + O(X^{-1/2} (\log X)^{c'(k)}), \tag{9}$$

where $a(k)$, $c'(k)$, and the O-constant are functions of k.

It may be noted that, in view of the case $k=1$, the error term in (9) is, essentially, the best possible. The sharpest result before was the estimate of Lavrik [3] with the exponent $-1/4$ instead of $-1/2$.

The estimate (7) gives also mean-value and density results for L-functions with real characters. It is our intention to discuss these questions in a forthcoming paper. We only mention the following result: *there exists a function* $f(x, y)$, *defined for* $x > 0$, $y > 0$, *such that*

$$\lim_{\substack{(x, y) \to (0, 0), \\ x > 0, \ y > 0}} f(x, y) = 0,$$

and such that, for any numbers $\varepsilon_1 > 0$, $\varepsilon_2 > 0$, $X \geqslant 3$ *the region*

$$\sigma \geqslant \frac{1}{2} + f(\varepsilon_1, \varepsilon_2), \qquad |t| \leqslant X^{\varepsilon_1}$$

contains no zero of any L-function $L(s, \chi^{(d)})$ *with* $|d| \leqslant X$, *with a possible exception of* $\ll X^{1-\varepsilon_2}$ *fundamental discriminants d at most.*

4. Now we shall briefly sketch the proof of the estimate (7).

Define the function $\psi(x, u)$ to be the function $\psi(x)$ with the following choice of the parameters:

$$\Delta = X^{-1/2}, \qquad \alpha = \frac{1}{2}\Delta, \qquad \beta = Yu^{-1} + \frac{1}{2}\Delta = \beta(u),$$

where u is variable, and suppose that $X^{1/2} \leqslant Y \leqslant \frac{1}{4} X$, $\frac{1}{2} X \leqslant u \leqslant X$, $X > 100$. Now, for each fundamental discriminant d with $\frac{1}{2} X \leqslant |d| \leqslant X$, the sum

$$S^{(d)} = \sum_{1 \leqslant n \leqslant Y} \left(\frac{d}{n}\right)$$

can be written in the form (2), with $\psi\left(\frac{n}{k}\right)$ replaced by $\psi\left(\frac{n}{|d|}, |d|\right)$. In this way, the sum

$$\sum_{1/2X \leqslant |d| \leqslant X}^{*} |S^{(d)}|^2 \tag{10}$$

can be split up into three parts. These are expressed by a partial summation formula which allows one to appeal to the estimate (3). In comparison to the proof of (3), we now need considerably closer information on the function $\psi(x, u)$, namely the explicit expressions for the Fourier coefficients $a_{m,u}$ of $\psi(x, u)$, as well as estimates for the derivatives $\frac{d}{du}\psi\left(\frac{n}{u}, u\right)$ (in the cases $n \in [1, X^{1/2}]$, $n \in [Y, Y + X^{1/2}]$), and $\frac{d}{du} a_{m,u}$. After that the proof of (7) is a matter of calculation.

5. It would be of interest to generalize the estimate (7) to higher powers of character sums. Write for $k = 1, 2, \ldots$

$$S_k(X, Y) = \sum_{|D| \leqslant X}' \left| \sum_{1 \leqslant n \leqslant Y} \left(\frac{D}{n}\right) \right|^{2k}.$$

We state the following

Conjecture. *For all* $k = 1, 2, \ldots$ *and* $X \geqslant 3$, $Y \geqslant 1$, *the estimate*

$$S_k(X, Y) \leqslant c_1(k)\, X Y^k (\log X)^{c_2(k)}$$

holds with certain coefficients $c_1(k)$, $c_2(k)$, *depending on* k.

By (7) this holds for $k = 1$, but already the next case $k = 2$ seems to be difficult and worth investigation.

BIBLIOGRAPHY

1. M. Jutila, *On character sums and class numbers*, J. Number Theory 5 (1973), 203–214.

2. I. M. Vinogradov, *The method of trigonometrical sums in the theory of numbers*, Trudy Mat. Inst. Steklov. 23 (1947); rev. ed., "Nauka", Moscow, 1971; English transl., Interscience, New York, 1954. MR 10, 599; 15, 941.

3. A. F. Lavrik, *On the moments of the class number of primitive quadratic forms with a negative discriminant*, Dokl. Akad. Nauk SSSR 197 (1971), 32–35 = Soviet Math. Dokl. 12 (1971), 399–403. MR 43 #6166.

Trudy Mat. Inst. Steklov.
132 (1973)

Proc. Steklov Inst. Math.
132 (1973)

ON LARGE SIEVE INEQUALITIES AND THEIR APPLICATIONS*

E. BOMBIERI

Dedicated to Professor I. M. Vinogradov

1. Introduction

The large sieve was introduced by Linnik in 1941 in order to handle problems about the statistical distribution of residue class mod p in an arbitrary sequence. Let $n_1 < n_2 < \ldots < n_z$ be a sequence of Z integers in an interval $[M+1, M+N]$ and let

$$Z(p, a) = \sum_{n_i \equiv a(p)} 1.$$

The prime number p is said to be exceptional if the number of residue classes mod p not represented by the numbers n_j is $> \tau p$, where $0 < \tau < 1$ is a fixed positive number. Linnik proved that the number of exceptional primes $p \leqslant \sqrt{N}$ does not exceed

$$\frac{c_0 N}{\tau^2 Z}$$

where c_0 is an absolute constant. In particular, if $Z \gg N$ the number of exceptional primes is bounded. Linnik himself gave a number of applications of this result.

The next step forward was taken by Rényi. Trying to give a quantitative form of Linnik's result, Rényi was led to consider the sum

$$S_X = \sum_{p \leqslant X} p \sum_{h=0}^{p-1} [Z(p, h) - Z/p]^2 \tag{1}$$

as a kind of variance (in probabilistic sense) and proved the estimate

$$S_X \leqslant 2NZ \quad \text{for} \quad X = \left(\frac{N}{12}\right)^{1/3}. \tag{2}$$

Though this result gives information which cannot be extracted from Linnik's theorem, it has the defect that it considers only primes $p \leqslant N^{1/3}$ while Linnik's result gives a non-trivial information about primes $p \leqslant N^{1/2}$.

AMS (MOS) subject classifications (1970). Primary 10H30; Secondary 10H35, 10G05.

**Editor's note.* The body of this article is an unaltered reprinting of the original paper. Details of the bibliography have been checked and put in standard form, and MR citations have been added.

This form of the large sieve received a striking application when Rényi proved that every large even integer is the sum of a prime and an almost prime with a bounded number of factors, that is

$$2n = p + p_1 p_2 \ldots p_c$$

for some absolute constant c, a result which was out of the reach of the small sieve techniques of Brun, Buchstab and Selberg.

The main point of Rényi's proof was an inequality of the type

$$\sum_{q \leqslant Q} \mu^2(q) \max_{(a,\, q)=1} \left| \psi(x;\, q,\, a) - \frac{x}{\Phi(q)} \right| \ll \frac{x}{(\log x)^A} \tag{3}$$

with $Q = x^{\eta - \varepsilon}$, for some small $\eta > 0$. Actually rather than having $\psi(x;\, q,\, a) - \frac{x}{\Phi(q)}$ he had other averages of $\Lambda(n) - 1$ but the result was almost equivalent to the one stated here.

It soon became clear that a result of the previous type would have been much more useful if one could get it with $\eta = {}^1/_2$. Various explicit values for η were proposed, including $\eta = {}^3/_8$ by Barban [1]. Quite independently and with different methods, essential progress was obtained in 1965 by Roth [2], A. I. Vinogradov [3] and Bombieri [4]. Roth [2] improved Rényi estimate (2) to the range $X = C\,(N/\log N)^{1/2}$, a result which, at least in principle, would suffice to prove (3) with $Q = x^{1/2 - \varepsilon}$; A. I. Vinogradov [3] instead, proved a result on the distribution of zeros of L-functions from which he could obtain (3) with $Q = x^{1/2 - \varepsilon}$ and Bombieri [4] proved that

$$\sum_{q \leqslant Q} \max_{(a,\, q)=1} \left| \psi(x;\, q,\, a) - \frac{x}{\Phi(q)} \right| \ll \frac{x}{(\log x)^A} \tag{4}$$

for $Q = x^{1/2}/(\log x)^B$ and some $B = B(A)$. The improved range for Q (i. e. replacing $x^{-\varepsilon}$ by $(\log x)^{-B}$) has some useful applications. The proof of (4) has been much simplified through the work of Davenport-Halberstam [5] and Gallagher [6].

Today the large sieve inequality (2) is understood in a more general setup, namely:

(A) *inequalities for trigonometric sums*

$$\sum_{r=1}^{R} |S(x_r)|^2$$

where

$$S(x) = \sum_{M+1}^{M+N} a_n e^{2\pi i n x};$$

(B) *inequalities for sums of characters*

$$\sum_{\omega \in \Omega} \left| \sum_{M+1}^{M+N} a_n \omega(n) \right|^2$$

where Ω *is a set of multiplicative characters.*

Inequalities of type (A) can be applied directly to a sequence and give the classical large sieve, while inequalities of type (B) are more useful if combined with analytical methods.

2. The inequalities of the large sieve

The general form of (A) as obtained by Davenport and Halberstam is

Theorem 1. *Let* x_r, $r=1, 2, \ldots, R$, *be real numbers,* $\|x_r - x_s\| \geqslant \delta$ *for* $r \neq s$ *where* $\|\ \|$ *denotes the distance from the nearest integer. Then*

$$\sum_{r=1}^{R} |S(x_r)|^2 \leqslant \left(N + \frac{2}{\delta}\right) \sum_{M+1}^{M+N} |a_n|^2.$$

The proof of this is so simple that it is worth giving (see [7]).

Lemma (Selberg). *Let* H *be a complex Hilbert space,* $\varphi_1, \varphi_2, \ldots, \varphi_R$, $f \in H$. *Then*

$$\sum_{i=1}^{R} \frac{|(f, \varphi_i)|^2}{\sum_{j=1}^{R} |(\varphi_i, \varphi_j)|} \leqslant \|f\|^2.$$

Proof. For $\xi_i \in \mathbf{C}$ we have

$$\|f - \sum \xi_i \varphi_i\|^2 \geqslant 0.$$

Therefore

$$\|f\|^2 - 2\,\mathrm{Re} \sum_{r=1}^{R} \xi_i \overline{(f, \varphi_i)} + \sum_{i, j=1}^{R} \xi_i \bar{\xi}_j (\varphi_i, \varphi_j) \geqslant 0,$$

but since

$$|\xi_i \bar{\xi}_j| \leqslant \frac{1}{2}(|\xi_i|^2 + |\xi_j|^2)$$

we obtain

$$\|f\|^2 - 2\,\mathrm{Re} \sum_{r=1}^{R} \xi_i \overline{(f, \varphi_i)} + \sum_{i=1}^{R} |\xi_i|^2 \left\{\sum_{j=1}^{R} |(\varphi_i, \varphi_j)|\right\} \geqslant 0.$$

Taking

$$\xi_i = \frac{(f, \varphi_i)}{\sum_{j=1}^{R} |(\varphi_i, \varphi_j)|}$$

we get what we want.

Proof of Theorem 1. We take H the space l^2 of sequences $\{a_n\}$, $n \in \mathbf{Z}$ with scalar product

$$(a, b) = \sum_{n \in \mathbf{Z}} a_n b_n.$$

Now let

$$f = \begin{cases} a_n, & |n| \leqslant N, \\ 0, & |n| > N, \end{cases}$$

$$\varphi_j = \begin{cases} e^{-2\pi i n x_j}, & |n| \leqslant N, \\ \left(\frac{N+L-|n|}{N}\right)^{1/2} e^{-2\pi i n x_j}, & N < |n| \leqslant N+L, \\ 0, & |n| > N+L. \end{cases}$$

We have

$$(f, \varphi_j) = \sum_{-N}^{N} a_n e^{2\pi i n x_j},$$

$$(\varphi_h, \varphi_j) = \frac{1}{L}\left\{\left[\frac{\sin \pi (N+L)(x_j - x_h)}{\sin \pi (x_j - x_h)}\right]^2 - \left[\frac{\sin \pi N (x_j - x_h)}{\sin \pi (x_j - x_h)}\right]^2\right\}.$$

Therefore

$$|(\varphi_h, \varphi_j)| \leqslant \begin{cases} 2N+L, & h=j, \\ L^{-1}\sin^{-2}\pi(x_j - x_h), & h \neq j. \end{cases}$$

Since

$$\|x_h - x_j\| \geqslant \delta \quad \text{for} \quad h \neq j$$

we have

$$\sum_{h=1}^{R} |(\varphi_h, \varphi_j)| \leqslant 2N + L + \frac{1}{L}\sum_{\substack{h=1 \\ h\neq j}}^{R} \frac{1}{\sin^2 \pi (x_j - x_h)} \leqslant$$

$$\leqslant 2N + L + \frac{2}{L}\sum_{k=1}^{1/(2\delta)} \frac{1}{\sin^2(\pi k \delta)} \leqslant 2N + L + \frac{1}{2L}\sum_{k=1}^{\infty}\frac{1}{k^2\delta^2} \leqslant 2N + L + \frac{\pi^2}{12}\frac{1}{L\delta^2}$$

and taking $L = \left[\frac{\pi}{2\sqrt{3}\,\delta}\right] + 1$ we get

$$\sum_{h=1}^{R} |(\varphi_h, \varphi_j)| \leqslant 2N + \frac{\pi}{\sqrt{3}\delta} + 1.$$

Theorem 1 now follows from Selberg's lemma.

Among the applications of Theorem 1 I will mention that many results previously obtained using Selberg's sieve now can be proved using (4); see [8] and [9] for details.

We turn to inequalities of type (B), which in many cases lie much deeper than those of type (A).

Let Ω be a set of distinct multiplicative characters, let $L^2(\mathcal{N})$ and $L^2(\Omega)$ be the spaces of sequences on $\mathcal{N} = [1, N]$ and Ω, with the usual scalar product. The linear operator

$$D : L^2(\mathcal{N}) \to L^2(\Omega)$$

defined by

$$D\{a_n\}_{n\in\mathcal{N}} = \left\{\sum_{\mathcal{N}} a_n \omega(n)\right\}_{\omega\in\Omega}$$

will be called the Dirichlet series operator. If $\|D\|_2$ is the norm of D we have the inequality of type (B)

$$\sum_{\Omega}\left|\sum_{\mathcal{N}} a_n\omega(n)\right|^2 \leqslant \|D\|_2^2 \sum_{\mathcal{N}} |a_n|^2$$

and the main problem is that of estimating $\|D\|_2$.

The first useful estimates valid for general characters $\omega(n)=\chi(n)\,n^s$ and not just for Dirichlet characters $\chi(n)$ have been obtained by Montgomery [10] using an ingenious trick of Halász. An analysis of Montgomery's paper shows that Halász' trick can be replaced by

$$\|D\|_2 = \|D^*\|_2$$

where

$$D^*: L^2(\Omega) \to L^2(\mathcal{N})$$

is the adjoint operator

$$D^*\{\alpha_\omega\}_{\omega\in\Omega} = \left\{\sum_{\Omega}\alpha_\omega\bar{\omega}(n)\right\}_{n\in\mathcal{N}}.$$

It turns out that $\|D^*\|_2$ is much easier to estimate than $\|D\|_2$. In order to see what one obtains, let us consider only characters

$$\omega(n)=\chi(n)\,n^{it}, \qquad t \text{ real}$$

and define

$$\|\omega\| = q(|t|+1)$$

where q is the modulus of χ, and

$$D = \max_{\omega,\,\omega'}\|\bar{\omega}\omega'\|, \qquad T = \max_{\omega,\,\omega'}|t-t'|+1.$$

We say that the characters $\omega\in\Omega$ are well-spaced if for $\omega, \omega'\in\Omega$, $\omega\neq\omega'$ we have that either $\bar{\chi}\chi'$ is non-principal or $|t-t'|>2(\log D)^2$. We also define $\mu(\sigma)$ as the lower bound of the exponents such that

$$|L(\sigma+it,\chi)| \ll [q(|t|+1)]^{\mu(\sigma)+\varepsilon}$$

uniformly in t and q.

Theorem 2. *Let Ω be a set of $|\Omega|$ well-spaced characters. Then for $\varepsilon>0$ and $\delta<1/2$ we have*

$$\sum_{\Omega}\left|\sum_{N}^{2N} a_n\omega(n)\right|^2 \ll \left\{N + N^{\delta}D^{\frac{1}{2}-\delta+\varepsilon}|\Omega| + N^{\delta}D^{\mu(\delta)+\varepsilon}\max_{\chi^*\in\Omega}\sum_{\chi\in\Omega}N(1-\delta,\ T;\ \chi\bar{\chi}^*)\right\}\sum_{N}^{2N}|a_n|^2$$

(the $\max_{\chi^*\in\Omega}\sum_{\chi\in\Omega}N(1-\delta,\ T;\ \chi\bar{\chi}^*)$ may also be replaced by $|\Omega|$), where $\chi\in\Omega$ means that Ω contains a character $\omega(n)=\chi(n)\,n^{it}$ and where $N(\alpha,\ T;\ \chi)$ is the number of zeros of $L(s,\chi)$ in the rectangle $\sigma>\alpha$, $|t|\leqslant T$.

Since Theorem 2 can be used to give non-trivial estimates for $\max_{\chi^*\in\Omega}\sum_{\chi\in\Omega}N(\alpha,\ T;\ \chi\bar{\chi}^*)$ it can be shown that in many cases there is $\delta_0>0$ such that

$$\sum_{\Omega}\left|\sum_{N}^{2N} a_n\omega(n)\right|^2 \ll \left\{N + N^{\delta_0}D^{\frac{1}{2}-\delta_0+\varepsilon}|\Omega|\right\}\sum_{N}^{2N}|a_n|^2,$$

a result which previously could be obtained only using the Lindelöf hypothesis $\mu(\sigma)=0$ for $\sigma>1-\delta_0$.

Theorem 2 leads to the best known bounds for $N(\alpha, T; \chi)$. I will give here an example of what can be obtained in this way for $N(\alpha, T)$.

Theorem 3 (M. N. Huxley). *We have*

$$N(\alpha, T) \ll T^{\frac{12}{5}(1-\alpha)} (\log T)^4$$

for $1/2 \leqslant \alpha \leqslant 1$. *Hence we have the bound*

$$p_{n+1} - p_n \ll p_n^{7/12+\varepsilon}$$

for the gaps between consecutive primes.

Theorem 4. *The density hypothesis*

$$N(\alpha, T) \ll T^{2(1-\alpha)+\varepsilon}$$

is true for $\alpha > 13/16$.

The result of Theorem 2 and many other new inequalities will appear in a forthcoming paper by M. Forti and C. Viola, two young researchers in Number Theory at the University of Pisa.

BIBLIOGRAPHY

1. M. B. Barban, *The "large sieve" method and its application to number theory,* Uspehi Mat. Nauk **21** (1966), no. 1 (127), 51–102 = Russian Math. Surveys **21** (1966), no. 1, 49–103. MR **33** #7320.

2. K. F. Roth, *On the large sieves of Linnik and Rényi,* Mathematika **12** (1965), 1–9. MR **33** #5589.

3. A. I. Vinogradov, *On the density conjecture for Dirichlet L-series,* Izv. Akad. Nauk SSSR **29** (1965), 903–934; correction, ibid. **30** (1966), 719–720; English transl., Amer. Math. Soc. Transl. (2) **82** (1969), 9–46. MR **33** #2607; #5579.

4. E. Bombieri, *On the large sieve,* Mathematika **12** (1965), 201–225. MR **33** #5590.

5. H. Davenport and H. Halberstam, *The values of a trigonometrical polynomial at well spaced points,* Mathematika **13** (1966), 91–96; correction, ibid. **14** (1967), 229–232. MR **33** #5592; **36** #2569.

6. P. X. Gallagher, *Bombieri's mean value theorem,* Mathematika **15** (1968), 1–6. MR **38** #5724.

7. E. Bombieri, *A note on the large sieve,* Acta Arith. **18** (1971), 401–404. MR **44** #3982.

8. E. Bombieri and H. Davenport, *On the large sieve method,* Number Theory and Analysis (Papers in Honor of Edmund Landau), Plenum Press, New York, 1969, pp. 9–22. MR **41** #5327.

9. H. L. Montgomery, *A note on the large sieve,* J. London Math. Soc. **43** (1968), 93–98. MR **37** #184.

10. ———, *Mean and large values of Dirichlet polynomials,* Invent. Math. **8** (1969), 334–345. MR **42** #3029.

Trudy Mat. Inst. Steklov.
132 (1973)

Proc. Steklov Inst. Math.
132 (1973)

ON SOME PROBLEMS OF PRIME NUMBER THEORY CONNECTED WITH I. M. VINOGRADOV'S METHOD

A. A. KARACUBA

ABSTRACT. Results are presented for problems connected with a sequence of shifted prime numbers, i.e. a sequence of integers of the form $p + k$, where $k \geqslant 1$ are natural numbers and $p = 2, 3, 5, \cdots$.

Bibliography: 6 items.

The linear additive problems with prime numbers solved by Vinogradov's method have become widely known thanks to the fact that one such problem is the Goldbach problem. However, the Goldbach problem can be derived from the extended Riemann hypothesis or even from weaker theorems concerning the distribution of primes. But there exists a whole class of additive problems with primes and a series of other number theory problems which can be solved by Vinogradov's method and cannot be derived from the strongest hypotheses concerning the distribution of primes, in particular, the extended Riemann hypothesis.

Without touching upon additive problems with primes, one of which is presented in I. M. Vinogradov's new book [1], I shall dwell on problems connected with a sequence of shifted primes, i.e. a sequence of the form $p + k$, where $k \geqslant 1$ is a natural number and $p = 2, 3, 5, \cdots$. This sequence will be denoted by P_k.

Let us introduce the concept of quadratic residue and nonresidue with respect to a certain modulus. Consider a number q, which we shall assume to be prime, and let us divide all the integers into two classes V and N. The class V will contain those integers which, after dividing by q, yield the same remainders as their squares. The integers in this class are called quadratic residues mod q. The class N will contain all other integers, which are called quadratic nonresidues mod q.

Consider the natural segment $[1, x]$, and let $V(x)$ and $N(x)$ be the number of integers of V and N respectively in this segment.

One of the first questions concerning these concepts is the following: *When is it possible to find an asymptotic formula for $N(x)$ and $V(x)$?*

The very first attempts to answer this question show that the problem can be solved if x is large in comparison with q, for example, of order q, and that the problem is unusually difficult if x is small, for example, of order q^{ϵ}.

AMS (MOS) subject classifications (1970). Primary 10H35, 10H25; Secondary 10G05.

In 1918 Vinogradov obtained asymptotic formulas for $V(x)$ and $N(x)$ when $x > \sqrt{q}\log q$. The remainder in these asymptotic formulas cannot be significantly improved. In 1958 Burgess proved that the asymptotic formula for $V(x)$ and $N(x)$ is valid when $x > q^{1/4}\log q$.

Problems similar to those about the magnitudes of $V(x)$ and $N(x)$ can be posed concerning the quadratic residues and nonresidues [2] of a sequence P_k lying in $[1, x]$.

Let $V_k(x)$ and $N_k(x)$ be respectively the number of quadratic residues and nonresidues $\bmod q$ of the sequence P_k lying in $[1, x]$.

It is natural to expect that $V_k(x)$ and $N_k(x)$ are asymptotically equal and approximately equal to half of all the integers of P_k in the segment $[1, x]$, i.e. $\pi(x)/2$, where $\pi(x)$ is the number of primes in $[1, x]$. But whereas the asymptotic formula for $V(x)$ and $N(x)$ was obtained in 1918 for $x > \sqrt{q}\log q$ and it follows from the extended Riemann hypothesis for $x > q^{\epsilon}$, the asymptotic formula for $V_k(x)$ and $N_k(x)$ follows from the extended Riemann hypothesis only for $x > q\log q$.

Therefore it was completely unexpected when Vinogradov [3], using his own method, obtained an asymptotic formula for $V_k(x)$ and $N_k(x)$ when $x > q^{3/4+\epsilon}$. Developing Vinogradov's method, in 1969 I succeeded in obtaining an asymptotic formula for $V_k(x)$ and $N_k(x)$ when $x > q^{1/2+\epsilon}$ (see [4]).

I shall go into the main features of the proof of this result in more detail.

Let us consider the Legendre symbol

$$\left(\frac{a}{q}\right)=\begin{cases} +1, & \text{if } (a, q)=1,\ a\equiv x^2 \pmod q,\\ -1, & \text{if } (a, q)=1,\ a\not\equiv x^2 \pmod q,\\ 0, & \text{if } a \text{ is divisible by } q.\end{cases}$$

The problem of finding an asymptotic formula for $V_k(x)$ and $N_k(x)$ reduces to the problem of finding a nontrivial estimate for the following sum:

$$T(x)=\sum_{p\leqslant x}\left(\frac{p+k}{q}\right).$$

We assume that $(k, q) = 1$. The following theorem holds concerning this sum:

THEOREM. *Let* ω *be any number in the interval* $(0, 1/2)$, *and let* $q^{1/2+\omega} \leqslant x \leqslant q^2$. *Then*

$$T(x)\ll xq^{-0.01\omega^2},$$

where the constant in the $\ll$ *sign depends only on* ω.

The proof of this theorem consists of the following parts:

1. The reduction of $T(x)$—a sum over primes—to a double sum W.

2. An estimate for W.

I shall begin with part 2. The following sum will be called the double sum for the case under consideration:

$$W = \sum_{U<u\leqslant 2U} \sum_{V<v\leqslant 2V} \theta_1(u)\theta_2(v)\left(\frac{uv+k}{q}\right),$$

where the numbers u and v are completely arbitrary and vary independently in the indicated intervals; $\theta_1(u)$ and $\theta_2(v)$ are arbitrary complex-valued functions of u and v which, in the problem at hand, do not exceed u^ϵ and v^ϵ in modulus. We shall assume that u and v are greater than $U^{1-\epsilon}$ and $V^{1-\epsilon}$. Then, in order for the sum W to have a nontrivial estimate, and in order to have the rough estimate $W \ll (UV)^{1+\epsilon}$, it is sufficient that

$$U > q^{1/2+\varepsilon_1}, \qquad V > q^{\varepsilon_2}.$$

The estimate for W is derived by Vinogradov's method as follows. It is obvious that

$$W < U^\varepsilon \sum_{U<u\leqslant 2U} \left| \sum_{V<v\leqslant 2V} \theta_2(v)\left(\frac{uv+k}{q}\right)\right|.$$

The outer summation over u can be performed over the whole interval. This only strengthens the inequality. Now, in order for the summation over u to be made the inner sum, both sides of the inequality must be raised to some even power $2r$, $r \geqslant 1$; The larger the number r, the smaller the number V, i.e. the shorter the internal sum.

Apply Hölder's inequality, we obtain the following inequality:

$$|W|^{2r} \leqslant U^{2r\varepsilon} U^{2r-1} \sum_{v_1,\ldots,v_{2r}} \theta_2(v_1)\ldots\theta_2(v_{2r}) \sum_{U<u\leqslant 2U} \left(\frac{(uv_1+k)\ldots(uv_{2r}+k)}{q}\right),$$

where we have used the multiplicativity of the Legendre symbol, and the summation over u has been made the inner sum. The number u runs through the whole interval, i.e. the sum over u can be estimated by applying Weil's estimates. The estimate with respect to u will be trivial if the product $(uv_1+k)\cdots(uv_{2r}+k)$ is a square. This will be the case, roughly speaking, when $v_1,\cdots,v_{2r}$ are pairwise equal, i.e. $v_1,\cdots,v_{2r}$ are not greater than V^r in order. In the remaining cases the estimate with respect to u yields $\sqrt{q}\log q$, i.e.

$$|W|^{2r} \ll (UV)^{2r\varepsilon} U^{2r-1}(V^r U + V^{2r}\sqrt{q}\log q) = (UV)^{2r(1+\varepsilon)}(V^{-r} + U^{-1}\sqrt{q}\log q),$$

and

$$W \ll (UV)^{1+\varepsilon}(V^{-r} + U^{-1}\sqrt{q}\log q)^{1/2r}.$$

The conditions on U and V for the estimate of W to be nontrivial are now clear.

Thus the dual sums have been estimated. The transition to a summation over the whold interval involves Vinogradov's idea of smoothing. I note that this idea of smoothing was used by Vinogradov in his solution of Waring's problem and in his method for estimating Weil sums.

I shall now turn to part 1, i.e. the reduction of $T(x)$ to W. Consider a number $H < x^{1/3}$ and the number

$$\mathbf{P} = \prod_{p \leqslant H} p.$$

By using a sieve method, $T(x)$ can be reduced to a small number of sums of the form

$$W_s = \sum_{d_1 \backslash \mathbf{P}} \cdots \sum_{\substack{d_s \backslash \mathbf{P} \\ d_1 \ldots d_s m_1 \ldots m_s \leqslant x}} \sum_{m_1 > 0} \cdots \sum_{m_s > 0} \mu(d_1) \ldots \mu(d_s) \left(\frac{d_1 \ldots d_s m_1 \ldots m_s + k}{q} \right).$$

The following difficulty arises here. Let $d_2 = \cdots = d_s = 2$ and $m_1 = \cdots = m_s = 1$, i.e. the summation runs over $d_1 \backslash \mathbf{P}$. How can the sum be estimated in this case?

This difficulty is overcome with the help of a deep and beautiful lemma of Vinogradov. This lemma, together with the ideas of dual sums and smoothing, is fundamental in Vinogradov's method of estimating trigonometric sums with primes.

LEMMA. *Let* $0 < \sigma \leqslant 1/3$ *and let* $\mathbf{P}$ *be the product of all primes not exceeding* x^σ, *i.e.*

$$\mathbf{P} = \prod_{p \leqslant x^\sigma} p.$$

Then all divisors d *of* $\mathbf{P}$ *not exceeding* x *can be distributed among less than* x^ϵ *sets with the following properties:*

a) *The numbers* d *belonging to one set have the same number* β *of prime factors and, therefore, the same value* $\mu(d) = (-1)^\beta$.

b) *One of the sets, called the simplest, consists of one number* $d = 1$. *To each of the remaining sets there corresponds a* φ *such that all the numbers of this set satisfy the condition* $\varphi < d \leqslant \varphi^{1+\epsilon_1}$, *where* $\epsilon_1 = \epsilon_1(\epsilon)$.

c) *For any set different from the simplest, for any* U *such that* $0 \leqslant U < \varphi$ *there exist two sets of numbers d–numbers* d' *and* d'' *with corresponding numbers* φ' *and* φ''*–which satisfy the conditions* $U < \varphi' \leqslant Ux^\sigma$ *and* $\varphi'\varphi'' = \varphi$, *such that for some natural number* B *we obtain all the numbers of the chosen set* B *times each, if from all the products* $d'd''$ *we choose only those that satisfy the condition* $(d', d'') = 1$.

Thus if the numbers d belong to some set with $\varphi > x^{\epsilon_0}$, then these d can be represented in the form $d = d'd''$, where d' runs through some interval of values whose magnitude can be regulated (this magnitude was denoted by the letter U in the lemma).

Let us turn to our multiple sum W_s. By the lemma, all divisors $d_1, \cdots, d_s$ which appear in this sum can be divided into fewer than x^ϵ classes; the sum W_s can be divided into fewer than $x^{s\epsilon}$ sums of the form

$$W'_s = \pm \sum_{d_1} \cdots \sum_{\substack{d_s \\ d_1 \cdots d_s m_1 \cdots m_s \leqslant x}} \sum_{m_1 > 0} \cdots \sum_{m_s > 0} \left(\frac{d_1 \cdots d_s m_1 \cdots m_s + k}{q} \right).$$

We can immediately assume that $d_1 \cdots d_s m_1 \cdots m_s > x^{1-\epsilon_0}$ for some $\epsilon_0 > 0$, since otherwise a trivial estimate yields a reduction of $x^{-\epsilon_0}$. Now if for at least one of the classes the corresponding $\varphi > x^{\epsilon_1}$, then, by representing d, according to the lemma, in the form $d = d'd''$ with the necessary limits of variation of d', i.e. with the necessary value U, we obtain a product of the form

$$d'd'' \ldots m_1 \ldots m_s = uv,$$

where $u = d'$ and $v = d'' \cdots m_1 \cdots m_s$, i.e. we obtain the dual sum which I mentioned above. The corresponding choice of U allows us to estimate this dual sum nontrivially provided $x > q^{1/2+\omega}$. The condition $(d', d'') = 1$ is easily overcome with the help of a well-known property of the Möbius function.

But if all $\varphi < x^{\epsilon_1}$ and all $m_i < x^{\epsilon_1}$, then the product $d_1 \cdots d_s m_1 \cdots m_s$ can again be represented in the form uv with the necessary limits of variation of u and v.

But if one of the m_i varies in limits bigger than $q^{1/4+\epsilon_1}$, then we can use Burgess' estimate [5]. Difficulties arise if all the m_i vary in limits less than $q^{1/4+\epsilon_1}$, i.e. in limits for which there are now no nontrivial estimates. These difficulties can be overcome with the help of arguments which I applied in 1968 to the problem of estimating sums of characters in finite fields [6]. This proves the theorem.

It is clear that a verbatim repetition of these arguments allows us to estimate nontrivially the sum of the values of a nonprinciple character mod q with respect to a sequence $p + k$ if the length of this sequence is greater than $q^{1/2+\epsilon}$. This in turn allows us to solve problems concerning the distribution of power residues and non-residues, primitive roots, indices in P_k, and a number of other problems.

But if we carry out these arguments with a linear sum over primes of the form

$$\sum_{p \leqslant x} e^{2\pi i \alpha p},$$

where in this case the summation over the whole interval is a summation of terms of a geometric progression, we obtain an estimate which implies an asymptotic formula for the number of representations for odd numbers by a sum of three primes and, in particular, the Goldbach problem.

If we consider the more complicated sum

$$\sum_{p \leqslant x} e^{2\pi i f(p)},$$

where f is a polynomial, then Vinogradov's method of estimating Weil sums should be added to the previous arguments. Here we also obtain an estimate from which nonlinear additive problems with primes follow.

I note that an asymptotic formula follows for the sum of the fractional parts of $f(p)$, i.e. the sum $\Sigma\ \{f(p)\}$, which appears in the emblem of our conference. If the degree of $f \geqslant 2$, this asymptotic formula cannot be obtained even from the extended Riemann hypothesis.

BIBLIOGRAPHY

1. I. M. Vinogradov, *The method of trigonometrical sums in the theory of numbers*, Trudy Mat. Inst. Steklov. **23** (1947); rev. ed., "Nauka", Moscow, 1971; English transl., Interscience, New York, 1954. MR **10**, 599; **15**, 941.

2. ————, *The distribution of quadratic residues and nonresidues of the form* $p + k$ *with respect to a prime modulus*, Mat. Sb. **3 (45)** (1938), 311–319. (Russian)

3. ————, *Improvement of an estimate for the sum of the values* $\psi(p, k)$, Izv. Akad. Nauk SSSR Ser. Mat. **17** (1953), 285–290. (Russian) MR **15**, 855.

4. A. A. Karacuba, *On character sums with primes*, Dokl. Akad. Nauk SSSR **190** (1970), 517–518 = Soviet Math. Dokl. **11** (1970), 135–137. MR **41** #5284.

5. D. A. Burgess, *The distribution of quadratic residues and non-residues*, Mathematika **4** (1957), 106–112. MR **20** #28.

6. A. A. Karacuba, *Character sums and primitive roots in finite fields*, Dokl. Akad. Nauk SSSR **180** (1968), 1287–1289 = Soviet Math. Dokl. **9** (1968), 755–757. MR **37** #5188.

Translated by R. LENET